TESTING 1-2-3

TESTING 1–2–3

Experimental Design with Applications in Marketing and Service Operations

JOHANNES LEDOLTER AND

ARTHUR J. SWERSEY

STANFORD BUSINESS BOOKS

AN IMPRINT OF STANFORD UNIVERSITY PRESS

Stanford, California 2007

Stanford University Press
Stanford, California

Printed in the United States of America on acid-free, archival-quality paper

Library of Congress Cataloging-in-Publication Data

Ledolter, Johannes.
   Testing 1–2–3 : experimental design with applications in marketing and service operations / Johannes Ledolter and Arthur J. Swersey.
        p.   cm.
   Includes bibliographical references and index.
   ISBN-13: 978-0-8047-5612-9 (cloth : alk. paper)
   1. Marketing research.   2. Service industries—Research.   3. Experimental design.   I. Swersey, Arthur J. II. Title.   III. Title: Testing one, two, three.
HF5415.2.L385   2007
658.8′3—dc22

                                                                                        2006029738

Typeset by Newgen–Austin in 10/13.5 Minion

Special discounts for bulk quantities of Stanford Business Books are available to corporations, professional associations, and other organizations. For details and discount information, contact the special sales department of Stanford University Press. Tel: (650) 736-1783, Fax: (650) 736-1784.

To Lea Vandervelde, my wife and inspiration (JL)
To my brother, Burt Swersey (AJS)

CONTENTS

PREFACE

Our interest in writing this book began about 10 years ago when in our own work we started to explore the applications of experimental design methods to problems outside of manufacturing. We recognized as others had before that these powerful approaches were valuable tools for marketing problems. We also discovered that beyond marketing applications there were other important questions outside of manufacturing for which experimental design methods could be usefully applied. For example, in education much research has been directed at determining the relationship between student learning as measured by standardized tests and class size. Large-scale tests have been carried out, but researchers have missed the opportunity to use experimental design methods that would allow the experimenter to simultaneously and efficiently test other variables such as textbook, use of computers, level of parental involvement, and amount of homework. We also observed that existing books on experimental design focused almost exclusively on industrial applications. Recognizing this, we have written a book that aims to fill this large gap in the literature by emphasizing marketing, service operations, and general business problems.

We have written this book for both academic and practitioner audiences. It can be used effectively in MBA courses in quality management and marketing research and in undergraduate and graduate engineering courses in design of experiments. It is also well suited for self-study by quality professionals, management consultants, and other practitioners.

We assume that readers have had a basic undergraduate course in statistics or an introductory statistics course at the MBA level. Chapter 2 provides a review of the basic statistical concepts that we use throughout the book. In subsequent chapters, material that is more mathematically advanced (review of regression using basic matrix algebra) is included in appendixes. We have included this material for the sake of mathematical rigor and completeness and to give those with more mathematical backgrounds the opportunity to delve deeper into the basic methodology.

In teaching statistical methods we have found that students learn best if they see the relevance of the material, learn clearly how to apply the tools, and understand the underlying statistical concepts. People learn about design of experiments best by solving exercises, analyzing real data sets, and designing and carrying out their own experiments. Each chapter includes many exercises ranging from straightforward "drill-type" problems to more challenging ones that test tools and concepts. The 13 cases involving real-world applications are a key and unique part of the book. Some of these cases describe how experiments were conducted and give readers the opportunity to analyze and interpret the results, while others are written so that students can develop their own designs and compare their approaches to what was actually done.

Each chapter ends with an important section of notes titled "Nobody Asked Us, But . . ." Including these notes allowed us to focus on the basic concepts in the main text and then to elaborate on them at the end of the chapter. Doing so gives the reader the opportunity to first learn the basics without being bogged down with too many details and then through the notes to build upon these core concepts. The title of this section comes from a well-known column of the same title written by the late New York sportswriter Jimmy Cannon.

The development of statistical computer software has made it easier to design experiments and analyze and interpret the results. We have not tied the book to a specific computer program, but discuss computer output from several packages, in particular, Minitab and JMP.

Instructors can use the book in a number of ways. The entire book can be covered in a full semester course on experimental design that would include most of the cases in the case study appendix, as well as an experimental design project that would combine methodology with real-world practice. The book can also be used for a section on experimental design in a course on quality management. To do so, the instructor would assign Chapters 1, 4, and 5, along with several of the cases. In addition, the instructor might assign selected sections of Chapter 2, which reviews basic statistical concepts, and Chapter 3.

Many people contributed to this book. We begin by acknowledging several people who greatly influenced our thinking and learning. We gratefully acknowledge George Box, Norman Draper, and the late Bill Hunter who taught courses on design of experiments and statistical modeling when one of the authors (JL) was a graduate student at the University of Wisconsin–Madison. The very lively "Monday Night Beer Seminars" in George Box's basement had a profound impact as these discussions showed the importance of well-designed experiments for learning and also provided a strategy for implementing these methods in real-world settings. We also pay tribute to the late Sebastian B. Littauer, a distinguished professor at Columbia University and a recipient of the American Society for Quality's Shewhart Medal, who was a mentor to one of us (AJS). He was an expert in statistical methods who influenced many by his extraordinary teaching of statistical quality control not only as a set of problem-solving tools but as the conceptional foundation of a quality management philosophy.

At Stanford University Press, a number of people made important contributions. We are especially grateful to Martha Cooley, our editor, for her encouragement, insights, and suggestions, and to Jared Smith who carefully checked and organized the manuscript in preparation for its production. We are grateful to the production services team at Newgen–Austin, including Andy Sieverman, who oversaw the production process from start to finish, and Teresa Berensfeld, whose excellent copy editing improved the presentation.

We wish to thank several people who helped us develop the cases and examples in this book. Mark Wachen, CEO of Optimost, provided the data for the PhoneHog case in Section 8.2 and shared with us his modeling insights. Optimost (www.optimost.com) is a technology and services company specializing in comprehensive real-time testing and conversion rate marketing. We also thank Phil Nadel, CEO of Gulfstream Internet (the parent company of PhoneHog), for carefully reviewing the case and allowing us to use it. Jay Harris, publisher of *Mother Jones*, was instrumental in the development of the Mother Jones (A) and (B) cases, providing access to his organization and contributing many helpful ideas as the experiment at *Mother Jones* was designed and carried out. Alexander Dean, president of David Brooks Company, was very generous with his time and expertise. The broken pots example that we introduce in Chapter 4 and discuss further in Chapter 5 was written based on many discussions with Alex and describes a simplified version of the production process his company uses in the making of clay pots.

We thank Elsevier Publishing Company for allowing us to include the article [Bell, G. H., Ledolter, J., and Swersey, A. J.: "Experimental Design on the Front Lines of Marketing: Testing New Ideas to Increase Direct Mail Sales," *International Journal of Research in Marketing*, Vol. 23 (2006)] as Case 9 of the case study appendix.

We are also grateful to Ronald Snee, Soren Bisgaard, and Barry O'Neil for their helpful suggestions.

There are several people who deserve special mention. We are pleased to acknowledge the contributions of Jullie Chon who over the course of a summer produced a comprehensive and very useful review of the experimental design literature. It was a pleasure working with her. We are also extremely grateful to Berton Gunter, a leader in teaching, applying, and writing on experimental design. Bert provided a detailed, excellent review of an earlier version of the manuscript. His input was invaluable and we have incorporated many of his suggestions.

Ken McLeod was the original editor of our book at Stanford University Press and was instrumental in the birth of this project. We will always remember and be grateful for his enthusiasm for the project, intelligence in understanding what we were trying to do, the constant encouragement he gave us, and his personal warmth. We are saddened that he did not live to see the book completed and are indebted to him for his contributions.

In acknowledging the contributors to this book, we have saved for last our deep gratitude to the most important contributor, Gordon Bell, president of the consulting firm LucidView. Gordon made major contributions to our book by providing us with cases and chapter examples that are based on his expertise and extensive experience helping firms apply experimental design methods. He contributed Case 2 (Magazine Price Test) and Case 5 (Office Supplies E-mail Test). A simplified version of that case is used in Section 5.8. Gordon also co-authored (with the authors of this book) two other cases: Case 8 (Experiments in Retail Operations: Design Issues and Application), and Case 9 (Experimental Design on the Front Lines of Marketing: Testing New Ideas to Increase Direct Mail Sales). Cases 8 and 9 are based on Gordon's exceptional consulting work. Parts of Case 9 are also used as a case example in Sections 4.5 and 6.3. One of the greatest benefits to us in writing this book has been the interactions, both professional and personal, that we have had with Gordon.

We thank him for contributing so much to this book and look forward to continuing our collaborations with him in the future.

We also thank the many students who took our classes at the University of Iowa, the Vienna University of Economics and Business Administration, and Yale University. We treasure the interactions we have had with our students and value all we have gained from them. Finally, we could not have completed this book without the encouragement of our families and closest friends. Writing a book is inevitably more time consuming than anticipated, and we will always be thankful for the patience and support we received from those nearest to us.

We welcome comments from readers. Our e-mail addresses are johannes-ledolter@ uiowa.edu and arthur.swersey@yale.edu. Throughout the book we have tried to convey our passion for the subject of experimental design and to share with readers our strongly felt beliefs in the power of these methods and their practical value. The success of this book will depend in large part on the experiments carried out in the future by those who read it.

Johannes Ledolter
Arthur J. Swersey

1 | INTRODUCTION

1.1 THE THEME OF THE BOOK

This book is about the power of statistical experiments. In the increasingly competitive global economy, firms are constantly under pressure to reduce costs, increase productivity, and improve quality. Testing or experimentation in the business world is commonplace, and the usual approach is to change one factor at a time while holding other factors constant. To some, this approach seems logical, simple, and therefore appealing. But as we will show, it is highly inefficient, and it may fail to identify important factors and lead to wrong conclusions. The better method is to test all factors simultaneously. Doing so not only reduces the costs of experimenting but, as we will demonstrate, also provides the experimenter with more and better information.

Elementary courses in statistics that cover topics such as probability, hypothesis testing, confidence intervals, and regression analysis often appear abstract; and although they are illustrated with numerous examples, they typically seem far removed from practical issues. In this book we use and build on basic statistical concepts to explore approaches for solving real-world problems. Although our focus is on practice, it is important to keep in mind that statistics is a science, and science is based on theory. While computer software has made the implementation of statistical methods much easier, there is a danger in relying on a cookbook approach in which the user fails to understand the underlying concepts. In contrast, this book's presentation combines theory and practice, and focuses on strengthening the reader's understanding of fundamental statistical ideas.

Our goal in writing this book is to share our passion for the subject and to provide students, practitioners, and managers with a set of highly relevant, interesting, and valuable tools. In the past, in the area of experimental design, nearly all the attention was focused on manufacturing rather than services. In contrast, most of the applications and examples in this book will involve marketing and service operations. In the next section, we give a brief introduction to some of the cases that are included.

1.2 A PREVIEW OF CASES

Throughout the book we illustrate concepts with practical examples. In addition, we include a group of real cases based on the actual implementation of experimental design methods. In this section, we discuss the highlights of a number of these cases.

In the marketing area, consumer testing is an important and widely used tool. But most marketing professionals hold firmly to the approach of changing only one variable at a time, which is often called "split-run testing" (also referred to as A/B splits, test-control, or champion-challenger testing). Only recently have marketing managers begun to embrace multifactor techniques that simultaneously test marketing variables. These experimental design methods are particularly well suited to product testing in supermarkets.

In one significant marketing application, and one of the cases in the book, a major magazine publisher sought to increase sales of its popular magazine in a chain of supermarkets. The firm identified 10 factors to test, including a discount on multiple copies (no or yes), an additional display rack in the snack food area (no or yes), and an on-shelf advertisement (no or yes). After considering a number of alternatives, the publisher implemented a 24-run Plackett-Burman experimental design (see Chapter 6). Each run consisted of a particular combination of settings of each of the 10 factors.

A key part of the experiment was to decide how many stores to include and how long to test, in order to achieve statistically significant results. A total of 48 stores were included, and the experiment ran for two weeks. As a result, the firm identified several changes that increased sales by 20%, and equally important, it gained insights into which changes would have a negative effect or no effect.

Direct mail is a common marketing channel, and firms use it for a wide range of products including credit cards, clothing, and magazines. Typically, response rates are very low, and a small increase in response can mean large financial benefits. *Mother Jones* magazine had extensive experience in direct mail testing aimed at increasing their subscription rates. Their protocol was to test only one change, such as the color of the envelope, in each mailing to potential subscribers. Using a fractional factorial design (Chapter 5) the firm was able to test seven factors simultaneously in a single mailing, gaining valuable and immediate insights that led to large increases in response. Moreover, the results were attained with a sample size (the number of people receiving the mailing) that was much smaller than would have been needed if the seven factors had been tested one factor at a time.

A leading office supplies retailer designed and implemented an e-mail test targeted at small business customers, a group the retailer wanted to attract to their stores and Web site. The retailer identified 13 factors that it wanted to include in this experiment, with each factor having two possible values. The factors included the background color of the e-mail (white or blue), a discount offer (normal price or 15% discount), a free gift (no gift or a pen-and-pencil set), and products pictured (few or many). Testing all possible combinations of the 13 factors would have required $2^{13} = 8,192$ different e-mail designs! But using a fractional factorial design, a methodology that we discuss in Chapter 5, the firm was able to successfully test all 13 factors with just 32 different designs.

Peak Electronics, a manufacturer of printed circuit boards, was faced with a recurring problem. In the circuit board production process, most of the holes on each board are plated with a thin layer of copper so that current can flow from one side to the other. Some holes, however, are not meant to be plated and instead are *tented*, meaning that they are protected by a thin layer of photographic film. During the manufacturing process, a significant number of these tents were breaking, and their holes were being plated. The result was the number-one cause of rework at the firm, because the copper in these holes had to be scraped out.

At the time, Peak was using film supplied by Dupont. The sales representative of Hercules, a competing filmmaker, suggested that Peak perform an experiment using the Hercules film to test the effect on broken tents of a number of key manufacturing variables. The sales representative designed the test and helped Peak analyze the results.

With the explosive growth of the Internet, Web site design has become an important issue, as firms attempt to attract a greater number of people to visit their sites and order their products or services. PhoneHog is a subscription-based service through which consumers get free long-distance phone calls. Participants sign up for the program and earn phone minutes by visiting Internet sites, entering sweepstakes, or trying new products and services. The PhoneHog case in Chapter 8 describes how experimental design can be used to improve a Web site to obtain more customers. In this case there were 10 factors to be tested with the number of variations, or levels, for each factor ranging between 2 and 10. For example, the top image on the Web page had four possible designs: (1) photos of five people talking on the phone with the PhoneHog logo on the right, (2) a cartoon image of a pig peeking through the O in the PhoneHog logo on a blue background, (3) the same image of the pig on a white background, and (4) the photos of the five people talking on the phone with a different PhoneHog logo on the right. If every possible combination of factor levels were included in the experiment, a total of 1,658,880 test Web pages would have been required. In fact, the experiment consisted of just 45 different Web pages (each page a combination of factor levels), with each person arriving to the site randomly assigned to one of them. The number of visitors to the site and the number of visitors who click on an icon to request additional information were recorded. As a result of this experiment, the click-through rate, which is the number of clicks divided by the number of visitors, increased by 35%.

1.3 A BRIEF HISTORY OF EXPERIMENTAL DESIGN

The field of experimental design began with the pioneering work of Sir Ronald Fisher, whose classic book, *The Design of Experiments*, was published in 1935. Fisher was responsible for statistical analysis at an agricultural experiments station in England, and his early work on experimental design was applied to improving crop yields and solving other agricultural problems. Over the years, applications of experimental design to industrial problems have been widespread, with particular attention given to problems in the chemical industry, such as maximizing chemical yields and assays. In 1978, George E. P. Box, William G. Hunter, and J. Stuart Hunter published *Statistics for Experimenters* (second edition, 2005), a book that became, and still is, a standard text in the field.

Beginning about 1980, U.S. manufacturing firms, faced with competitive challenges, especially from Japanese companies, took a renewed interest in quality management and design of experiments. This period spurred renewed interest among U.S. manufacturers in experimental design, and in the 1980s the American Society for Quality (ASQ) and many other organizations started to offer numerous seminars on experimental design. However, little or no attention was given to the application of experimental design to service organizations.

More recently that has slowly begun to change, and several articles have appeared showing that multivariable experimental design techniques provide powerful approaches to service problems. "The New Mantra: MVT" (*Forbes*, March 11, 1996) discussed the experimental design applications to services by a quality consulting firm, while "Tests Lead Lowe's to Revamp Strategy" (*Wall Street Journal*, March 11, 1999) explained how that firm helped Lowe's improve its advertising policy. The article, "Boost Your Marketing ROI with Experimental Design" (Almquist and Wyner, *Harvard Business Review*, October 1, 2001), told how another consulting firm used experimental design to improve marketing decisions. In short, business leaders are beginning to realize that experimental design has widespread applications to management decision making, particularly in service organizations.

1.4 OUTLINE OF THE BOOK

Chapter 2 presents important basic concepts of probability and statistics. It is meant to be a concise review and provides the common language and notation that we use throughout the book. While writing it, we assumed that readers have previously had an exposure to most of the material covered in the chapter. We discuss a number of important distributions such as the binomial, normal, t-, and F-distributions. In subsequent chapters, they are used extensively. We also discuss useful tools for displaying data such as dot plots, histograms, and scatter diagrams. Chapter 2 also shows how confidence intervals and tests of hypotheses are constructed based on sample information and are used to make inferences about a population mean or the difference in two population means. These important statistical tools are applied in later chapters to identify statistically significant factors.

In Chapter 3, we extend the discussion in Chapter 2 and focus on comparing more than two population means. For example, we might want to compare the effectiveness of three different advertising strategies by testing them in a number of stores. We present two statistical models: the completely randomized design and the randomized block design. In Chapter 3, we emphasize two important ideas, randomization and blocking, that are used throughout the book.

The heart of the book begins with Chapter 4, where we focus on so-called 2-level factorial designs. In these designs, there are k factors to be tested, and each factor is studied at two different values (levels). For example, in a Web site test, one factor might be the banner headline (version 1 vs. version 2), while another might be the image under the headline (product photo vs. happy user). In the full factorial design, the experimenter tests all combinations of factors and levels, with each combination called a run. With k factors, there are 2^k runs. Thus, testing two factors requires $2^2 = 4$ runs, testing three factors requires $2^3 = 8$ runs, and so forth. The main effect of a factor is the difference in response at one level of the factor versus the other. For example, for the image under the banner headline, the main effect of that factor is the difference in response if the happy user image is employed rather than the product photo. In some instances, there may be an interaction between factors. For example, the difference in response between version 1 and version 2 of the banner headline may depend on which image under the banner is used. In Chapter 4, we show how main and interaction effects are estimated and discuss the various approaches for determining which effects are statistically significant.

The focus of Chapter 5 is on 2-level fractional factorial designs. Full factorial designs are useful for experimenting with relatively few factors. As the number of factors increases, the number of runs required in a full factorial design increases dramatically. In fact, the inclusion of each additional factor in a full factorial design doubles the number of runs required, with 4 factors requiring $2^4 = 16$ runs, 5 factors requiring $2^5 = 32$ runs, and so forth. If full factorials were the only option, the experimental design approach would have limited value. In a fractional factorial design, the experiment requires only a fraction of the number of runs needed for a full factorial design. For example, a full factorial design with seven factors requires $2^7 = 128$ runs, or separate experiments. But as we shall see, it is possible to construct a fractional design requiring only 16 runs that provides nearly as much information as in a full factorial design. In some instances, a fractional experiment may produce results that are difficult to interpret. We show how a follow-up experiment can be designed and executed to resolve these ambiguities.

In Chapter 6, we discuss Plackett-Burman designs. The number of runs required in a fractional factorial design is a power of 2. Thus, the number of runs would be 8, 16, 32, 64, and so forth. In a Plackett-Burman design, the number of runs required is a multiple of 4, so the number of runs would be 4, 8, 12, 16, and so forth. For example, in a particular situation, if the experimenter were limited to fractional factorial designs, she might have to choose between a design of 16 runs and a design of 32 runs. There is a rather large gap between allowable run sizes. The Plackett-Burman designs give the experimenter additional options that may be advantageous. We discuss the characteristics of Plackett-Burman designs and illustrate their use with several case examples.

The designs in Chapters 4 through 6 are all 2-level designs, with each factor being set at one value or another. In Chapter 7, we extend the analysis to include designs in which factors may be at more than two levels. We show how regression analysis can be used to estimate effects, and we discuss the construction and analysis of simple fractional designs that include factors at more than two levels.

The last chapter of the book, Chapter 8, is devoted mainly to the most advanced topic. The designs discussed in earlier chapters have an important property called *orthogonality*. In an orthogonal design, effects are estimated independently of one another. That means that the particular estimate of one effect is not influenced by the estimated value of another. In Chapter 8, we consider nonorthogonal designs involving many factors and several levels. We show how regression analysis can be used to analyze these designs, and we illustrate the approach with the PhoneHog case, which was described earlier in this chapter. Chapter 8 ends with a discussion of experimental design software focusing on two software products, Minitab and JMP.

1.5 NOBODY ASKED US, BUT . . .

R. A. Fisher, The Life of a Scientist, is an interesting biography of Sir Ronald Fisher written by his daughter, Joan Fisher Box (1978). Fisher is one of the statisticians included in *The Lady Tasting Tea: How Statistics Revolutionized Science in the Twentieth Century*, by David Salsburg (2001). The title of the book comes from a paper that Fisher wrote, which is included in Fisher's *The Design of Experiments*. As the story goes, a lady claimed that by tasting it she could tell whether milk or tea was put into the cup first. Fisher designed an experiment to test her claim. Salsburg's book has stories of other great statisticians including William Gossett, famous for the t-distribution, which we discuss in Chapter 2. The online (and free) encyclopedia Wikipedia (www.wikipedia.org) has interesting biographical information on Fisher, Gossett, and many other important figures in the world of statistics.

The NBC television white paper "If Japan Can, Why Can't We?" which was broadcast in 1980, was a milestone that marked the beginning of a quality revolution in manufacturing in the United States. W. Edwards Deming was featured on the program, and he castigated American firms for shoddy quality. Deming, a statistician with a Ph.D. in physics, gave a series of lectures in Japan in 1950 that greatly influenced that country's quality efforts. The Deming prize, the highest award for quality in Japan, is named in his honor. Deming's (1982) book, *Out of the Crisis*, is a good source for learning about his quality management ideas.

Not long after that NBC program, the work of Genichi Taguchi, a Japanese consultant and former professor, began receiving widespread attention from manufacturers in the United States, particularly in the automobile industry. Taguchi methods became a familiar buzzword for his approaches to experimental design. Statisticians have often criticized Taguchi's statistical methods, but there is general agreement that his engineering ideas are very useful. He is probably best known for two concepts: robust design and the Taguchi loss function. *Robust design* means designing a product or process that is insensitive to environmental factors. For example, a robust cake recipe would produce a good cake even with considerable variation in baking time and oven temperature. The *Taguchi loss function* is an appealing alternative to the traditional approach to determine whether a product or process meets customer specifications. For example, to illustrate the traditional approach, suppose the plating thickness in millimeters of a printed circuit board is acceptable if it falls within certain upper and lower specification limits. So, a board having thickness just below the upper specification would be judged acceptable, whereas a board whose thickness was just above that limit would be classified as defective. In reality, there is a target that is ideal, and the closer each board comes to that target, the better. In contrast, under Taguchi's loss function, the loss associated with an individual board would be equal to a constant times the squared deviation between the board's thickness and the target. With this function, doubling the distance from the target would quadruple the loss. In the traditional approach, where each board is either in or out of specifications, two processes might have the same fraction of boards meeting specifications but, in reality, very different quality levels. One process might have most of its acceptable boards with thicknesses close to the target, whereas the other might have a more uniform distribution with board thicknesses evenly spread within the window defined by the specification limits. This process would have much lower quality than the other, but under the traditional approach, the quality of products produced under the two systems would be judged as equal.

In recent years, the Six Sigma approach to quality has been embraced by numerous organizations. Six Sigma was originally developed at Motorola in the mid-1980s and refined first by Allied Signal and more recently by General Electric. Six Sigma has many similarities to total quality management (TQM) and other programs in the past, but it also has some distinctive characteristics. One is its focus on defining and responding to customer needs. In doing so, it takes a broader view of quality management compared to some more narrowly focused programs of the past, better integrating quality activities into all areas of the organization and aligning these activities with the strategic goals of the firm. In addition, Six Sigma programs have been more widely applied to service processes, including many implementations in hospitals and other health care organizations. The design of efficient experiments is an important component of the Six Sigma approach. One of the many books on Six Sigma is *The Six Sigma Way: How GE, Motorola, and Other Top Companies are Honing Their Performance*, by Peter S. Pande, Robert P. Neuman, and Roland R. Cavanagh (2000).

EXERCISES

Exercise 1 Search the Web for the work of Sir Ronald Fisher on experimental design, including his earliest efforts performing agricultural experiments at the Rothamsted Experimental Station in the United Kingdom.

Exercise 2 Read *Mother Jones* (Case 3 in the case study appendix) and *Peak Electronics: The Broken Tent Problem* (Case 4). Both of these cases describe a company's first exposure to experimental design methods.

(a) *Mother Jones*: Suppose the organization wanted to test each of the seven factors in a separate mailing. What specific shortcomings would this approach have compared to the approach in the case?

(b) *Peak Electronics*: Suppose the company did not use experimental design to examine and solve the broken tent problem. Imagine how they would have approached the problem instead. What difficulties would they have likely encountered? Would it have been possible to identify interactions between factors? If so, how?

Exercise 3 Pick a Web site on the Internet. Suppose you were designing an experiment for increasing visitors' response to a product or service offered on the site. What seven factors do you think would be most important to test? In each case, if possible, specify two levels (values) for each of the factors.

2 | A REVIEW OF BASIC STATISTICAL CONCEPTS

2.1 INTRODUCTION

This chapter reviews basic concepts that we use in the remainder of this book. Section 2.2 reviews discrete and continuous probability distributions including two important special cases, the binomial and normal distributions. In Section 2.3, we focus on the graphical display and numerical summary of information. Topics covered include bar and pie charts for categorical data; dot diagrams; histograms and scatter plots for continuous data; and summary measures including the mean, median, standard deviation, and correlation coefficient. In Section 2.4, we discuss sampling and random sampling, and in Section 2.5, we review the basics of statistical inference. We discuss confidence intervals and hypothesis tests for a single mean and a single proportion, and determine the sample size that is required for estimates to achieve a given level of precision. We also address the comparison of two populations, using data from the completely randomized as well as the randomized block experiment. A case study on the effectiveness of two advertising strategies completes the chapter.

2.2 PROBABILITY DISTRIBUTIONS

The world is uncertain, and measurements on products and processes vary. Probability distributions describe the variability among the measurements.

Random variables are variables whose outcomes are uncertain. For example, the purchasing response of a customer who receives a catalog or an e-mail offer can be "yes" or "no"—or in coded form, 1 for "yes," and 0 for "no." Similarly, the soldering quality of a circuit board, expressed in terms of the number of flaws, is a random variable. The board may have zero flaws, exactly one error, two errors, and so on.

Random variables with a discrete number of possible outcomes (in the first example, 0 and 1; in the second example, 0, 1, 2, . . .) are called *discrete random variables*. We use discrete probability distributions to describe the uncertainty. Later in this chapter, we discuss the binomial distribution, the most important discrete distribution.

Variables such as the length or the width of a product, the amount spent on purchases, the commuting time to work, the gas mileage of a car, or the yield of a process are continuous in nature. Here, any number—obviously within a certain interval—is a feasible outcome. We call such random variables *continuous random variables*, and we use continuous distributions to characterize the variability. The normal distribution, the t-distribution, and the F-distribution are important examples.

2.2.1 Discrete Random Variables

The distribution of a discrete random variable is described by the

- collection of possible distinct outcomes, and
- their associated probabilities. Probabilities are numbers between 0 and 1, and the sum of the probabilities over all possible outcomes must be 1. The probabilities may represent prior beliefs, come from previous studies, or be implied by a theoretical model.

It is standard and useful notation to use capital letters to denote the random variable $(X, Y, Z, \dots)$, and lowercase letters $(x, y, z, \dots)$ to denote the possible outcomes. The notation $P[Y = y]$ stands for the probability that the random variable Y takes on the value y.

Example 1 The random variable Y describes a customer's purchasing decision. Possible outcomes are 1 (purchase) and 0 (no purchase). Based on historical data, it is estimated that 5% of customers will place an order. Thus, $P[Y = 1] = 0.05$, and hence $P[Y = 0] = 1 - 0.05 = 0.95$.

Example 2 The random variable Y is the number of flaws on a circuit board produced on an assembly line. The possible outcomes are

$$y = 0 \text{ (no flaw)}$$

$$y = 1 \text{ (exactly one flaw)}$$

$$y = 2 \text{ (exactly two flaws), and so on}$$

The following probabilities are given: $P[Y = 0] = 0.90$, $P[Y = 1] = 0.08$, and $P[Y = 2] = 0.02$. This probability distribution implies that producing an item having three or more flaws is impossible.

Example 3 Let the random variable Y be the number showing on a thrown die. The possible outcomes are $y = 1, 2, 3, 4, 5, 6$. Assuming the die is fair, the outcomes are equally likely. Hence $P[Y = 1] = P[Y = 2] = \dots = P[Y = 6] = 1/6$.

Example 4 Let the random variable Y be the number of times a customer orders from a catalog during a specified time period. The possible outcomes are $y = 0, 1, 2, 3$, with $P[Y = 0] = 0.2$, $P[Y = 1] = 0.5$, $P[Y = 2] = 0.2$, $P[Y = 3] = 0.1$. Note that the probabilities sum to 1. Notice also that ordering four or more times has zero probability; it cannot occur.

We can easily calculate the probabilities of various events. For example, the probability of at most two orders is given by $P[\text{at most } 2] = P[Y \leq 2] = P[Y = 0 \text{ or } Y = 1 \text{ or } Y = 2] =$

$P[Y = 0] + P[Y = 1] + P[Y = 2] = 0.2 + 0.5 + 0.2 = 0.9$. Similarly, the probability of at least one order is $P[Y \geq 1] = P[Y = 1 \text{ or } Y = 2 \text{ or } Y = 3] = P[Y = 1] + P[Y = 2] + P[Y = 3] = 0.5 + 0.2 + 0.1 = 0.8$. Alternatively, $P[Y \geq 1] = 1 - P[Y < 1] = 1 - P[Y = 0] = 1 - 0.2 = 0.8$.

Mean of a Discrete Distribution

The *mean* of a discrete distribution (also called its *expected value*) with outcomes y and probabilities $P[Y = y]$ is given by

$$\mu = \sum_{y} y P[Y = y]$$

We use the Greek letter μ to denote the mean. It is the weighted sum of the possible outcomes, with each outcome weighted by its probability. The mean or expected value is the long-run average.

Example 1 $\mu = (0)(0.95) + (1)(0.05) = 0.05$. The mean is 0.05.

Example 2 $\mu = (0)(0.90) + (1)(0.08) + (2)(0.02) = 0.12$. The expected number of flaws is 0.12.

Example 3 $\mu = (1)(1/6) + (2)(1/6) + \cdots + (6)(1/6) = 3.5$. The mean is 3.5.

Example 4 $\mu = (0)(0.2) + (1)(0.5) + (2)(0.2) + (3)(0.1) = 1.2$. The company expects, on average, 1.2 orders per customer. Of course, the number of orders can only be an integer; however, in the long run (i.e., over many customers) the number of orders averages to 1.2.

Variance of a Discrete Distribution

$$\sigma^2 = \sum_{y} (y - \mu)^2 P[Y = y]$$

The variance, denoted by the Greek letter sigma squared (σ^2) is a measure of spread. It is the weighted sum of squared deviations from the mean, with squared deviations weighted by their probability of occurrence.

Standard Deviation of a Discrete Distribution

$$\sigma = \sqrt{\sum_{y} (y - \mu)^2 P[Y = y]}$$

The standard deviation is equal to the square root of the variance. If the units of the random variable are, say, dollars, the variance will be in units of dollars squared. That makes the variance difficult to interpret. Taking the square root of the variance to obtain the standard deviation expresses the spread of the distribution in the same units as the random variable— in this case, dollars.

Example 1 $\sigma = [(0 - 0.05)^2(0.95) + (1 - 0.05)^2(0.05)]^{0.5} = [0.0475]^{0.5} = 0.218$.

Example 2 $\sigma = [(0 - 0.12)^2(0.90) + (1 - 0.12)^2(0.08) + (2 - 0.12)^2(0.02)]^{0.5} = [0.1456]^{0.5} = 0.382$ (flaws).

Example 3 $\sigma = [(1 - 3.5)^2(1/6) + (2 - 3.5)^2(1/6) + \cdots + (6 - 3.5)^2(1/6)]^{0.5} = [2.9167]^{0.5} = 1.71$.

Example 4 $\sigma = [(0 - 1.2)^2(0.2) + (1 - 1.2)^2(0.5) + (2 - 1.2)^2(0.2) + (3 - 1.2)^2 (0.1)]^{0.5} = [0.76]^{0.5} = 0.87$ (orders).

The Binomial Distribution

The binomial is the most important discrete probability distribution. A binomial situation is one that is analogous to repeatedly tossing a coin (not necessarily a fair one) and counting the number of heads. The purchasing decision of buying or not buying, or the outcomes of a pass/fail inspection, can be viewed as the outcomes of such coin tosses.

The assumptions are as follows:

- Each individual experiment (also called a trial) can result in only one of two outcomes. We refer to the outcomes as success (S) and failure (F). We assume that the probability of a success is $P(S) = \pi$ for each trial, and hence the probability of a failure is $P(F) = 1 - \pi$.

- There are n such independent trials. Independence means that the outcome of any trial does not affect the outcome of any other trial.

- The random variable Y represents the number of successes in n independent trials.

The random variable Y has outcomes $y = 0, 1, 2, \ldots, n$. The probabilities associated with these $n + 1$ outcomes are given by the binomial formula

$$P[Y = y] = \frac{n!}{y!(n - y)!}\pi^y(1 - \pi)^{n-y} \quad \text{for } y = 0, 1, 2, \ldots, n$$

Here y factorial is defined as $y! = (1)(2) \ldots (y - 1)(y)$. For example, $3! = (1)(2)(3) = 6$, and $5! = (1)(2)(3)(4)(5) = 120$. By definition, $0! = 1$. The number of trials n and the probability of success in a single trial π are called the *parameters of the binomial distribution*.

It can be shown that the mean of the binomial distribution is given by

$$\mu = n\pi$$

The standard deviation is given by

$$\sigma = \sqrt{n\pi(1 - \pi)}$$

The binomial distribution is tabulated in statistics textbooks. Also, its probabilities can easily be determined using functions in computer packages such as Excel or Minitab.

Example   Assume that a production process is characterized by a 10% defect rate. That is, the probability of producing a defective item is $P[\text{defective}] = 0.10$; the probability of producing a good item is $P[\text{good}] = 0.9$. Assume also that the quality of each item (defective or not) is independent of the quality of every other item.

Assume that as part of a sampling inspection program, $n = 10$ items are selected at random. The distribution of the number of defectives Y in a sample of size $n = 10$ items is binomial with parameters $n = 10$ and $\pi = 0.1$. The mean number of defectives is $\mu = (10)(0.1) = 1$; there will be one defective, on average. The standard deviation is $\sigma = \sqrt{10(0.1)(0.9)} = 0.9487$. Individual probabilities such as

$$P[Y = 2] = \frac{10!}{2!8!}(0.1)^2(0.8)^8 = 0.1937$$

can be calculated either from the expression above, or from the binomial function of readily available computer programs. Note that the probabilities, summed over the possible outcomes, add to 1. Computer programs also calculate the cumulative probabilities such as $P[Y \le 1] = P[Y = 0] + P[Y = 1] = 0.3487 + 0.3874 = 0.7361$, or in general

$$P[Y \le y] = \sum_{i=0}^{y} P[Y = i] \quad \text{for } y = 0, 1, 2, \ldots, n$$

The Excel function BINOMDIST(y, n, π, FALSE) returns $P[Y = y]$, if n is the number of trials and π is the probability of success. Replacing FALSE with TRUE returns the cumulative probability, $P[Y \le y]$. In Minitab, the calculations are carried out by using the convenient pull-down menu "Calc > Probability Distributions > Binomial."

Probabilities such as

$$P[1 \le Y \le 3] = P[Y = 1] + P[Y = 2] + P[Y = 3] = P[Y \le 3] - P[Y \le 0]$$
$$= 0.9872 - 0.3487 = 0.6385$$

can be calculated by summing the individual probabilities, or as the difference of two cumulative probabilities.

2.2.2 Continuous Random Variables

A continuous random variable Y is described by its *probability density function f(y)*, which is nonnegative. For any density function, the area under the curve described by the density function is equal to 1. The probability that the random variable falls between two constants, a and b, is the area under the density curve between a and b. That is,

$$P[a \le Y \le b] = \int_{a}^{b} f(y)\,dy = P[Y \le b] - P[Y \le a]$$

Percentiles of the distribution are defined by cumulative probabilities. The $(100p)$th percentile is given by y_p, the value of the random variable for which the area under the curve from $-\infty$ to y_p equals p; that is, $p = P[Y \le y_p]$.

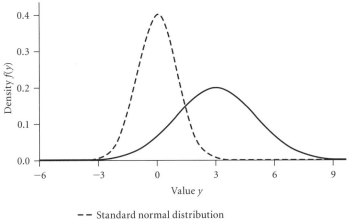

Figure 2.1   Densities of Two Normal Distributions

The probability that a continuous random variable is exactly equal to a particular value is zero; that is, $P[Y = a] = 0$, for any a. Hence,

$$P[a \le Y \le b] = P[a \le Y < b] = P[a < Y \le b] = P[a < Y < b]$$

The Normal Distribution

The normal distribution is the most important distribution in statistics. It is characterized by two parameters: its mean μ and standard deviation σ. The distribution is symmetric around the mean and bell-shaped. The standard deviation σ determines the spread of the distribution.

Densities of two normal distributions are shown in Figure 2.1: the so-called standard normal distribution with mean 0 and standard deviation 1, and the normal distribution with mean 3 and standard deviation 2. For any normal distribution about 68% of the values will fall within 1 standard deviation of the mean, about 95% of the values will fall within 2 standard deviations of the mean, and 99.7% of the values will fall within 3 standard deviations of the mean.

A random variable that follows a *standard normal distribution* (mean 0 and standard deviation 1) is denoted by the capital letter Z. The probability density of the standard normal distribution, $f(z)$, is shown in Figure 2.2. Cumulative probabilities can be looked up in the table of the standard normal distribution (the "z-table"), or they can be obtained by pushing certain buttons on advanced calculators or by executing appropriate functions of statistical computer software. The Excel function NORMSDIST(z) returns the cumulative probability, the area under the standard normal curve below the value z. For example:

$$P[Z \le 0] = 0.5$$

$$P[Z \le -1] = 0.1587$$

$$P[Z \le -0.6] = 0.2743 \quad \text{and} \quad P[Z \ge -0.6] = 1 - P[Z \le -0.6] = 1 - 0.2743 = 0.7257$$

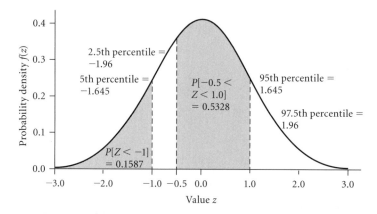

Figure 2.2 Density of the Standard Normal Distribution

$$P[Z \leq 0.7] = 0.7580 \quad \text{and} \quad P[Z \geq 0.7] = 1 - P[Z \leq 0.7] = 1 - 0.7580 = 0.2420$$

$$P[-0.5 \leq Z \leq 1.0] = P[Z \leq 1.0] - P[Z \leq -0.5] = 0.8413 - 0.3085 = 0.5328$$

$$P[0.4 \leq Z \leq 1.2] = 0.8849 - 0.6554 = 0.2295$$

Important percentiles of the standard normal distribution are

2.5th percentile $z_{0.025} = -1.96$ and 97.5th percentile $z_{0.975} = 1.96$

5th percentile $z_{0.05} = -1.645$ and 95th percentile $z_{0.95} = 1.645$

Suppose Y has a normal distribution with mean μ and standard deviation σ, and that for any value a, we want the probability that Y is less than or equal to a. We convert the probability statement about Y into an equivalent statement about Z. We have

$$P[Y \leq a] = P\left[\frac{Y - \mu}{\sigma} \leq \frac{a - \mu}{\sigma}\right] = P\left[Z \leq \frac{a - \mu}{\sigma}\right]$$

Note that on either side of the inequality we subtract the mean μ and divide by the standard deviation σ. The random variable $Z = (Y - \mu)/\sigma$ follows a standard normal distribution, and the probability in the above equation can be looked up in the z-table. Similarly,

$$P[a \leq Y \leq b] = P\left[\frac{a - \mu}{\sigma} \leq \frac{Y - \mu}{\sigma} \leq \frac{b - \mu}{\sigma}\right] = P\left[\frac{a - \mu}{\sigma} \leq Z \leq \frac{b - \mu}{\sigma}\right]$$

$$= P\left[Z \leq \frac{b - \mu}{\sigma}\right] - P\left[Z \leq \frac{a - \mu}{\sigma}\right]$$

Example The weight of toothpaste Y in a 2.7 ounce tube follows a normal distribution with mean $\mu = 2.8$ and standard deviation $\sigma = 0.05$. The fraction of underfilled tubes is

$$P[Y \leq 2.7] = P\left[\frac{Y - 2.8}{0.05} \leq \frac{2.7 - 2.8}{0.05}\right] = P[Z \leq -2.00] = 0.0228$$

Statistical software allows us to obtain cumulative probabilities for any normal random variable directly, without the conversion to the standard normal. The Excel function NORMDIST(y, μ, σ) returns the cumulative probability, that is, the area under the density curve below y, for a normal random variable with mean μ and standard deviation σ.

Some percentiles of the normal distribution with mean μ and standard deviation σ are the following:

50th percentile $y_{0.50} = \mu$

5th percentile $y_{0.05} = \mu - (1.645)\sigma$ and 95th percentile $y_{0.95} = \mu + (1.645)\sigma$

2.5th percentile $y_{0.025} = \mu - (1.96)\sigma$ and 97.5th percentile $y_{0.975} = \mu + (1.96)\sigma$

99th percentile $y_{0.99} = \mu + (2.326)\sigma$

The Excel function NORMINV(p, μ, σ) returns the $(100p)$th percentile of a normal distribution with mean μ and standard deviation σ. For example, suppose a random variable Y has a normal distribution with mean 100 and standard deviation 20. Then NORMINV (0.99, 100, 20) = 146.53, and $P[Y \leq 146.53] = 0.99$. In Minitab, cumulative probabilities and percentiles (referred to as "inverse" cumulative probabilities) of a normal distribution are obtained with the pull-down menu "Calc > Probability Distributions > Normal."

The t-Distribution

The t-distribution has one parameter $\nu > 0$, called its degrees of freedom. In most statistical applications, the parameter ν is a positive integer.

The t-distributions are very similar to the standard normal distribution. They are symmetric around mean 0, and their densities resemble the bell-shaped curve of the normal. The only difference is that the tails of t-distributions are slightly heavier than those of the standard normal (as explained later). The standard deviation of the t-distribution with $\nu > 2$ degrees of freedom is given by $\sigma = \sqrt{\nu/(\nu - 2)}$.

Figure 2.3 compares the densities of the t-distributions with 3 and 10 degrees of freedom to the density of the standard normal. Notice that for very large or small values y of the random variables, the densities and hence the tail areas are larger for the t-distributions compared to the normal. This gives t-distributions a somewhat larger chance to generate large deviations from the mean. The t-distribution converges to the standard normal as the degrees of freedom approach infinity.

Percentiles and cumulative probabilities of the t-distribution can be calculated using Excel or any other statistical software. With Excel, percentiles are found using the TINV function. The user specifies α, the area in *both* tails of the distribution, and the number of degrees of freedom ν. Thus $\alpha/2$ is the upper tail probability and $1 - (\alpha/2)$ is the corresponding cumulative probability. For example, for a t-distribution with 3 degrees of freedom, TINV(0.10, 3) returns the value 2.3534, which is the 95th percentile of the distribution. Other examples are

95th percentile of $t(10)$: $t_{0.95}(10) = 1.8125$

95th percentile of $t(\infty)$, the standard normal: $t_{0.95}(\infty) = 1.645$

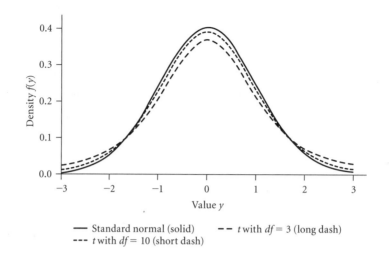

— Standard normal (solid)   – – t with df = 3 (long dash)
--- t with df = 10 (short dash)

Figure 2.3   Densities of the Standard Normal and Two t-Distributions

97.5th percentile of $t(3)$: $t_{0.975}(3) = 3.1824$

97.5th percentile of $t(10)$: $t_{0.975}(10) = 2.2181$

97.5th percentile of $t(\infty)$, the standard normal: $t_{0.975}(\infty) = 1.96$

The Chi-Square Distribution

The chi-square distribution is a skewed distribution with values from 0 to ∞. It has one parameter ν, its degrees of freedom, which is a positive integer. We write the distribution as $\chi^2(\nu)$, with the symbol χ denoting the Greek lowercase letter chi.

Figure 2.4 shows the densities of three chi-square distributions, with 3, 6, and 10 degrees of freedom. The mean of the chi-square distribution is the same as its degrees of freedom: $\mu = \nu$. The standard deviation is $\sigma = \sqrt{2\nu}$.

The F-Distribution

The F-distribution takes on values from 0 to ∞, and it is skewed to the right. It has two parameters, its degrees of freedom $\nu_1 > 0$ and $\nu_2 > 0$, which in most statistical applications are positive integers. We use the notation $F(\nu_1, \nu_2)$, to describe an F-distribution with ν_1 and ν_2 degrees of freedom.

The mean of the $F(\nu_1, \nu_2)$ distribution, $\mu = \nu_2/(\nu_2 - 2)$, depends only on ν_2, and it is always slightly larger than 1. The standard deviation depends on both parameters, ν_1 and ν_2.

Figure 2.5 shows densities of four F-distributions: $F(4, 10)$ and $F(4, 20)$, and $F(8, 10)$ and $F(8, 20)$. Percentiles and cumulative probabilities can be calculated with standard statistics software. For example, the 95th percentiles of these four F-distributions are

$$F_{0.95}(4, 10) = 3.4780 \quad F_{0.95}(4, 20) = 2.8661$$

$$F_{0.95}(8, 10) = 3.0717 \quad F_{0.95}(8, 20) = 2.4471$$

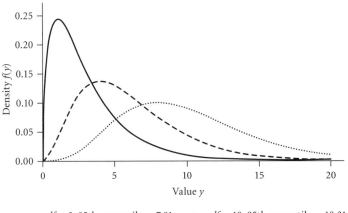

— $df = 3$: 95th percentile = 7.81   ⋯⋯ $df = 10$: 95th percentile = 18.31
-- $df = 6$: 95th percentile = 12.59

Figure 2.4   Densities of Three Chi-Square Distributions

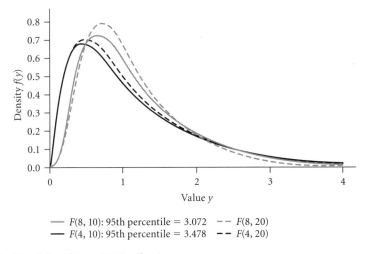

— $F(8, 10)$: 95th percentile = 3.072   -- $F(8, 20)$
— $F(4, 10)$: 95th percentile = 3.478   -- $F(4, 20)$

Figure 2.5   Densities of Four F-Distributions

2.3 DESCRIBING DATA

In this book, we focus on methods for designing experiments and analyzing the resulting data. As part of this process, simple graphical displays such as data plots and histograms, and summary measures such as the mean, median, and standard deviation, provide extremely useful complements to the more formal statistical methodology. In this section, we discuss these simple tools for displaying, summarizing, and analyzing data. In most cases the data are a sample from a larger underlying population. Occasionally, if the population is small, the data will consist of all of its elements.

Categorical data are observations that are grouped into qualitative categories. Examples are marital status (single, married, divorced, widowed), advertising media (radio, television, print), and type of real estate (residential, commercial). For categorical data, we can calculate relative frequencies of observed outcomes. For example, it may be that among 500 clients who received an advertising message, 20 made a purchase and 480 did not. Then the (sample) proportion of clients who purchased is $p = 20/500 = 0.04$ (4%), and the proportion of clients who did not is $1 - p = 480/500 = 0.96$ (96%). We can display these proportions in a *bar chart* or a *pie chart*. For more than two outcomes, there are more proportions (adding up to 1), more bars in the bar chart, and more pie slices in the pie chart.

Continuous data, on the other hand, reflects measurements that can be any (possibly rounded) value within a certain interval. We display continuous measurement data using *dot diagrams* and *histograms*. In dot diagrams, each measurement is displayed as a dot on a line graph (the x-axis). In a histogram, the observations are binned into nonoverlapping equal-width intervals on the x-axis and the frequencies (either absolute or relative) are displayed on the y-axis.

Statistical software makes it easy to construct bar and pie charts for categorical data and dot diagrams and histograms for continuous measurement data. Illustrative examples are shown at the end of this section.

Summary statistics are useful for describing data sets. The center (or location) of a data set is measured by the mean or median, while its variability is described best by the standard deviation or the interquartile range. Assume that we have a sample of n observations $y_1, y_2, \ldots, y_n$, such as the dollar purchases of n customers or the annual donations to a college made by n alumni. The arithmetic *mean* (average) is given by

$$\bar{y} = [y_1 + y_2 + \cdots + y_n]/n = \frac{\sum_{i=1}^{n} y_i}{n}$$

The *median* is the "middle" observation in rank. First, order the observations according to their size $y_{(1)} \leq y_{(2)} \leq \cdots \leq y_{(n)}$; the numbers in parentheses are the ranks. The median is the observation with rank $(n + 1)/2$. If this "middle" rank is not an integer, then the median is the average of the two observations with ranks adjacent to $(n + 1)/2$.

The *percentile of order p*, where p is a number between 0 and 1, is the observation with rank $(n + 1)p$. If this is not an integer, we take the average of the two observations with adjacent ranks. $100p\%$ of the observations are smaller than the percentile, while $100(1 - p)\%$ of the observations are larger.

The *range* is defined as the difference between the largest and the smallest observation:

$$\text{Range} = y_{(n)} - y_{(1)}$$

The *interquartile range* is the difference between the 75th percentile (the third quartile) and the 25th percentile (the first quartile):

$$\text{IQR} = y_{((n+1)0.75)} - y_{((n+1)0.25)}$$

The range is very sensitive to extreme observations. The interquartile range covers the middle 50% of the observations and is less sensitive to extreme values.

The *sample standard deviation* is the most commonly used measure of variability. For a sample of n observations, it is defined as

$$s = \sqrt{\frac{\sum_{i=1}^{n} (y_i - \bar{y})^2}{n - 1}}$$

The sample standard deviation is nonnegative; it is zero only if there is no variability and all observations are the same. The standard deviation approximates the "average" distance of the observations from their mean. In many data sets (reasonably symmetric and bell-shaped), the cumulative probabilities of the normal distribution will apply approximately and about 95% of the observations will fall within two standard deviations from the mean, while about 2/3 of the observations will fall within one standard deviation.

The square of the standard deviation results in the *sample variance.*

$$s^2 = \frac{\sum_{i=1}^{n} (y_i - \bar{y})^2}{n - 1}$$

The numerator, the sum of the squared deviations from the sample average, is referred to as the sum of squares, corrected for the mean. The denominator, $n - 1$, reflects the degrees of freedom of the sum of squares. The degrees of freedom of a sum of squares are the number of "independent" components that are needed for its calculation. The sum of the deviations from a sample average $\sum_{i=1}^{n}(y_i - \bar{y})$ is always zero, and consequently specifying any $n - 1$ deviations determines the final deviation; the value of the last deviation must equal the negative of the sum of the others. The division of the sum of squares by its degrees of freedom $n - 1$, instead of the number of observations n, results in a better estimate of the population variance σ^2. This issue is discussed further in our end-of-chapter notes. Of course, the division by $n - 1$ instead of n usually won't make a difference, provided, of course, that n is reasonably large.

Scatter diagrams display relationships between two measurement variables, and *correlation coefficients* measure the degree of their linear association. Assume that the data set contains n pairs of observations; for example, the family income (x_i) and the amount that is being donated to the college that was attended (y_i), for $i = 1, 2, \ldots , n$. The correlation coefficient

$$r = \frac{1}{n - 1} \sum_{i=1}^{n} \left(\frac{x_i - \bar{x}}{s_x} \right) \left(\frac{y_i - \bar{y}}{s_y} \right)$$

is always between -1 and $+1$. Its sign indicates the direction of the linear association. For positive values of r, above-average values on y tend to occur with above-average values on x. The absolute value of r indicates the strength of the linear association. A correlation of $+1$ occurs if the observations plotted on a scatter diagram lie on a straight line with positive slope. A correlation of -1 occurs if the observations plotted on a scatter diagram lie on a straight line with negative slope. A high correlation does not necessarily imply causality; this is especially relevant if one analyzes data from observational studies (as compared to data from designed experiments). Statistical software such as Excel and Minitab can be used to carry out the calculations. The Excel function CORREL(*array1*, *array2*) returns the correlation coefficient. The user enters the n pairs of observations into two columns of the spreadsheet, with *array1* being the cell range for one variable and *array2* being the cell range for the other.

2.3.1 Example: Alumni Donations

The file *contribution* (available on our Web site) summarizes the 2004 contributions received by a selective private liberal arts college in the Midwest. The college has a very large endowment and, like all private colleges, keeps detailed records on alumni donations. Here, we analyze the 2004 contributions of five graduating classes (the cohorts who have graduated in 1957, 1967, 1977, 1987, and 1997). The data set consists of $n = 1,635$ individuals. In addition to donations in 2004 and class, the data set includes several other variables such as donations made in previous years, gender, marital status, college major, subsequent graduate work, and whether alumni have attended a fundraising event. Not all variables are used in this example.

The summary statistics are shown next. The variable "donation" is categorical, with two outcomes: no donation and donation. The overall proportion of alumni donating to the college is given by 570/1,635 = 0.349, or 34.9%. Bar and pie charts are shown in Figure 2.6. The proportions of donors for the five cohorts are given in Table 2.1, and a bar chart of this information is shown in Figure 2.6.

We show a dot plot of the 570 donation amounts in Figure 2.7, where each dot represents up to seven individual measurements. It is useful to aggregate the information in the form of a histogram. The dot diagram shows that the distribution is skewed to the right with a very long right tail. The largest contribution is $14,655. To display the majority of the donations more clearly, we redraw the histogram for donations that are $2,000 or less. Furthermore, we stratify the histogram and show separate histograms for each of the five graduating classes. To bring out differences, we have drawn these five histograms on the same scale. The y-axes on these histograms represent relative frequencies.

Summary statistics are given in Table 2.1. For each year separately, we show the summary statistics for all donations, but also for donations that are less than $2,000, and donations that are $2,000 or more. The means are influenced by occasional large contributions. The comparisons of the five groups may be more meaningful after omitting these large donations that are difficult to predict, and focusing on donations that are less than $2,000. Alternatively, we can compare the cohorts in terms of their medians, which are not affected by rare large donations.

First and third quartiles of the donation amounts are calculated for each graduating year, and the information is displayed in Figure 2.7 through comparative box plots. Box plots have a box around the middle 50% of the observations (i.e., the observations between the first and third quartiles) and lines added that point to the extremes. These plots, as well as the information in Table 2.1, show that both the proportion of donors and the magnitude of the donations increase with the time since graduation.

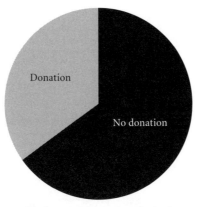

Pie chart: 2004 donations (no/yes)

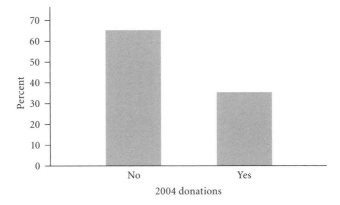

Bar chart: 2004 donations (no/yes)

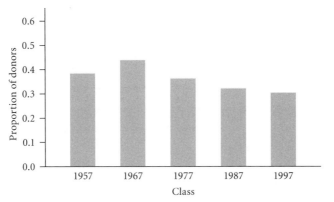

Bar chart: proportion of donors by class

Figure 2.6   Bar and Pie Charts of a Categorical Variable

TABLE 2.1
Summary Statistics of 2004 College Donations

PROPORTION OF ALUMNI DONATING

	1957	1967	1977	1987	1997	All
No donation	157	159	215	234	300	1,065
Donation	95	120	119	108	128	570
Percentage	37.7	43.0	35.6	31.6	29.9	34.9

PROPORTION OF ALUMNI DONATING AND THE MAGNITUDE OF ALUMNI DONATIONS

| | | | | ALL DONATIONS | | | | DONATIONS | | | | |
| | | | | | | | | <$2,000 | | | | ≥$2,000 | |
Year	Nu	NuDo	%Do	Mean	Median	StDev	Max	Nu	Mean	Median	StDev	Nu	Mean
1957	252	95	37.7	606	158	1,480	11,506	89	298	158	416	6	5,168
1967	279	120	43.0	559	158	1,804	14,655	113	211	152	238	7	6,156
1977	334	119	35.6	356	120	879	6,500	113	181	100	268	6	3,640
1987	342	108	31.6	246	90	470	2,716	106	201	80	341	2	2,608
1997	428	128	29.9	73	48	110	1,000	128	73	48	110	0	(no data)

PROPORTION OF ALUMNI DONATING AND EVENT ATTENDANCE

| | DONATIONS | | | |
	No	Yes	Total	Percentage Donating
No prior attendance	647	182	829	21.95
Prior attendance	418	388	806	48.14
All	1,065	570	1,635	34.86

MAGNITUDE OF ALUMNI DONATIONS AND EVENT ATTENDANCE

| | | | | | DONATIONS | | | | |
| Atten- dance | ALL DONATIONS | | | <$2,000 | | | ≥$2,000 | | |
	Nu	Mean	Median	Nu	Mean	Median	Nu	Mean	Median
No	182	134	50	182	134	50	0	No data	
Yes	388	460	100	367	210	100	21	4,820	3,000

We investigate whether attendance at alumni fund-raising events affects donations. It is reasonable to suppose that people who attend college functions are more likely to give. The information in Table 2.1 shows that almost 50% of those who attend fund-raising events are donating, while the proportion of donors among nonattending alumni is only 22%. Also, the magnitude of the donation increases if alumni attend such events.

A scatter plot of 2004 donations against 2003 donations is shown in Figure 2.7. We consider the $n = 410$ alumni who have given donations of $1,000 or less in both years. The scatter plot and the correlation coefficient $r = 0.812$ indicate that the magnitudes of donations in different years are strongly related.

2.4 SAMPLING ISSUES

An important objective of statistical analysis is to generalize findings that are based on a *sample* to the *population* from which the sample was drawn. The population consists of all

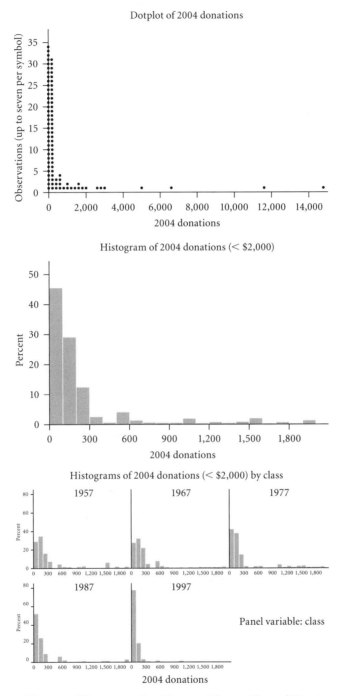

Figure 2.7 Dot Diagrams, Histograms, Box Plots, and Scatter Plots of Continuous Variables

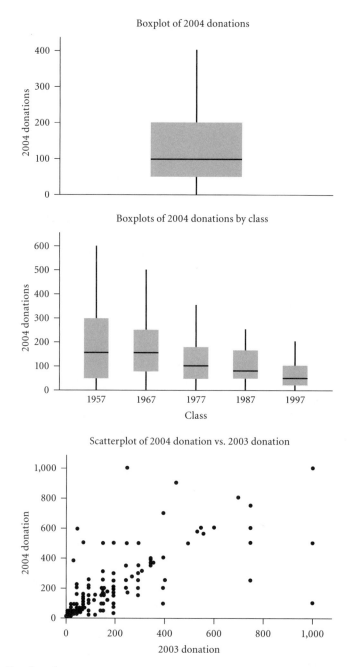

Figure 2.7 Continued

elements, whereas the sample consists of a subset of the population. It is important that the sample be representative of the population, otherwise reliable inferences about characteristics of the population would not be possible. Characteristics of the population are usually referred to as *parameters*, and summaries that are calculated from the sample are referred to as *sample statistics*.

Consider the population of all graduating seniors at State University. There are about one thousand each year; we denote the population size by $N = 1,000$. We may be interested in population characteristics such as the average grade point average (GPA), the average number of weekly study hours, and the proportion of smokers (a percentage) among graduating seniors at State University. These characteristics can be determined without any uncertainty if we are willing and able to collect information on *all* graduating seniors. We call this a *census*. Of course, asking students about this information may be subject to error; some respondents may not tell the truth.

If the population is large, a census is not feasible, and sampling becomes an alternative. The sample size is denoted by n; usually it is much smaller than the population size N. A *random sampling* method guarantees that the sample results are "representative." In this case, one can use statistical tools to assess the likely size of the resulting sampling error. Random sampling guarantees that each possible sample has the same likelihood of being selected. For a large population size N and a small sample size n, many samples are possible; in fact, there are $\binom{N}{n} = \dfrac{N!}{n!(N-n)!}$ different samples. Under random sampling, each of these samples is equally likely to be selected.

How is a random sample drawn from a population of elements? A simple approach is to prepare slips of paper—one for each element in the population (slips with numbers 1 through N), put them into a box, mix them thoroughly, and draw n items one after the other without replacement. Obviously, this method would only be practical for sampling from very small populations. In all other cases, a numbered list of all elements in the population (the sampling frame) and computer-generated random numbers would be used. Minitab's function "Calc > Random Data > Sample From Columns" makes it very easy to select n items at random and without replacement from a column containing a list of N distinct items.

In experiments in which two or more groups are compared, several independent random samples may have to be drawn. Assume that we want to study the effectiveness of two different online experimental design tutorials, which we identify as A and B. A group of State University seniors will complete the tutorial and then take an exam. Suppose we want 30 subjects in each group and want the two groups to be different (i.e., no overlap). As already noted, there are 1,000 graduating seniors at the school. Assume that at State University all seniors graduate, so that 1,000 students are available to take the tutorial during the school year. We enter the names of the students into a column of length 1,000 and select 60 of them at random and without replacement by executing the Minitab command "Calc > Random Data > Sample From Columns." The first 30 students in the sample become the students for tutorial A, and the second group of 30 students use tutorial B.

Assume that gender plays a role. A sampling strategy such as the one just discussed may not be optimal because it could lead to an unbalanced gender composition in the sample. The student body at State includes about the same number of men and women. However, it could be—by bad luck of the draw—that the first sample for A includes only 40% women, while the second for B includes 65%. It is better to take *stratified random samples*. From the 500 women, select at random 30 and randomly divide the 30 into two groups of 15 to receive A or B. The same is done with the 500 men.

2.5 STATISTICAL INFERENCE

2.5.1 Central Limit Effect for Averages

Suppose we have a very large population, and the random variable of interest Y is continuous in nature and varies around a certain unknown mean μ, with standard deviation σ. Our objective is to estimate the unknown mean μ from the results of a random sample of size n.

Many different samples of size n from N elements are possible, and each one results in a particular sample mean $\bar{y} = (y_1 + y_2 + \cdots + y_n)/n$. Random sampling, which gives each of the samples the same probability of being selected, induces a *sampling distribution* for the sample average $\overline{Y}$. This sampling distribution has a certain mean $\mu_{\overline{Y}}$ and standard deviation $\sigma_{\overline{Y}}$. The sampling distribution has the following characteristics:

- The mean of the sampling distribution of $\overline{Y}$ is given by μ. That is,

$$\mu_{\overline{Y}} = \mu$$

  The sampling distribution of $\overline{Y}$ is centered at the population mean μ. Repeated sample averages fluctuate around μ. Averages of some samples are smaller, and averages of others are larger; however, the mean of sample averages from repeated samples will be μ.

- The standard deviation of the sampling distribution of $\overline{Y}$ is given by $\sigma_{\overline{Y}} = \sigma/\sqrt{n}$, and its variance is

$$\sigma_{\overline{Y}}^2 = \sigma^2/n$$

  Averaging reduces the variability, with averages varying less than individual population values. Sample results from a single observation ($n = 1$) fluctuate around the population mean with standard deviation σ. Averages of n observations fluctuate around the same mean with standard deviation $\sigma_{\overline{Y}} = \sigma/\sqrt{n}$. If n becomes large, the sampling variability approaches zero, and μ can be estimated perfectly. But, of course, taking a very large sample would in most cases be prohibitively expensive.

- For reasonably large sample sizes, the distribution of $\overline{Y}$ is approximately normal, regardless of the distribution of Y.

The bulleted paragraphs are consequences of the central limit theorem, one of the most important results in statistics.

2.5.2 Confidence Intervals for a Population Mean

A random sample of size n is taken from a process with mean μ and standard deviation σ. The sample average $\bar{y}$ provides a point estimate of the population mean μ. Suppose the process (population) standard deviation σ is known. The standard deviation of the sample average $\sigma_{\overline{Y}} = \sigma/\sqrt{n}$ quantifies the estimation error; it tells us how far the estimate could be from the true population mean. Then a 95% confidence interval for the population mean is given by the interval

$$\bar{y} \pm 1.96\sigma_{\bar{y}} \quad \text{or} \quad \bar{y} \pm 1.96\sigma/\sqrt{n}$$

Suppose σ is unknown. The sample standard deviation $s = \sqrt{\sum_{i=1}^{n}(y_i - \bar{y})^2/(n-1)}$ provides an estimate of σ. Replacing σ in $\sigma_{\bar{y}} = \sigma/\sqrt{n}$ by its estimate s gives us an estimated standard deviation of a sample average. We refer to it as the *standard error* of the sample average, and we write it as $se_{\bar{y}} = s/\sqrt{n}$. For a reasonably large sample size, an approximate 95% confidence interval for the population mean is given by the interval

$$\bar{y} \pm 1.96 se_{\bar{y}} \quad \text{or} \quad \bar{y} \pm 1.96 s/\sqrt{n}$$

The factor 1.96 follows from the central limit effect and the approximating normal distribution; it is the 97.5th percentile of the standard normal distribution. Using the factor 2 rather than 1.96 results in a close approximation.

For small sample sizes, and under the additional assumption that the distribution of Y in the population is normal, we replace the factor 1.96 with the 97.5th percentile of the t-distribution with $n-1$ degrees of freedom. Then the 95% confidence interval is given by

$$\bar{y} \pm [t_{0.975}(n-1)]s/\sqrt{n}$$

where $t_{0.975}(n-1)$ is the 97.5th percentile of the t-distribution with $n-1$ degrees of freedom. For sample sizes larger than 30, the difference between percentiles of the t- and normal distribution is small, and it does not matter which distribution is used.

Thus, 95% confidence intervals cover the true population mean in 95% of repeated samples. Intervals with other coverages, such as 90% or 99% confidence intervals, can be obtained by using different percentiles, such as $t_{0.95}(n-1)$ for a 90% or $t_{0.995}(n-1)$ for a 99% confidence interval.

Example A random sample of 60 customers selected from among all customers who have ordered from a catalog in 2005 showed an average purchase amount of $\bar{y} = 125$ dollars, with a sample standard deviation of $s = 24$ dollars. The standard error of the average is $se_{\bar{y}} = 24/\sqrt{60} = 3.098$, and a 95% confidence interval for the mean purchase amount in the population is given by

$$125 \pm (1.96)(3.098) \quad \text{or} \quad (118.9 \text{ to } 131.1)$$

2.5.3 Central Limit Effect for Proportions

Assume that we are interested in estimating an unknown proportion π, such as the proportion of smokers among State University graduating seniors. The sample proportion

$$p = (\text{number of successes})/n = \bar{y} = (y_1 + y_2 + \cdots + y_n)/n$$

is an average of n sample responses. Each response is the outcome of a discrete random variable Y with possible values 0 or 1 (smoker), and associated probabilities $1 - \pi$ and π. The variable Y follows a binomial distribution from a single trial and with success probability π. Section 2.2.1 shows that its mean is π, and its standard deviation is $\sqrt{\pi(1-\pi)}$. Applying the central limit effect (Section 2.5.1) to the sample proportion $P = \overline{Y}$, we find that for reasonably large samples, the sampling distribution of a proportion can be approximated by a normal distribution with mean π and standard deviation $\sigma_P = \sqrt{\pi(1-\pi)/n}$. Sample proportions fluctuate around the population proportion π, and their standard deviation decreases with the square root of the sample size.

The sample size needs to be large for the central limit theorem to take effect—certainly much larger than when averages of continuous measurement data are considered. Sample sizes of 100 or more will be sufficient as long as the population proportion is not too close to 0 or 1. If π is close to the boundary (0 or 1), the distribution of the sample proportion will be skewed (and not normal) even for large values of n.

2.5.4 Confidence Intervals for a Population Proportion

A random sample of size n is taken from a population. The resulting sample proportion p provides an estimate of the population proportion π. The substitution of this estimate into the standard deviation of the sample proportion $\sqrt{\pi(1-\pi)/n}$ provides the standard error $se_p = \sqrt{p(1-p)/n}$. The standard error quantifies the estimation error, telling us how far the estimate can be from the true population proportion. An approximate 95% confidence interval for the population proportion π is given by the interval

$$p \pm 1.96 se_p \quad \text{or} \quad p \pm 1.96\sqrt{p(1-p)/n}$$

Example A random sample of 400 customers selected at random from all our catalog customers found that 108, or 27%, are repeat customers; in other words, 27% is our best estimate for the proportion of repeat buyers in the population of all our customers. A 95% confidence interval for the population proportion is

$$0.27 \pm (1.96)\sqrt{(0.27)(0.73)/400}$$

The interval extends from 0.226 to 0.314.

Comment. The term "margin of error" is often used in reporting the results of political and other polls. For example, a report might say that 43% favored candidate A with a margin of error of 3 percentage points. The margin of error is half the width of a 95% confidence interval, which for the population proportion is approximately $(2)\sqrt{p(1-p)/n}$.

2.5.5 Statistical Tests of Hypotheses

Prior to collecting data, the decision maker often has a certain hypothesis about the population characteristic of interest. For example, she may be interested in the *mean* purchasing amount of catalog customers and hypothesize that it is larger than 115 dollars. Or, she may be interested in the *proportion* of repeat buyers and hypothesize that it is less than 30%. Or, she may be interested in whether or not two advertising strategies affect the mean purchasing amount. Suppose experiments are conducted to learn about the validity of these hypotheses. A sample from the customer base is taken, and the average purchasing amount and the sample proportion of repeat customers are calculated. An experiment with two different advertising strategies is also conducted, and the average sales response for each group is calculated.

Hypotheses address unknown population characteristics. The research hypothesis (i.e., the hypothesis we put forward as the hypothesis to be tested) is called the *alternative hypothesis*, H_1. The opposite of the research hypothesis becomes the *null hypothesis*, H_0. It is the status quo or the fallback hypothesis in case we cannot show that the research hypothesis is more appropriate. In our first example, H_0: $\mu \leq 115$ and H_1: $\mu > 115$. In the second example H_0: $\pi \geq 0.30$ and H_1: $\pi < 0.30$. In the third example, H_0: $\mu_1 - \mu_2 = 0$ and H_1: $\mu_1 - \mu_2 \neq 0$.

The burden of proof always lies on the research (i.e., the alternative) hypothesis. If our sample or experiment does not provide enough evidence against the null hypothesis, we will not embrace the research hypothesis and will retain the status quo. We are aware that sample information may not always give an accurate picture of the population, as sample statistics are fraught with sampling error. We want to be reasonably confident that we do not reject the null hypothesis (the status quo) in error. That is, if in fact the null hypothesis is correct, we want to fix the error of rejecting it at a certain low value; say, 5%. This value is referred to as the *significance level* of the test.

The test of the two hypotheses such as H_0: $\mu \leq 115$ and H_1: $\mu > 115$ proceeds as follows. A random sample is taken, and from that sample we calculate the sample statistics $\bar{y}$ and s. The test statistic is the difference between the sample average and the hypothesized value, that is, $\bar{y} - 115$. If the difference is positive and large, we reject H_0: $\mu \leq 115$ and conclude H_1: $\mu > 115$; otherwise, we retain H_0. But the sample mean is subject to sampling variability, and its standard error, $se_{\bar{y}} = s/\sqrt{n}$, must be taken into account and used to standardize the difference. This results in the standardized test statistic $TS = \dfrac{\bar{y} - 115}{se_{\bar{y}}} = \dfrac{\bar{y} - 115}{s/\sqrt{n}}$. If this test statistic is large, larger than what could be expected under the null hypothesis, we reject the null hypothesis. Under the null hypothesis that the population mean μ is 115, the standardized test statistic follows a t-distribution with $n - 1$ degrees of freedom (or a standard normal distribution, if n is large). The probability that the t-distributed random variable exceeds the computed test statistic can be found in t-tables or by using certain functions in statistical software packages. For example, one can use the Excel function TDIST(t, $n - 1$, 1), where t is the value of the standardized test statistic, $n - 1$ is the number of degrees of freedom, and 1 indicates that the user wants the upper tail probability (replacing the 1 with a 2 would return the probability in both tails of the distribution). We call this the *probability value*,

$$probability\ value\ =\ P\left[t(n - 1) \geq \frac{\bar{y} - 115}{se_{\bar{y}}} \right]$$

A small probability value indicates that under the null hypothesis it would be unlikely to observe such a large sample test statistic. In this case, we reject H_0 in favor of the alternative H_1. The significance level 0.05 is taken as the cutoff value. On the other hand, a large probability value (larger than the significance level 0.05) makes it plausible that the sample test statistic resulted from the null hypothesis, and therefore we would retain H_0.

Example 1 The sample average from purchases of 60 customers is $\bar{y} = 125$, with sample standard deviation $s = 25$. We wish to test a research hypothesis about the mean purchasing amount of our catalog customers, and we hypothesize that it is larger than 115 dollars. That is, $H_1: \mu > 115$ and $H_0: \mu \leq 115$. The standardized test statistic is

$$TS = \frac{125 - 115}{25/\sqrt{60}} = 3.10$$

This statistic is quite large; certainly larger than 2, which is a reasonable cutoff, because it is close to the 97.5th percentile (1.96) of the standard normal distribution. The probability value

$$probability\ value = P[t(59) \geq 3.10] = 0.0015$$

is very small, which makes the null hypothesis highly unlikely. We reject the null hypothesis in favor of the alternative that the population mean is in fact larger than 115.

Example 2 We are interested in the *proportion* of repeat buyers, and we want to test the research hypothesis that it is less than 30%. Here we test $H_0: \pi \geq 0.30$ against $H_1: \pi \leq 0.30$. We reject the null hypothesis if the sample proportion p is much smaller than the hypothesized value of 0.30. Under the null hypothesis, the standard deviation of p in repeated samples of size n is $\sqrt{0.3(1 - 0.7)/n}$; see Section 2.5.3. The standardized test statistic becomes

$$TS = \frac{p - 0.30}{\sqrt{0.3(1 - 0.7)/n}}$$

Suppose a random sample of 400 catalog customers found that 108, or 27%, were repeat customers. The value of the standardized test statistic, $TS = -1.31$, is not extreme and within the range ± 2 that we associate with a normal distribution. The

$$probability\ value = P[Z \leq -1.31] = 0.0951$$

is larger than the standard significance level, and therefore we retain the null hypothesis. There is not enough evidence to say that the proportion of repeat buyers is less than 30%.

2.5.6 Determination of the Sample Size

Sample statistics vary around the true population characteristics that they estimate. The standard deviation of the sampling distribution (i.e., the standard error) indicates the margin of error, and we learned that it decreases with the sample size. How large must the sample size be if we want to be reasonably confident that our estimate is within a certain distance from the true value? Determining the required sample size is very important, because we need to know whether a certain sample size is sufficient for estimating a population characteristic to the desired accuracy.

Estimating a Mean

Assume that we want to estimate an unknown population mean μ, and suppose that we want to be 95% confident that the estimate is within $\pm B$ units of the true value. How large a sample size is needed? The standard deviation of the sample average is $\sigma/\sqrt{n}$, and a 95% confidence interval is given by $\bar{y} \pm 1.96\sigma/\sqrt{n}$. For simplicity, replacing 1.96 with 2, the quantity $\pm 2\sigma/\sqrt{n}$ must equal $\pm B$. Solving the equation $B = 2\sigma/\sqrt{n}$ leads to the required sample size

$$n = \left(\frac{2\sigma}{B}\right)^2$$

A prior estimate of the standard deviation of individual measurements is needed. One could argue that it would be unreasonable to know σ if μ were unknown. But often one has access to prior data and experiments that looked at similar issues. In this case, an estimate of σ from these prior studies would be used. Alternatively, if no previous estimates were available, we could first take a small preliminary sample of (say) 50 observations and use it to estimate σ.

Example Assume that we want to estimate the mean GPA for undergraduate students at the Central University. Similar studies on GPA may have been conducted at other comparable schools, and we may even have access to estimates of the variability in GPA at Central University for previous years. Suppose these studies indicate that a good planning value for the standard deviation among individual GPAs is $\sigma = 0.8$.

Suppose that we want to be 95% confident that our estimate is within ± 0.15 of the true population mean. How large must the sample be? Using the equation just given, we find that

$$n = \left(\frac{2(0.8)}{0.15}\right)^2 = 113.8 \approx 114$$

Estimating a Proportion

Assume that we want to estimate an unknown proportion π, and suppose that we want to be 95% confident that our estimate is within $\pm B$ units of the true population proportion. How large a sample do we need?

The standard deviation of the sample proportion is $\sqrt{\pi(1-\pi)/n}$, and an approximate 95% confidence interval for the population proportion is $p \pm (2)\sqrt{\pi(1-\pi)/n}$. Solving the equation $B = (2)\sqrt{\pi(1-\pi)/n}$ leads to the required sample size

$$n = \frac{4\pi(1-\pi)}{B^2} \leq \frac{1}{B^2}$$

The value $1/B^2$ is an upper bound on n, the required sample size. It results from setting $\pi = \frac{1}{2}$. The function $\pi(1-\pi)$ resembles a half-dome shape, with a maximum value of $\frac{1}{4}$ when $\pi = \frac{1}{2}$. If we had prior knowledge about the proportion π, we could substitute this value into the equation just given. Previous studies with similar objectives would help with this selection. On the other hand, we could substitute $\frac{1}{2}$ and use the safe upper bound

$n = 1/B^2$ if no prior guess on π is available. Setting $\pi = 0.5$ is frequently used in determining the sample sizes in political polls.

Example We know from past studies that two-party elections are close, with the probability of the candidate of the incumbent party winning, π, at around 0.5. Usually there is much interest in "calling" an upcoming close election, and we want to estimate the proportion of votes for the candidate of the incumbent party from a random sample of likely voters. We want to be 95% confident that our estimate is within ± 0.02 (i.e., 2 percentage points) of the true value. How large a sample is needed? The above equation implies that we should take a sample of size

$$n = \frac{1}{(0.02)^2} = 2,500$$

Although this is not an overly large number, the challenge of sampling is in making sure that a true random sample is taken, and that each possible sample from the population of interest is given the same chance of being selected. We need to be certain that our sample does not exclude voters that are difficult to reach, nor do we want to include in our sample people who will not be eligible or willing to vote at election time.

How does the sample size change if we want to be 99%, or 90% confident? For that we need to replace the factor 2 (which is roughly the 97.5th percentile of the standard normal distribution) with the 99.5th percentile (which is 2.576), or the 95th percentile (which is 1.645), and solve for n.

2.5.7 Confidence Intervals and Tests of Hypotheses: Comparing Means of Two Independent Samples

We may be interested in whether or not two advertising strategies (A and B) affect the mean purchasing amounts of catalog customers. Suppose we have no prior opinion on whether one strategy is better than the other. We merely want to test our research hypothesis that they are different. In this case $H_0: \mu_A - \mu_B = 0$ and $H_1: \mu_A - \mu_B \neq 0$, where μ_A and μ_B are the average purchase amounts of customers exposed to advertising strategies A and B, respectively.

Assume we conduct the following experiment. One subset of n_1 customers is drawn randomly from our regular customer base (the population) and sent advertisement A. A second, and different, randomly selected subset of size n_2, is sent advertisement B. Purchases over the next 6 months are monitored. Suppose in this particular experiment we selected $n_1 = n_2 = 30$ customers in each group, and found that $\bar{y}_A = 132$ and $s_A = 20$, and $\bar{y}_B = 141$ and $s_B = 25$. Is this enough evidence to conclude that the effects of the two strategies differ?

Here we base our decision on the difference between the two sample means, $\bar{y}_A - \bar{y}_B$. However, one must realize that many different independent pairs of samples could have been drawn, and that the difference of the resulting means would have changed with each pair of samples. What is the sampling variability of the difference of sample averages from two independent random samples? Another version of the central limit effect implies the following:

- The mean of the sampling distribution of $\overline{Y}_A - \overline{Y}_B$ is given by $\mu_A - \mu_B$, which says that the sampling distribution is centered at the difference of the population means.
- The standard deviation of the sampling distribution of $\overline{Y}_A - \overline{Y}_B$ is given by

$$\sigma_{\overline{Y}_A - \overline{Y}_B} = \sqrt{\frac{\sigma_A^2}{n_1} + \frac{\sigma_B^2}{n_2}}$$

The sample standard deviations s_A and s_B can be substituted for the unknown population standard deviations. $se_{\overline{y}_A - \overline{y}_B} = \sqrt{\frac{s_A^2}{n_1} + \frac{s_B^2}{n_2}}$ is referred to as the *standard error* of the difference of two sample averages.

- For reasonably large sample sizes, the distribution of $\overline{Y}_A - \overline{Y}_B$ is approximately normal.

Consequently, an approximate 95% confidence interval for $\mu_A - \mu_B$ is given by

$$\overline{y}_A - \overline{y}_B \pm 1.96 se_{\overline{y}_A - \overline{y}_B} \quad \text{or} \quad \overline{y}_A - \overline{y}_B \pm 1.96\sqrt{\frac{s_A^2}{n_1} + \frac{s_B^2}{n_2}}$$

If the sample sizes are small (smaller than 20 to 30), the percentile of the normal distribution should be replaced by the percentile of a t-distribution. Available computer programs calculate the appropriate degrees of freedom automatically, using an approximation due to Welch (1937).

A test of $H_0: \mu_A - \mu_B = 0$ and $H_1: \mu_A - \mu_B \neq 0$ is based on the standardized test statistic

$$TS = \frac{(\overline{y}_A - \overline{y}_B) - 0}{\sqrt{\frac{s_A^2}{n_1} + \frac{s_B^2}{n_2}}}$$

We reject $H_0: \mu_A - \mu_B = 0$ in favor of the two-sided alternative $H_1: \mu_A - \mu_B \neq 0$ if the test statistic is a large positive or large negative value, with ± 2 being a good cutoff value. In addition, we can calculate the probability value

$$probability\ value = P[Z \geq |TS|] + P[Z \leq -|TS|] = 2P[Z \geq |TS|]$$

Z follows the standard normal distribution, and the probability can be looked up in the z-table. Because of the two-sided nature of the alternative hypothesis we must double the tail probability. This was not needed in the one-sided alternative of the two previous examples. We reject the null hypothesis in favor of the alternative hypothesis if the probability value is smaller than the significance level 0.05. Equivalently, for a 5% significance level, we reject the null hypothesis if a 95% confidence interval fails to include the value zero.

Example In our experiment we considered $n_1 = n_2 = 30$, and found that $\overline{y}_A = 132$ and $s_A = 20$, and $\overline{y}_B = 141$ and $s_B = 25$. The standard error of the difference of the two averages is $se_{\bar{y}_A - \bar{y}_B} = \sqrt{\dfrac{20^2}{30} + \dfrac{25^2}{30}} = 5.85$, and the 95% confidence interval for $\mu_A - \mu_B$ is $(132 - 141) \pm (1.96)(5.85)$. The confidence interval extends from -20.46 to 2.46. The value zero is within this interval, which indicates that the "no difference" hypothesis cannot be rejected with this data.

The identical conclusion is reached with the probability value. The test statistic for H_0: $\mu_A - \mu_B = 0$ and H_1: $\mu_A - \mu_B \neq 0$ is $(132 - 141)/5.85 = -1.54$, with probability value $2P[Z \geq 1.54] = 2(0.0618) = 0.1236$. Since it is larger than the significance level 0.05, we find no reason to reject the null hypothesis. The effects of the two advertising strategies are about the same.

2.5.8 Inference in the Blocked Experiment: Comparing Means of Two Dependent Samples

In comparative experiments, treatments need to be assigned to experimental units. For example, an experiment comparing the yield of two corn hybrids must assign hybrids A and B to each of several small test fields within a larger experimental plot. In many industrial studies, experiments are conducted sequentially in time, and the assignment of the treatments A and B to the available time slots needs to be addressed. The same issue arises in medical trials for evaluating the effectiveness of new drugs, where the treatments are the drugs that are tested and the subjects are the experimental units.

One design approach is to *randomize* the assignment of treatments to the experimental units. The experimenter would list the experimental units—the test fields, the available times, or the available subjects—and randomly assign treatments to units. Randomization is important and certainly better than a nonrandom arrangement, as it spreads the existing variability among the experimental units fairly across all treatments. However, the experimenter can do considerably better if the experimental units can be grouped into groups or blocks, such that the units are homogeneous within the same block but differ across blocks. For example, test fields close together are more similar than fields far apart. Or, experiments run on the same day benefit from more homogeneous conditions than experiments that are conducted on different days. Or, a within-subject comparison of the effectiveness of a drug is exposed to fewer interfering variables than a comparison across subjects. In *randomized block experiments*, one randomizes the assignment within each block. For example, if 20 experiments need to be carried out over 5 days, the experimenter would randomize the order of two A and two B experiments on each day. Or, instead of assigning a certain blood pressure medication to 50 patients and "no treatment" to 50 others and comparing the blood pressure readings of these two groups after a period of 3 months, a better approach would be to establish the initial blood pressure (the no-treatment group) on all 100 patients, then put all patients on the new medicine, and analyze changes after 3 months.

Example In Table 2.2, we report results of a blood pressure experiment on 10 patients. Initial blood pressures (x) and blood pressures after 3 months on the new drug (y) are listed. The table also lists the summary statistics (mean and standard deviation) of the initial blood

TABLE 2.2
Initial Blood Pressure and Blood Pressure After 3 Months: 10 Subjects

Patient	Initial Blood Pressure (x)	Blood Pressure after 3 Months (y)	Reduction ($x - y$)
1	190	181	9
2	221	211	10
3	212	200	12
4	232	218	14
5	200	185	15
6	178	175	3
7	186	169	17
8	220	212	8
9	204	191	13
10	196	187	9
Mean	203.9	192.9	11.0
Standard deviation	17.22	16.69	4.06

pressure and the blood pressure after 3 months. Furthermore, it lists the changes for each patient, $d_i = x_i - y_i$, the average change $\bar{d} = \sum_{i=1}^{n} d_i / n$ and the standard deviation of the changes $s_d = \sqrt{\sum_{i=1}^{n}(d_i - \bar{d})^2/(n-1)}$.

Comparative dot plots of the initial blood pressures and blood pressures after 3 months, shown on the same scale, are given in Figure 2.8. We notice considerable variability among the blood pressures, initially as well as after 3 months. In other words, there is considerable variation in blood pressure levels across patients. If we treated the two groups (initial, after 3 months) as independent, it would be difficult to conclude that the medication has made a difference. A two-sample test treating the two samples as independent fails to show any improvement due to the medication. The test statistic

$$\frac{203.9 - 192.9}{\sqrt{[(17.22)^2/10] + [(16.69)^2/10]}} = 1.45$$

and its probability value $P[Z \geq 1.45] = 0.0735$ are inconclusive and do not allow us to reject the null hypothesis $H_0: \mu_{\text{Initial}} - \mu_{\text{After}} = 0$. Note that we used the one-tail probability, because our research hypothesis specifies an improvement $H_1: \mu_{\text{Initial}} - \mu_{\text{After}} > 0$. Also, observe that we used the normal distribution; we could have used the t-distribution with the appropriate degrees of freedom, but the results would have been very similar, and our conclusions would not have changed.

The assumption that the two samples in this experiment are independent is incorrect. Two blood pressure readings (x and y) are taken on the same person. If the initial reading on one subject is high compared to all other subjects, we would expect that also his or her reading after 3 months would be high compared to the other patients. Each subject acts as his or her own block. The variability between the two readings from the same subject is small, certainly much smaller than the variability across patients. It is the differences in the blood pressure readings that need to be analyzed. Taking differences eliminates the subject variability, which constitutes a large part of the variability that we see in Figure 2.8.

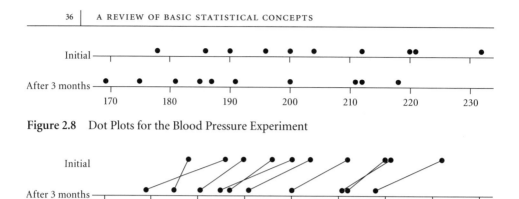

Figure 2.8 Dot Plots for the Blood Pressure Experiment

Figure 2.9 Blood Pressure Experiment. Measurements for the Same Subject are Connected

We have redrawn the information in Figure 2.9, but we have connected the observations that come from the same subject. It is obvious from this graph that the type of treatment makes a big difference. In all subjects, blood pressure is reduced by the medication.

The correct test procedure in this blocked (paired) experiment is to consider the differences and test whether $H_0: \mu_\delta = \mu_{\text{Initial}} - \mu_{\text{After}} = 0$ against $H_1: \mu_\delta = \mu_{\text{Initial}} - \mu_{\text{After}} > 0$. The appropriate test statistic is $\dfrac{\bar{d} - 0}{s_d/\sqrt{n}} = \dfrac{11}{4.06/\sqrt{10}} = 8.57$. Its probability value $P[t(9) > 8.57] = 0.00001$ is essentially zero. Hence, there is very strong evidence that the medication has lowered the blood pressure. The average reduction is 11 units; the 95% confidence interval for the reduction is given by $\bar{d} \pm t_{0.975}(9)s_d/\sqrt{n}$, or $11 \pm (2.2622)(4.06)/\sqrt{10}$. The interval extends from 8.10 to 13.9.

Comment. Here we assess whether a particular drug "works." Of course, one should be concerned that the observed effect is a combination of two effects: the real effectiveness of the drug and a placebo effect due to the person's belief of being given something useful. Apart from much higher sample sizes, FDA-approved drug studies usually compare a new experimental drug to the currently available "best-practice" drug. The best-practice drug could be a placebo. In such a study, one would divide patients into two groups (preferably, at random) and conduct the experiment discussed in this example with both groups. This would result in two sets of blood pressure differences (final readings minus initial readings), one set for each group. The procedure in Section 2.5.7 for comparing the means of two independent samples can be applied to test whether the mean effectiveness of these two drugs is different.

2.6 CASE STUDY: ADTEL

The following discussion is adapted from a Harvard Business School case reported in Chapter 5 of Clarke (1987). In the past, the Barrett Foods Company had enjoyed a market leadership position for its peanut butter, but recently was faced with a declining market share for this product. The company commissioned AdTel, a marketing research company, to assess the impact of a dramatically increased advertising budget and determine the potential payoff of a \$6 million television advertising campaign versus the current \$2 million strategy. Management had estimated that a 15% sales increase (established with 90% confidence or higher) would be required to justify the added expense.

AdTel maintained a 2,000-family panel. It also employed a dual-cable television system to determine the sales effect of television advertising alternatives. AdTel had two separate cable circuits. Television sets owned by half of the test-families were wired to cable A, while those of the other half were wired to cable B. The panels were carefully balanced according to demographic characteristics and shopping preferences. By the push of a button, AdTel was able to block the commercial broadcast on one side of the cable and simultaneously cut in the desired test commercial, while the other side carried the regular program. The panel families recorded their purchases in weekly diaries.

The basic study covered a period of 18 months. The first 6 months represented a control period, where both circuits received the same advertising at the level of the $2 million campaign. The next 12 months represented the test period where advertising for panel A tripled. To avoid distortions by families joining and dropping the panel during the test, a static sample was created that only included those families returning at least 80% of their diaries. Panel A contained 829 families, while panel B comprised 922. The average monthly volumes per family and the monthly market shares of Barrett's peanut butter for the 18 months (6 pretest and 12 test periods) are shown in Table 2.3.

Time sequence graphs of average sale volumes for Panels A and B are shown in Figure 2.10. Time series graphs of market shares for Barrett's peanut butter are given in Figure 2.11.

The pretest data (weeks 1–6) show that there is no appreciable difference between the two panels. The graphs also show convincingly that sales and market shares—for both panels A and B—change with the reporting period. Hence *period* is an important blocking variable, and the analysis needs to be conducted with the monthly differences between A and B.

T A B L E   2.3
Volume and Market Share for Barrett's Peanut Butter

Period (month)	Pretest and Test	Volume Panel A	Volume Panel B	Volume A − B	Market Share Panel A	Market Share Panel B	Market Share A − B
1	Pretest	43	41	2	50.0	50.0	0.0
2	Pretest	22	23	−1	30.0	30.0	0.0
3	Pretest	31	31	0	40.0	39.5	0.5
4	Pretest	17	18	−1	23.0	24.0	−1.0
5	Pretest	29	25	4	45.0	44.0	1.0
6	Pretest	31	25	6	39.0	35.0	4.0
7	Test	22	22	0	20.0	20.0	0.0
8	Test	21	23	−2	23.0	26.0	−3.0
9	Test	29	25	4	33.0	30.0	3.0
10	Test	29	32	−3	27.0	33.0	−6.0
11	Test	46	42	4	44.0	43.5	0.5
12	Test	40	35	5	32.0	30.0	2.0
13	Test	38	29	9	29.0	27.0	2.0
14	Test	53	38	15	38.0	33.0	5.0
15	Test	47	34	13	41.0	38.0	3.0
16	Test	45	26	19	43.0	34.0	9.0
17	Test	65	38	27	55.0	54.0	1.0
18	Test	40	47	−7	51.0	57.0	−6.0

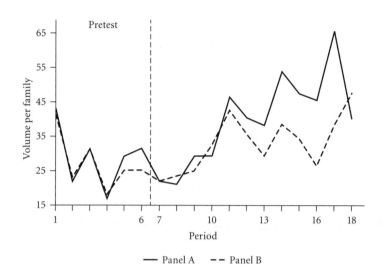

Figure 2.10   Volume per Family: Panels A and B

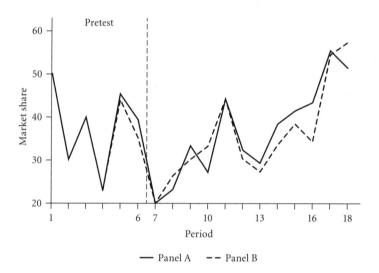

Figure 2.11   Market Shares: Panels A and B

Dot plots of monthly differences of A and B for volume and market share are shown in Figure 2.12.

We consider the test period (months 7–18) and test H_0: $\mu_\delta = \mu_A - \mu_B = 0$ against H_1:

$\mu_\delta = \mu_A - \mu_B > 0$. The test statistics are $\dfrac{\bar{d} - 0}{s_d/\sqrt{n}} = \dfrac{7}{9.98/\sqrt{12}} = 2.43$ for volume, and

$\dfrac{\bar{d} - 0}{s_d/\sqrt{n}} = \dfrac{0.88}{4.32/\sqrt{12}} = 0.71$ for market share. The probability values are $P[t(11) \geq 2.43] = 0.0167$ and $P[t(11) \geq 0.88] = 0.1988$, respectively.

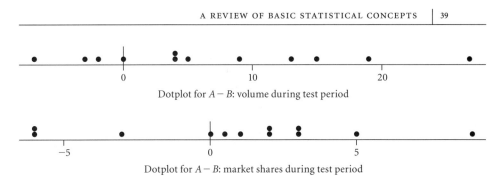

Dotplot for $A - B$: volume during test period

Dotplot for $A - B$: market shares during test period

Figure 2.12   Dot Plots of Monthly Differences in Volume and Market Share

There is evidence that the increased advertising has increased the volume. The average increase of seven units is statistically significant; a 90% confidence interval for the mean increase extends from $7 - (1.7959)(9.98)/\sqrt{12} = 1.83$ to $7 + (1.7959)(9.98)/\sqrt{12} = 12.17$.

An increase of seven units over the average for panel B with standard marketing (which is 32.58 units) represents a 21.5% increase in sales. However, the lower limit of a 90% confidence interval for the percent increase in sales amounts to only $100(1.83/32.58) = 5.6\%$. It appears from the graph in Figure 2.10 that the extra advertising has done very little during the first 6 months of the test period. It is only during periods 13 through 17 that we notice appreciable differences. The last period is also quite remarkable, in that the benefit of the extra advertising has disappeared completely. In summary, while we see some increase in volume due to the increased advertising, it is doubtful that this strategy meets management's goal of a 15% sales increase that can be established with minimum 90% confidence.

A conclusion that increased advertising has affected market share is even less convincing; the small average increase of 0.88 percentage points is not statistically significant.

2.7 NOBODY ASKED US, BUT . . .

What is now called the normal distribution first appeared in 1733 in a paper by the French mathematician Abraham de Moivre. (For a discussion of the paper, see Anders Hald (1986), *History of Probability and Statistics and Their Applications Before 1750.*) At the time, games of chance such as tossing coins or rolling dice were very popular, and both gamblers and mathematicians were interested in knowing the probabilities of various outcomes. The binomial distribution was well known, but calculating binomial probabilities was extremely difficult computationally if the number of trials n was fairly large, and impossible if n was very large. In his paper, de Moivre derived the equation that would later be called the normal density function as an approximation to the binomial when the number of trials is very large. Later, Laplace in 1783 and Gauss in 1809 made important contributions by developing theoretical arguments to support the normal distribution as a model of errors of measurement, in particular for errors in the observations of heavenly bodies. Over time, interest in this probability distribution continued to grow with noteworthy contributions made by the Belgian social statistician Adolphe Quetelet (1796–1874) who used the normal distribution to describe data on measurements of physical characteristics. Quetelet used the normal distribution to measure variations about the "average man." The name *normal* was first applied to the distribution in the 1870s by Galton and several others, reflecting the fact that the distribution described the *normal* or natural variation in many observed phenomena. (See Chapter 22 of Stigler, 1999, *Statistics on the Table*, for an interesting discussion of how the normal got its name.)

In 1908, in the paper "The Probable Error of a Mean," William Gossett derived the probability distribution that became known as the *t*-distribution. Gossett, a young chemist and statistician, was studying quality problems at the Guinness brewery in Dublin. He was interested in calculating the probability that a population mean was within a specified distance of a sample average. The approach had been to use $s/\sqrt{n}$ as an estimate of $\sigma/\sqrt{n}$, the standard deviation of the distribution of sample averages, and to calculate the probabilities using the normal distribution. Gossett knew this worked well for large samples where s would be close to σ. But he realized that when n was small, calculated values of s would vary greatly and therefore so would the estimate $s/\sqrt{n}$. As a consequence, errors in the calculated probabilities would be large. This led him to derive a theoretical density function for the random variable $T = \dfrac{\overline{Y} - \mu}{s/\sqrt{n}}$ and to compare it to the density function for $Z = \dfrac{\overline{Y} - \mu}{\sigma/\sqrt{n}}$, the density function for the standard normal distribution. He showed that because of the variability in the estimate $s/\sqrt{n}$, his new distribution was more likely than the normal to take on values in the tails. Gossett's employers at Guinness viewed their quality efforts as proprietary, and as a result, he published his papers under the name Student. His distribution became known as Student's *t*-distribution.

Students in introductory statistics courses are often puzzled by the fact that the sample variance $s^2 = \sum_{i=1}^{n}(y_i - \overline{y})^2/(n - 1)$ divides the sum of squares by $n - 1$, and not by n. The division by $n - 1$ is a consequence of having to calculate the sum of squares around the sample mean $\overline{y}$ instead of the unknown population mean μ. If the population mean μ were known, the quantity $\sum_{i=1}^{n}(y_i - \mu)^2/n$ would be an unbiased estimate of the population variance σ^2. It is easy to show that the sum of squares $\sum_{i=1}^{n}(y_i - \mu)^2$ is smallest if $\mu = \overline{y}$. Hence the sum of squares $\sum_{i=1}^{n}(y_i - \overline{y})^2$ is always smaller than the sum of squares around the unknown μ, and the estimate of σ^2 that divides $\sum_{i=1}^{n}(y_i - \overline{y})^2$ by n would be too small. Dividing the sum of squares by $n - 1$ compensates for this and eliminates the bias.

Based on the central limit theorem, we stated that for reasonably large sample sizes, the distribution of the sample average will be approximately normal regardless of the population distribution of the individual values. How large does the sample size n have to be? Many introductory textbooks specify or at least suggest that n should be at least 30. But in most cases of practical interest that figure is too high. For example, in statistical process control, sample averages are plotted on X-bar control charts, sometimes called Shewhart charts for their originator, Walter Shewhart. Typically, samples of size 4 or 5 are used, and the control limits are based on sample averages following a normal distribution. In his classic book, *Economic Control of Quality of Manufactured Product*, originally published in 1931, Shewhart presented the results of his experiments taking 1,000 sample averages of size 4 from populations that were rectangular (uniform) and triangular. In both cases, the sample averages were well approximated by a normal distribution. As Shewhart said, "The closeness of fit is striking and illustrates the rapid approach of the distribution to normality as the sample size is increased. Such evidence . . . leads us to believe that in almost all cases in practice we may establish sampling limits for averages of four or more upon the basis of normal law theory." In some instances, sample sizes greater than 30 may be needed if the distributions of population values are highly skewed with long tails (sampling from an exponential distribution would be one example). But in the experiments we consider in this book, these situations would be highly unlikely.

As we discussed in this chapter, for confidence intervals and hypothesis tests on the population proportion π, large samples are typically needed before the central limit theorem takes effect. In these cases the normal distribution is approximating the binomial distribution, which for $\pi = 0.5$ is symmetric. As the value of π moves away from 0.5, the binomial distribution becomes increasingly skewed, and larger samples are needed before the normal approximates it well. A useful rule of thumb is that the normal distribution is a good approximation if $n\pi > 10$ for $\pi \leq 0.5$, and $n(1 - \pi) > 10$ for $\pi \geq 0.5$. With this rule, a sample size of 100 will be large enough (for the normal to be a good approximation) as long as π is no smaller than 0.1 or no larger than 0.9.

SAMPLE SIZE DETERMINATION IN A COMPARATIVE EXPERIMENT

In many comparative studies we evaluate the success of a new strategy or method through the resulting change in a proportion. For example, we may have two different advertising strategies (1 and 2) and may be interested in whether or not strategy 2 increases the proportion of people who buy a certain product. Under the null hypothesis H_0: $\pi_1 = \pi_2 = \pi$, the distribution of the difference of the two sample proportions $p_2 - p_1$ is normal with mean 0 and variance $2\pi(1 - \pi)/n$, where n is the size of the first (and second) sample. For a test with significance level α, we reject the null hypothesis in favor of the one-sided alternative H_1: $\pi_2 - \pi_1 = \delta > 0$ whenever $p_2 - p_1 > z_{1-\alpha}\sqrt{2\pi(1 - \pi)/n}$; $z_{1-\alpha}$ is the $100(1 - \alpha)$ percentile of the standard normal distribution.

We are looking for a test with power $1 - \beta$, which implies probability β of falsely accepting the null hypothesis if the alternative ($\pi_2 - \pi_1 = \delta$) is actually true. This requirement implies the equality

$$\frac{z_{1-\alpha}\sqrt{\dfrac{2\pi(1 - \pi)}{n}} - \delta}{\sqrt{\dfrac{\pi(1 - \pi)}{n} + \dfrac{(\pi + \delta)(1 - \pi - \delta)}{n}}} = -z_{1-\beta}$$

The above equation can be solved for the sample size n, leading to

$$n = \frac{[z_{1-\alpha}\sqrt{2\pi(1 - \pi)} + z_{1-\beta}\sqrt{\pi(1 - \pi) + (\pi + \delta)(1 - \pi - \delta)}]^2}{\delta^2}$$

Setting $\delta = 0$ in the numerator leads to the approximation

$$n \cong \frac{2\pi(1 - \pi)[z_{1-\alpha} + z_{1-\beta}]^2}{\delta^2}$$

Example Consider the planning value $\pi = 0.03$ for the common success proportion, and assume that it is important to detect an increase of one-half of a percent ($\delta = 0.005$). For $\alpha = \beta = 0.05$ and $z_{0.95} = 1.645$, we must sample

$$n \cong \frac{2(0.03)(0.97)[1.645 + 1.645]^2}{(0.005)^2} = 25{,}200$$

in each group, for a total of 50,400 people for the two groups combined. Statistical computer software such as Minitab and JMP includes routines for such sample size calculations. Minitab, for example, returns this sample size when asking for the power/sample size in the two proportions case.

Comment. The result in Appendix 2.1 is pertinent to the design of *comparative* experiments that attempt to estimate the difference between two unknown success proportions. It shows how to select the two sample sizes such that a certain specified difference (δ) in the success proportions is detected with reasonably large power. A planning value for the common success proportion (π) and a meaningful detectable difference (δ) of the two success proportions need to be specified. Information on the success rate is usually available from prior experiments, and worthwhile changes are determined with economic considerations in mind. In our illustration, this has led us to the values $\pi = 0.03$ and $\delta = 0.005$.

The discussion of the sample size in Section 2.5.6 is different in two respects. First, it focuses on the *one sample* situation, not on comparative experiments. Second, it determines the sample size that is needed to achieve a certain precision of the estimate (either a single mean or a single proportion), but does not address the power of detecting a certain meaningful difference.

EXERCISES

Exercise 1 The file *contribution* summarizes the 2004 contributions to a selective private liberal arts college. Refer to Section 2.3 for a description of the data set.

  (a) Confirm the information in Table 2.1 and Figures 2.6 and 2.7. Use available computer software such as Excel or Minitab.

  (b) Consider some of the other factors that were not used in Section 2.3. In particular, assess the effect of gender, marital status, graduation status (graduated/not graduated), and major on the likelihood of donating and the donation amount.

Exercise 2 Consider the data in Section 2.6 on AdTel. Recreate the information in Figures 2.10 through 2.12 and the results of the hypothesis test discussed in this section.

Exercise 3 Search the Web for useful statistics applets. You can do this by searching for expressions such as "applets for central limit effect," "applets for confidence intervals," "applets for hypothesis testing," "applets for visualization of statistical concepts," or "applets for sample size." Experiment with these applets. These applets will reinforce the concepts discussed in Section 2.5. They will demonstrate the central limit effect by drawing repeated samples of a certain specified size. They show through simulations how the variability of sample statistics such as the sample mean decreases with increasing sample size. They show through simulations that 95% confidence intervals for a mean or a proportion cover the population (process) mean and proportion in 95% of the cases. Applets for the correlation coefficient illustrate the connection between scatter plots and correlation coefficients, and they show how the correlation changes if observations are changed.

Exercise 4 John, in charge of custodial services at the business school, installed brand new lightbulbs into the offices of the marketing faculty. He kept track of burned-out bulbs and the times when he had to replace them. After 12 months, he had to replace 25 of the 30 bulbs. The length of life (in weeks) for the 25 bulbs is given below:

33	19	11	22	22	15	37	5	7	10	38	19	46
20	23	50	30	22	10	15	37	15	22	40	22	

(a) Obtain a dot diagram and calculate the mean, median, and standard deviation of the 25 observations. Calculate the 90th percentile.

(b) We would like to obtain the mean and the median life length of all 30 lightbulbs. However, by the end of the 12 months, five bulbs had not yet burned out. Can you calculate the median of the 30 observations without waiting until the five remaining bulbs fail?

Can you calculate the mean of the 30 observations without waiting until the five remaining bulbs fail? If not, what can you say about this mean?

Exercise 5 The data set given below lists the annual 2005 salary (in $1,000) and the educational background for a sample of 25 employees at a large Midwestern manufacturing company. Educational background is measured by the number of years of formal schooling (12 refers to a high school graduate; 16 refers to a college graduate; 17 through 20 refer to college degree plus the number of years of graduate work).

Employee	Education	Salary	Employee	Education	Salary
1	16	52.3	14	17	49.4
2	12	43.7	15	16	45.4
3	12	39.5	16	13	41.3
4	16	47.8	17	12	37.6
5	18	53.0	18	12	33.3
6	15	49.0	19	19	64.8
7	11	33.7	20	16	50.7
8	12	32.1	21	17	54.5
9	11	9.8	22	16	27.3
10	20	37.7	23	12	14.8
11	15	26.3	24	16	21.7
12	16	22.0	25	16	33.8
13	16	27.0			

Construct a scatter diagram of salary against educational achievement. Calculate the correlation coefficient.

Exercise 6 In an NYT/CBS poll, 56% of 2,000 randomly selected voters in New York City said that they would vote for the incumbent in a certain two-person race. Calculate a 95% confidence interval for the population proportion. Discuss its implication. Carefully discuss what is meant by the population, how you would carry out the random sampling, and what other factors could lead to differences between the responses in the survey and the actual votes on the day of the election.

Exercise 7 A sample of $n = 50$ bread loaves is taken from the sizable production that left our bakery this morning. We find that the average weight of the 50 loaves is 1.05 pounds; the standard deviation is $s = 0.06$ pound.

(a) Obtain a 95% confidence interval for the mean weight of this morning's production.

(b) One of the employees claims that the current process produces loaves that are heavier than one pound, on average. Is there enough information in our sample to reject the null hypothesis that $\mu = 1.00$ lb, in favor of the alternative that $\mu > 1.00$ lb?

(c) Assume that the distribution of weights is normal; furthermore assume that the sample average and standard deviation are good estimates of the corresponding population characteristics μ and σ. Calculate the proportion of loaves that are underweight (i.e., weigh less than 1.0 pounds).

(d) Predict the weight of a single loaf from this morning's production. Obtain an approximate 95% prediction interval.

Exercise 8 Prior studies showed that the standard deviation among individual measurements of a certain air pollutant is 0.6 parts per million (ppm). You are planning on using the information from a random sample (i.e., the sample average) to estimate the unknown process (population) mean μ. How large do you have to select the sample size if you want to be 95% certain that your sample average is within plus or minus 0.2 ppm of the unknown process mean?

Exercise 9 Thirty lightbulbs were selected randomly from among a very large production batch, and they were put on test to determine the time until they burn out. The average failure time for these 30 bulbs was 1,080 hours; the sample standard deviation was 210 hours. The lightbulbs are advertised as having a mean life length of 1,200 hours. Test this hypothesis against the alternative that the mean life length of this batch is actually smaller than 1,200 hours.

Exercise 10 A new emergency procedure was developed to reduce the time that is required to fix a certain manufacturing problem. Past data under the old system were available ($n = 25$). The staff was trained under the new procedure, and the response times for the next 15 occurrences of this manufacturing problem were recorded.

Old Procedure

4.3	6.5	4.6	4.3	6.4	4.8	5.1	6.8	4.9	4.5	5.1	7.3	3.3
5.0	4.6	7.0	5.1	3.8	5.2	4.1	5.7	4.6	5.9	3.1	6.2	

New Procedure

6.2	4.0	3.3	4.5	2.3	3.0	3.2	6.0
3.7	4.5	5.3	4.0	5.4	4.3	3.8	

Compare the response times under the old and the new procedure. Are there differences in the mean responses? Discuss using appropriate graphs, summary statistics, and statistical tests. Would you switch to the new procedure?

Exercise 11 Two different fabrics are tested on a wear tester. A *wear tester* is a mechanical device that rubs the attached fabric against a fixed object. This particular machine has two separate attachments that allow us to compare two pieces of fabrics in the same run. The weight losses (in milligrams) from 8 runs are as follows:

				RUN				
Fabric	1	2	3	4	5	6	7	8
A	36	26	31	38	28	37	22	29
B	39	27	35	42	31	39	21	32

Analyze the data and determine whether the mean wear of fabric A is different from that of fabric B. If it is different, how does it differ?

Discuss why the design of assigning both fabrics to each run is preferable to a design that assigns fabric A to both positions of runs 1–4, and fabric B to both positions of runs 5–8.

Exercise 12 In the past, the sign-up rate for your credit card has been around 6%. Your marketing team wants to decide between two different sets of promotional materials that it plans to send to potential customers: a traditional set that is very similar to the one that has been used in the past, and a new, bolder set that is expected to increase the sign-up rate. Before switching to the new materials, your company wants to run a comparative experiment that evaluates the two sign-up rates. Assuming a significance level of 0.05, determine the common sample size for the two groups that can detect a 1% increase in the sign-up rate with power of 0.95. How does the sample size change if you require less power (say, 0.90 or 0.80)? How does the sample size change if you want to detect a difference of one-half of a percent? You may want to use computer software to carry out the calculations.

Exercise 13 Using a computer software of your choice, perform Shewhart's experiment of drawing random samples of size 4 from a continuous uniform distribution between 0 and 1. You may use the Minitab function "Calc > Random Data > Uniform" or the Excel command RAND(). Generate four columns of 1,000 random numbers, and calculate 1,000 sample averages from samples of size 4. Construct a histogram of the 4,000 individual observations, and calculate their mean and standard deviation. Construct a histogram of the 1,000 sample averages, and calculate their mean and standard deviation. Compare the two histograms and the two mean and standard deviation estimates. Are these results what you expected to see? Explain.

3 | TESTING DIFFERENCES AMONG SEVERAL MEANS: COMPLETELY RANDOMIZED AND RANDOMIZED COMPLETE BLOCK EXPERIMENTS

3.1 INTRODUCTION

In Section 2.5 we used sample information to test whether the means of two populations are equal. In this chapter, we extend the discussion to the comparison of more than two means. We discuss two designs for making this comparison: the *completely randomized experiment* and the *randomized complete block experiment*.

Internet experiments that present one of several advertising messages at random to users of search engines are examples of completely randomized experiments. There the k advertising messages, which may differ with respect to advertising text, background color, and font size, are offered to distinct Internet users at random; each user responds to one and only one advertising message. The response in such studies is the sales volume generated from each advertising message, or the "hit ratio" (the proportion of those who access a particular Web site in response to the message).

Consider another example. A firm wants to test three different in-store promotions for a major product, and identifies a group of 15 stores of similar size to participate in the experiment. Each store will test one and only one of the promotions for a certain period of time (say three weeks). The promotions are randomly assigned to the stores, with five different stores per promotion. In the language of experimental design, the three promotions are called treatments, and the 15 stores are called the experimental units. Since the treatments are assigned to the experimental units at random, we call this a *completely randomized experiment*.

An alternative design for comparing the three promotions is the *randomized complete block experiment*. Suppose the firm believes the 15 stores are not homogeneous and that possible store effects could introduce additional noise that would make it difficult to recognize differences among the treatments. Hence it may be better to observe each store under all three in-store promotions. We could divide the study period into three one-week periods and, for every store, assign each of the three promotions to a different week. In this design, each of the 15 stores acts as a *block*. Within each block, treatments are assigned to the three one-week periods at random. The design is called a complete block design because each of the three promotions is assigned to every block. The design is called a randomized complete block design as, within a given block, promotions are randomly assigned to weeks. In some situations, the number of treatments is greater than the size of the block and a complete block experiment is not possible. In those situations, which we will not discuss, it is possible to construct incomplete block experiments (see Box, Hunter, and Hunter, 2005).

The blocking approach has potential advantages. Suppose one particular store has some characteristic that would make its sales volume particularly high under all three promotions A, B, and C. In the earlier completely randomized design, a store is assigned to a single promotion, and it is due to chance whether the well-performing store becomes part of promotion A, B, or C. If the store were assigned to A, then A would benefit. However, this benefit is not due to the treatment, but due to the store effect. If the store happened to be assigned to group C, then C would benefit. Through "blocking," we control for the "luck of the draw" by assigning all three treatments to every store (block). In the randomized block design, we can focus on the relative changes within the block, thus canceling out possible block effects. Consequently, any differences in results will be due to the treatments, not the stores. If there is an actual block effect, the randomized complete block design will increase our ability to detect differences in the treatments.

By blocking on stores, we have eliminated a possible store effect. But sales might also be affected by a time (week) effect. The three 1-week periods might not be homogeneous; expected sales might vary from week to week. Randomizing the assignment of the promotions across the three 1-week periods of each block is important because it spreads a possible week's effect across the three treatments. But it is possible to do better, by blocking the experiment with respect to weeks as well as stores. Experiments that block on two factors (here, stores and weeks) are called *Latin square designs*. We discuss this type of design in Chapter 7 and in Case 11 of the case study appendix.

Our discussion of the randomized complete block experiment is a natural extension of the material covered in Chapter 2. There, in comparing two means, we discussed the difference between the completely randomized experiment in which two treatments are assigned to two groups of different experimental units, and the paired comparison (blocked experiment) in which each unit receives both treatments. We illustrated the paired comparison approach in a test of a blood pressure medication. We showed that measuring blood pressure on the same patient before and after treatment eliminated the variation in pressure among patients, increasing the precision of the test and hence its ability to detect differences in the two treatments. In this chapter, in the randomized complete block experiment, we will apply this idea to the comparison of more than two treatments.

3.2 THE COMPLETELY RANDOMIZED EXPERIMENT

A firm planned and executed a test of the effectiveness of three different product displays. Fifteen stores were available for the test, and each display was used in five different stores. To make the results comparable and to minimize bias, displays and stores were randomly assigned. Sales volume for the week during which the display was present was measured and

TABLE 3.1
Marketing Study: Results of a Test of Three Product Displays

	DISPLAY		
	1	2	3
	9.5	8.5	7.7
	3.2	9.0	11.3
	4.7	7.9	9.7
	7.5	5.0	11.5
	8.3	3.2	12.4
Sample size	5	5	5
Sample mean	6.64	6.72	10.52
Sample variance	6.82	6.28	3.43

TABLE 3.2
*Observations and Summary Statistics for k Treatment Groups
(the General Case)*

	TREATMENT GROUPS			
	1	2		k
	y_{11}	y_{21}	$\cdots$	y_{k1}
	y_{12}	y_{22}	$\cdots$	y_{k2}
	$\vdots$	$\vdots$	$\vdots$	$\vdots$
	y_{1n_1}	y_{2n_2}	$\cdots$	y_{kn_k}
Sample size	n_1	n_2	$\cdots$	n_k
Sample mean	$\bar{y}_1$	$\bar{y}_2$	$\cdots$	$\bar{y}_k$
Sample variance	s_1^2	s_2^2	$\cdots$	s_k^2

compared to the base sales of that store. Percentage changes were calculated, and they are given in Table 3.1.

In this particular example, the observations come from three treatment groups. In general, there are k treatment groups with observations y_{ij}. The first subscript denotes the treatment group ($i = 1, 2, \ldots, k$), while the second subscript j denotes the replication. A listing of the observations for the case of k treatment groups is shown in Table 3.2.

Note that the number of observations in the k treatment groups need not be the same. Let us denote the number of observations by $n_1, n_2, \ldots, n_k$, and the total number of observations by $N = \sum_{i=1}^{k} n_i$. We call the study *balanced* if the sample sizes in the k groups are the same. There are advantages to having equal (or nearly equal) sample sizes. Balanced designs allow us to estimate the treatment means with uniform precision, and they maximize the power of the test procedure (the F-test) that is discussed in this section.

Table 3.2 also lists summary statistics for the k treatment groups. The sample mean and the sample variance for treatment group i are given by

$$\bar{y}_i = \frac{\sum_{j=1}^{n_i} y_{ij}}{n_i} \quad \text{and} \quad s_i^2 = \frac{\sum_{j=1}^{n_i} (y_{ij} - \bar{y}_i)^2}{n_i - 1}$$

See Section 2.3. There are $k = 3$ groups in Table 3.1, with equal sample sizes $n_1 = n_2 = n_3 = 5$ and total sample size $N = 15$. You may want to check the sample means and the sample variances that are given in Table 3.1, using a calculator or a statistical software package.

We assume that the samples were randomly drawn from normal populations with possibly different means $\mu_1, \mu_2, \ldots, \mu_k$, but each with the same variance σ^2. Given the sample results, we wish to test the null hypothesis that the k population means are equal against the alternative that at least one of the means is different. Formally, we have

$$H_0: \mu_1 = \mu_2 = \cdots = \mu_k$$

$$H_1: \text{Not all population means are the same.}$$

In the following sections we will discuss a statistical test for this null hypothesis. Initially, we assume that the sample sizes in the k treatment groups are the same ($n_1 = n_2 = \cdots = n_k = n$), because this will make it easier for us to motivate the procedure. Later we will relax this assumption, considering the more general case when sample sizes are different.

3.2.1 Variation Within Samples

The sample variance of each treatment group is an estimate of the common population variance σ^2. The sample variance in group i is given by

$$s_i^2 = \frac{\sum_{j=1}^{n} (y_{ij} - \bar{y}_i)^2}{n-1} = \frac{SS_i}{n-1}$$

The numerator of the sample variance is the sum of the squared deviations of the observations from their mean, and we denote it by SS_i. The denominator $n - 1$ is the number of degrees of freedom that is associated with this sum of squares.

In our example of three groups ($k = 3$) of five observations each ($n = 5$), we have

$$s_1^2 = \frac{SS_1}{4}$$

$$= \frac{(9.5 - 6.64)^2 + (3.2 - 6.64)^2 + (4.7 - 6.64)^2 + (7.5 - 6.64)^2 + (8.3 - 6.64)^2}{4}$$

$$= \frac{27.27}{4} = 6.82$$

$$s_2^2 = \frac{SS_2}{4}$$

$$= \frac{(8.5 - 6.72)^2 + (9.0 - 6.72)^2 + (7.9 - 6.72)^2 + (5.0 - 6.72)^2 + (3.2 - 6.72)^2}{4}$$

$$= \frac{25.11}{4} = 6.28$$

$$s_3^2 = \frac{SS_3}{4}$$

$$= \frac{(7.7 - 10.52)^2 + (11.3 - 10.52)^2 + (9.7 - 10.52)^2 + (11.5 - 10.52)^2 + (12.4 - 10.52)^2}{4}$$

$$= \frac{13.73}{4} = 3.43$$

Each sample variance is an estimate of the common population variance σ^2, and this is true whether or not the population means are the same. The average of these three variances, $(s_1^2 + s_2^2 + s_3^2)/3 = (6.82 + 6.28 + 3.43)/3 = 5.51$, provides an even better pooled estimate of σ^2. It estimates the variation within samples.

In the general case with k treatments (groups) and varying sample sizes, the pooled estimate of the population variance σ^2 is given by

$$s_W^2 = \frac{(n_1 - 1)s_1^2 + (n_2 - 1)s_2^2 + \cdots + (n_k - 1)s_k^2}{(n_1 - 1) + (n_2 - 1) + \cdots + (n_k - 1)} = \frac{n_1 - 1}{N - k}s_1^2 + \frac{n_2 - 1}{N - k}s_2^2 + \cdots + \frac{n_k - 1}{N - k}s_k^2$$

It is a weighted average of the k individual within-sample variances; hence the subscript W (W for "within"). The weights are proportional to the degrees of freedom that are associated with each variance. In the case of equal sample sizes, the pooled estimate simplifies to the unweighted average of the individual estimates.

The pooled estimate is called the *within-sample (or within-treatment) estimate* of σ^2. Since $(n_i - 1)s_i^2 = SS_i$, we can write it as

$$s_W^2 = \frac{SS_1 + SS_2 + \cdots + SS_k}{N - k} = \frac{\sum_{i=1}^{k} \sum_{j=1}^{n_i} (y_{ij} - \bar{y}_i)^2}{N - k}$$

The numerator in this equation is called the *sum of squares within groups* (SSW). Its degrees of freedom are given by the sum of the degrees of freedom of the individual sums of squares, $(n_1 - 1) + (n_2 - 1) + \cdots + (n_k - 1) = (\sum_{i=1}^{k} n_i) - k = N - k$. The pooled estimate of the population variance s_W^2 is the ratio of the sum of squares SSW and its degrees of freedom $N - k$; it is called the *mean square error within groups*.

In our example, $N - k = 15 - 3 = 12$, and

$$s_W^2 = \frac{s_1^2 + s_2^2 + s_3^2}{3} = \frac{SS_1 + SS_2 + SS_3}{12} = \frac{27.72 + 25.11 + 13.73}{12} = 5.51$$

3.2.2 Variation Between Samples

We discussed the central limit effect for a sample average in Chapter 2. Suppose we take random samples of n observations from a population with mean μ and standard deviation σ. We learned in Section 2.5.1 that the sampling distribution of sample averages $\bar{y} = (y_1 + y_2 + \cdots + y_n)/n$ is approximately normal with mean μ and standard deviation $\sigma_{\bar{Y}} = \sigma/\sqrt{n}$. The variance of the sampling distribution is given by the square of the standard deviation, $\sigma_{\bar{Y}}^2 = \sigma^2/n$.

Assume that the sample sizes of the k treatment groups are the same and suppose that the null hypothesis $\mu_1 = \mu_2 = \cdots = \mu_k$ is true. Then the group averages $\bar{y}_i$ (for $i = 1, 2, \ldots, k$) are realizations from the same distribution with a common mean and variance σ^2/n. The sample variance of the k group averages $\bar{y}_1, \bar{y}_2, \ldots, \bar{y}_k$,

$$s_{\bar{y}}^2 = \frac{(\bar{y}_1 - \bar{y})^2 + (\bar{y}_2 - \bar{y})^2 + \cdots + (\bar{y}_k - \bar{y})^2}{k - 1} = \frac{\sum_{i=1}^{k}(\bar{y}_i - \bar{y})^2}{k - 1}$$

is an estimate of σ^2/n. Hence, a second estimate of the variance σ^2 is given by

$$s_B^2 = \frac{\sum_{i=1}^{k} n(\bar{y}_i - \bar{y})^2}{k - 1} = n s_{\bar{y}}^2$$

The numerator $\sum_{i=1}^{k} n(\bar{y}_i - \bar{y})^2$ measures the variation of the k group means from the grand mean $\bar{y} = \sum_i \sum_j y_{ij}/N$; it represents the variability between the groups and is called the *sum of squares between groups* (SSB). The denominator $k - 1$ is the number of degrees of freedom that is associated with this sum of squares; note that there are k means and one restriction. The estimate s_B^2 is called the *mean square between groups*; hence the subscript B.

In the example in Table 3.1 with $n = 5$, $\bar{y}_1 = 6.64$, $\bar{y}_2 = 6.72$, $\bar{y}_3 = 10.52$, $\bar{y} = (6.64 + 6.72 + 10.52)/3 = 7.96$,

$$s_{\bar{y}}^2 = \frac{(6.64 - 7.96)^2 + (6.72 - 7.96)^2 + (10.52 - 7.96)^2}{2} = 4.92$$

and

$$s_B^2 = \frac{\sum_{i=1}^{k} n(\bar{y}_i - \bar{y})^2}{k - 1} = (5)(4.92) = 24.6$$

3.2.3 Comparing the Within-Sample and the Between-Sample Estimates of σ^2

In the example, the between-sample estimate $s_B^2 = 24.6$ is much larger than the within-sample estimate $s_W^2 = 5.51$. What inferences can we draw from this discrepancy?

The within-sample estimate is an estimate of the population variance σ^2, and this is true whether or not the population means are equal. The between-sample estimate is also an estimate of σ^2, but only if the population means are the same. If they are not, the between-sample estimate is inflated, in that it also reflects the differences between the population means.

A test of the null hypothesis that the population means are equal examines the ratio of the between-sample and the within-sample variance estimates, $s_B^2/s_W^2 = 24.6/5.51 = 4.46$. Under the null hypothesis of equal population means, the two estimates of σ^2 will be similar in magnitude, and the ratio will be close to 1. If the null hypothesis is false, the numerator in this ratio will be larger than the denominator, and the ratio will be greater than 1.

How large does the ratio have to be before one can reject the null hypothesis that the population means are equal? The answer is given by the F-distribution, which was introduced in Section 2.2.2. The F-distribution is used to test the equality of two variances and arises in the following way. Suppose we take two independent samples from a normal distribution with variance σ^2: one sample of size n_1 and the other of size n_2. Then the ratio of the two sample variances $\dfrac{s_1^2}{s_2^2} = \dfrac{\sum_{j=1}^{n_1}(y_{1j} - \bar{y}_1)^2/(n_1 - 1)}{\sum_{j=1}^{n_2}(y_{2j} - \bar{y}_2)^2/(n_2 - 1)}$ follows an F-distribution with $n_1 - 1$ and $n_2 - 1$ degrees of freedom.

Applying this result to our problem, a test of the null hypothesis that the k population means are equal is given by the ratio of the between-sample and the within-sample variance estimates,

$$ F = \frac{s_B^2}{s_W^2} = \frac{SSB/(k - 1)}{SSW/(N - k)} = \frac{\sum_{i=1}^{k} n(\bar{y}_i - \bar{y})^2/(k - 1)}{\sum_{i=1}^{k}\sum_{j=1}^{n}(y_{ij} - \bar{y}_i)^2/(N - k)} $$

Under the null hypothesis of equal population means, this F-ratio follows an F-distribution with $k - 1$ and $N - k$ degrees of freedom. In our example, $F = s_B^2/s_W^2 = 24.6/5.51 = 4.46$. The numerator has $k - 1 = 3 - 1 = 2$ degrees of freedom, while the denominator has $N - k = 15 - 3 = 12$ degrees of freedom. The probability value is the probability of obtaining the value 4.46 or larger from this F-distribution. It is given by $P[F(2, 12) \geq 4.46] = 0.036$. This is smaller than the usually adopted 5% significance level and gives us reason to reject the null hypothesis. We conclude that there are differences among the three population means.

Excel or any other statistical software package can be used to find the probability value. For example, the Excel command FDIST(4.46, 2, 12) returns the probability value 0.036. Alternatively, we obtain the 95th percentile of the $F(2, 12)$ distribution and use it as the critical value for the test. The Excel command FINV(0.05, 2, 12) returns 3.89. Our test statistic exceeds this critical value.

3.2.4 The Analysis of Variance Table and the Output of Standard Computer Software

The earlier equation for the F-statistic assumes that the sample sizes are the same for all treatments groups. For the more general case with different sample sizes, the equation must be modified slightly. The only change is in the expression for the sum of squares between

TABLE 3.3

ANOVA Table for the Completely Randomized Experiment

Source of Variation	Sum of Squares SS	Degrees of Freedom df	Mean Squares MS	F-Ratio
Between groups	$\sum_{i=1}^{k} n_i(\bar{y}_i - \bar{y})^2$	$k - 1$	$SSB/(k-1)$	$F = \dfrac{SSB/(k-1)}{SSW/(N-k)}$
Within groups	$\sum_{i=1}^{k}\sum_{j=1}^{n_i}(y_{ij} - \bar{y}_i)^2$	$N - k$	$SSW/(N-k)$	
Total	$\sum_{i=1}^{k}\sum_{j=1}^{n_i}(y_{ij} - \bar{y})^2$	$N - 1$		

groups, $SSB = \sum_{i=1}^{k} n_i(\bar{y}_i - \bar{y})^2$. The common sample size n in the earlier expression is replaced by n_i. The F-statistic is given by

$$F = \frac{\sum_{i=1}^{k} n_i(\bar{y}_i - \bar{y})^2/(k-1)}{\sum_{i=1}^{k}\sum_{j=1}^{n_i}(y_{ij} - \bar{y}_i)^2/(N-k)}$$

The calculations are summarized in a convenient format in the *analysis of variance (ANOVA) table*; see Table 3.3. It can be shown that the total sum of the squared deviations of the observations from their common mean $SST = \sum_{i=1}^{k}\sum_{j=1}^{n_i}(y_{ij} - \bar{y})^2$ can be partitioned as

$$\underbrace{\sum_{i=1}^{k}\sum_{j=1}^{n_i}(y_{ij} - \bar{y})^2}_{\substack{SST \\ \text{Total Sum of Squares}}} = \underbrace{\sum_{i=1}^{k} n_i(\bar{y}_i - \bar{y})^2}_{\substack{SSB \\ \text{SS Between Groups}}} + \underbrace{\sum_{i=1}^{k}\sum_{j=1}^{n_i}(y_{ij} - \bar{y}_i)^2}_{\substack{SSW \\ \text{SS Within Groups}}}$$

The *total sum of squares* (SST) is shown in the last row of Table 3.3. It measures the variability of the observations around their common mean $\bar{y} = \sum_{i=1}^{k}\sum_{j=1}^{n_i} y_{ij}/N$, and its degrees of freedom are $N - 1$.

Programs for calculating the ANOVA table and for testing the hypothesis that the population means are the same are part of most statistical software packages. To illustrate, we use Minitab, a popular and useful statistical software package. Commands in Minitab carry out statistical analyses of data that are entered into columns and rows of a spreadsheet. In this example, one enters the response (percent changes) and the treatment identifier (1, 2, 3) into two columns, say, columns 1 and 2. There are 15 rows in each column because there are 15 stores. The first row has 9.5 in column 1 and 1 in column 2; the second row has 3.2 in column 1 and 1 in column 2; . . . ; the last row has 12.4 in column 1 and 3 in column 2. The consecutive arrangement of the 15 stores is arbitrary given that there is no particular order to the stores. The Minitab command "ANOVA > One-Way" provides the ANOVA table and the confidence intervals that you see at the bottom of Table 3.4. Other statistical software packages such as JMP and SPSS work in pretty much the same way.

TABLE 3.4
Minitab Output: Test of Three Product Displays

```
Source   DF      SS     MS     F      P
Display   2   49.17  24.58  4.46  0.036
Error    12   66.11   5.51
Total    14  115.28
                             Individual 95% CIs For Mean Based on
                             Pooled StDev
Display   N    Mean   StDev  ---+----------+----------+----------+------
1         5   6.640   2.611  (----------*----------)
2         5   6.720   2.505   (--------*---------)
3         5  10.520   1.853                       (--------*---------)
                             ---+----------+----------+----------+------
                              5.0        7.5       10.0       12.5

Pooled StDev = 2.347
```

Table 3.4 shows the *F*-statistic and the probability value that we had calculated earlier. The test result provides fairly strong evidence that the means are different. Once we have decided that there are differences among the population means, we look at how they differ. This can be done by displaying the data graphically. Dot diagrams, separate for each group but shown on the same scale, are very informative because they show differences in the levels as well as differences in the variability (which, for our test to work, should be about the same). The sample means and their 95% confidence intervals shown in Table 3.4 are also very informative. The mean square error within the groups, $s_W^2 = 5.51$, estimates the variance of individual observations by pooling the variability across the k groups. Its square root gives the pooled standard deviation $s_W = \sqrt{5.51} = 2.347$, which is also listed in the Minitab output. We use it to calculate confidence intervals of the population means. The 95% confidence interval for μ_i is given by $\bar{y}_i \pm (t)\dfrac{s_W}{\sqrt{n_i}}$, where t is the 97.5th percentile of the t-distribution with $N - k$ degrees of freedom, which are the degrees of freedom that are associated with the pooled estimate.

The confidence intervals show how the means differ. The third display is more effective than the first two. Also, there is not much difference between displays 2 and 3.

3.3 THE RANDOMIZED COMPLETE BLOCK EXPERIMENT

A large supermarket chain plans to test three different versions of an in-store promotion. The firm identifies 15 stores in one region to participate in the experiment. A test of a particular version of the promotion in a store will run for one week. The company wants to run the entire experiment over a consecutive 5-week period and plans to run three tests per week. Initially, the marketing director decided to use the following completely randomized design. Each promotion strategy would be randomly assigned to five stores. Then three of the 15 stores would be randomly chosen for the first week, three other stores would be randomly chosen for the second week, and so forth.

A young analyst in the marketing department, fresh from a course in experimental design, suggests an alternative. She realizes that in a completely randomized design there is a chance that all three stores selected for week 1 could have been assigned promotion

TABLE 3.5
Results of a Blocked Experiment with Three Different In-Store Promotions

| Treatment | BLOCK (WEEK) | | | | | Treatment Average |
	Week 1	Week 2	Week 3	Week 4	Week 5	
Version 1	52	47	44	51	42	47.2
Version 2	60	55	49	52	43	51.8
Version 3	56	48	45	44	38	46.2
Block (week) Average	56	50	46	49	41	48.4

TABLE 3.6
Responses from the Randomized Complete Block Experiment (the General Case)

| Treatment | BLOCK | | | | Treatment Average |
	1	2	...	b	
1	y_{11}	y_{12}	...	y_{1b}	$\bar{y}_1.$
2	y_{21}	y_{22}	...	y_{2b}	$\bar{y}_2.$
$\vdots$	$\vdots$	$\vdots$	$\vdots$	$\vdots$	$\vdots$
k	y_{k1}	y_{k2}	...	y_{kb}	$\bar{y}_k.$
Block Average	$\bar{y}._1$	$\bar{y}._2$	...	$\bar{y}._b$	$\bar{y}..$

(treatment) A, while in week 2, none of the stores would be testing A. She points out that the effectiveness of a promotion might depend on the week in which it is run, due to differences in the weather and other conditions that might vary from week to week. She explains, "In this case there are two sources of variation, the promotions themselves and the week. To eliminate the week as a source of variation, we should run the experiment in 'blocks' with each of the three versions of the promotion tested in every week." The marketing director is impressed and decides to follow the analyst's advice. The design and results are shown in Table 3.5. Based on the percentage increase in sales, the firm measures the effectiveness of a promotion on a scale from 0 to 100.

In the general randomized complete block design, there are k treatments and b blocks, and $n = bk$ observations. The observations y_{ij}, for $i = 1, 2, \ldots, k$ (treatments) and $j = 1, 2, \ldots, b$ (blocks), are arranged in Table 3.6. The observations in the second column represent the k treatment responses on the first block; the observations in the third column are the responses on the second block, and so forth; and the penultimate column contains the responses on block b. The treatment means (averaged over the blocks) are denoted by $\bar{y}_i.$, for $i = 1, 2, \ldots, k$. The block means (averaged over the k treatments) are given by $\bar{y}._j$, for $j = 1, 2, \ldots, b$. The symbol $\bar{y}..$ denotes the overall average.

Observe our dot notation for averages. A dot in place of an index expresses the fact that we have averaged the observations over that index. For example, $\bar{y}._j$ is the average for block j, averaging over all treatments in that block. The overall average $\bar{y}..$ is the result of averaging over both indexes i and j.

We model the observation y_{ij} in treatment group i and block j through an overall level and additive effects of treatment i and block j. Using the summary data in Table 3.6, we estimate the observation y_{ij} by $\bar{y}.. + (\bar{y}_i. - \bar{y}..) + (\bar{y}._j - \bar{y}..)$. The first term is an estimate of the grand mean; the second term is an estimate of the incremental effect of treatment i;

TABLE 3.7
ANOVA Table for the Randomized Complete Block Experiment

Source of Variation	Sum of Squares SS	Degrees of Freedom	Mean Squares MS	F-Ratio
Treatment	$SS(\text{TR}) = b \sum_{i=1}^{k} (\bar{y}_{i\cdot} - \bar{y}..)^2$	$k - 1$	$MS(\text{TR})$	$\dfrac{MS(\text{TR})}{MS(\text{error})}$
Block	$SS(\text{BL}) = k \sum_{j=1}^{b} (\bar{y}_{\cdot j} - \bar{y}..)^2$	$b - 1$	$MS(\text{BL})$	$\dfrac{MS(\text{BL})}{MS(\text{error})}$
Error	$SS(\text{error}) = \sum_{i=1}^{k} \sum_{j=1}^{b} (y_{ij} - \bar{y}_{i\cdot} - \bar{y}_{\cdot j} + \bar{y}..)^2$	$(k-1)(b-1)$	$MS(\text{error})$	
Total	$SST = \sum_{i=1}^{k} \sum_{j=1}^{b} (y_{ij} - \bar{y}..)^2$	$kb - 1$		

and the last term is an estimate of the incremental effect of block j. The difference between the observation and this estimate is the error component, $y_{ij} - [\bar{y}.. + (\bar{y}_{i\cdot} - \bar{y}..) + (\bar{y}_{\cdot j} - \bar{y}..)] = y_{ij} - \bar{y}_{i\cdot} - \bar{y}_{\cdot j} + \bar{y}...$

This model allows us to express the deviation of each observation from the grand mean as

$$y_{ij} - \bar{y}.. = (\bar{y}_{i\cdot} - \bar{y}..) + (\bar{y}_{\cdot j} - \bar{y}..) + (y_{ij} - \bar{y}_{i\cdot} - \bar{y}_{\cdot j} + \bar{y}..)$$

The first component on the right-hand side compares the treatment mean to the overall mean; the second component compares the block mean to the overall mean; and the last component measures the error after correcting the observation for its treatment average and block average.

Squaring the left- and right-hand sides of the equation, summing over both indexes i and j, and using the fact that all sums of cross-product terms are zero lead to the sum of squares decomposition:

$$\underbrace{\sum_{i=1}^{k} \sum_{j=1}^{b} (y_{ij} - \bar{y}..)^2}_{\substack{SST \\ \text{Total Sum of Squares}}} = \underbrace{b \sum_{i=1}^{k} (\bar{y}_{i\cdot} - \bar{y}..)^2}_{\substack{SS \\ \text{Treatment}}} + \underbrace{k \sum_{j=1}^{b} (\bar{y}_{\cdot j} - \bar{y}..)^2}_{\substack{SS \\ \text{Block}}} + \underbrace{\sum_{i=1}^{k} \sum_{j=1}^{b} (y_{ij} - \bar{y}_{i\cdot} - \bar{y}_{\cdot j} + \bar{y}..)^2}_{\substack{SS \\ \text{Error}}}$$

$$SST = SS(\text{TR}) + SS(\text{BL}) + SS(\text{error})$$

Similarly, the degrees of freedom can be partitioned as

$$(kb - 1) = (k - 1) + (b - 1) + (k - 1)(b - 1)$$

The sum of squares and their degrees of freedom become part of the ANOVA table in Table 3.7.

Here, we are testing whether the treatment means differ and whether the block means differ; thus, there are two F-ratios. The relevant statistic for testing whether there are statistically significant treatment effects is given in the last column. The F-statistic $\dfrac{MS(\text{TR})}{MS(\text{error})}$ needs to be compared to the percentiles of the $F(k - 1, (k - 1)(b - 1))$ distribution. A test

TABLE 3.8
*Minitab Output: Blocked Experiment with Three
Different In-Store Promotions*

Source	DF	SS	MS	F	P
Treatment	2	89.2	44.60	7.62	0.014
Block (Week)	4	363.6	90.90	15.54	0.001
Error	8	46.8	5.85		
Total	14	499.6			

at significance level 0.05 concludes that there are significant treatment effects if this F-statistic is larger than the 95th percentile of that distribution. The second F-statistic $\frac{MS(\text{BL})}{MS(\text{error})}$ tests for block effects and is compared to the 95th percentile of the $F(b - 1, (k - 1)(b - 1))$ distribution.

Our example considers $k = 3$ displays and $b = 5$ blocks (weeks). The observations in Table 3.5, together with the treatment and block means, can be used to calculate the sums of squares in Table 3.7. We have done this for illustration, even though statistical computer software will be used in practice.

$$SST = (52 - 48.4)^2 + (47 - 48.4)^2 + \cdots + (60 - 48.4)^2 + \cdots + (38 - 48.4)^2 = 499.6$$

$$SS(\text{TR}) = 5[(47.2 - 48.4)^2 + (51.8 - 48.4)^2 + (46.2 - 48.4)^2] = 89.2$$

$$SS(\text{BL}) = 3[(56 - 48.4)^2 + (50 - 48.4)^2 + (46 - 48.4)^2 + (49 - 48.4)^2 + (41 - 48.4)^2]$$

$$= 363.6$$

and

$$SS(\text{error}) = SST - SS(\text{TR}) - SS(\text{BL}) = 499.6 - 89.2 - 363.6 = 46.8$$

We use Minitab for the calculations. The ANOVA table in Table 3.8 was obtained with the Minitab command "ANOVA > Two-Way." The spreadsheet containing the data consists of three columns: column 1 contains the response (effectiveness rating); column 2, the treatment (promotion) indicators (1, 2, 3); and column 3, the blocking groups (weeks 1 through 5). There are 15 rows to each column. Row 1 contains 52 in column 1, 1 in column 2, and 1 in column 3. Row 2 contains 47 in column 1, 1 in column 2, and 2 in column 3, and so forth; the 15th row contains 38 in column 1, 3 in column 2, and 5 in column 3. The order in which the rows are entered is arbitrary.

The ANOVA table shows that the F-ratio for testing the null hypothesis of no treatment effect is $MS(\text{TR})/MS(\text{error}) = 44.60/5.85 = 7.62$. The 95th percentile of the $F(2, 8)$ distribution is 4.46. We conclude that there are treatment differences because the F-ratio is larger than this percentile. We come to the same conclusion when looking at the small probability value $P[F(2, 8) > 7.62] = 0.014$. There are differences among the three promotions, with promotion 2 scoring significantly higher than promotions 1 and 3, which are about the same.

Table 3.8 shows that there is considerable variability among the weeks (F-statistic = 15.54, with probability value 0.001), and that it is important to incorporate this variability into the analysis when comparing the three treatments. What would happen if the blocking effect were ignored? Combining $SS(\text{block}) + SS(\text{error}) = 363.6 + 46.8 = 410.4$ with

4 + 8 = 12 degrees of freedom, and calculating the test statistic that is appropriate for the completely randomized design in Section 3.2, would have led to the *F*-statistic

$$\frac{MS(\text{TR})}{MS(\text{error})} = \frac{89.2/2}{(363.6 + 46.8)/12} = 1.30$$

with probability value $P[F(2, 12) > 1.30] = 0.31$. An analysis that ignores the week effects would have made the error of accepting the null hypothesis and concluding that there are no differences among the treatments.

Note. The analysis in this section assumes that we have exactly one observation at each factor-block combination. The Minitab command "ANOVA > Two-Way" fails if there are any missing observations. In this situation, one needs to use the general linear (regression) model approach (Minitab command "ANOVA > General Linear Model") to analyze the data; this will be discussed in Chapter 8.

3.4 CASE STUDY

This case is adapted from Clarke (1987). Additional relevant details and further analyses are discussed in Case 11 of the case study appendix.

Researchers at the United Dairy Industry Association (UDIA) evaluated the results of a recent field experiment to test the impact of varying levels of advertising on the sales of cheese. The principal objective of the study was to measure the retail sales response (pounds of cheese sold) to varying levels of advertising. Four test markets, selected from different geographic regions, were used in this study. Executives determined the levels of advertising to be tested in the experiment. It was believed that the levels should be distinct enough to generate measurable differences in the results. They decided to test the impact of four levels of advertising: 0 cents (level A), 3 cents (B), 6 cents (C), and 9 cents (D), all expressed on a per-capita basis. The 6-cents per-capita level represents a national campaign costing approximately $12 million (in 1973). The principal medium for advertising was television, with point-of-purchase display materials in stores and newspaper ads playing a secondary role. Each of the four levels of advertising was implemented within each test market during a 3-month period between May 1972 and April 1973; see Table 3.9. The sequence in which the advertising levels were tested was selected so that each advertising level was used in only one test market during any one time period. You can check that each letter in Table 3.9 (A, B, C, D) appears only once in each column and each row. Such an arrangement is referred to as a *Latin square* design. In Case 11 of the case study appendix, we will discuss the analysis of observations that originate from a Latin square design, and we will illustrate that this design can be used to further isolate a possible time effect. However, for the purpose of this illustration, we ignore the time effect and assume that the observations are from a blocked experiment that studies four treatments (A–D) on each of four blocks (test markets).

Within each market, UDIA executives obtained the cooperation of approximately 30 supermarkets in obtaining quarterly audits of cheese sales. The average cheese sales (in pounds per store) during 3-month periods between May 1972 and April 1973 are listed in Table 3.9.

The ANOVA table and the treatment and block means are given in Table 3.10. The results show that there are large differences between test markets (with highest sales in Rockford

TABLE 3.9

UDIA Study: Test Markets, Treatments, Test Periods, and Results (Average Sales Per Store)

	TEST MARKETS			
Time	Binghamton	Rockford	Albuquerque	Chattanooga
May–July 72	A	C	B	D
Aug–Oct 72	B	D	A	C
Nov–Jan 73	C	A	D	B
Feb–Apr 73	D	B	C	A

Sales/Store Treatment	BLOCK (TEST MARKET)			
	Binghamton	Rockford	Albuquerque	Chattanooga
A	7,360	13,153	11,852	7,557
B	7,364	11,258	12,089	7,900
C	8,049	13,880	11,800	8,501
D	9,010	13,147	11,450	7,776

TABLE 3.10

ANOVA Table: UDIA Study

Source	DF	SS	MS	F	P
Treatment	3	1917416	639139	1.31	0.329
Block (Market)	3	79308210	26436070	54.31	0.000
Error	9	4380871	486763		
Total	15	85606498			

Treatment	Mean	Block	Mean
A	9980.5	Binghamton	7945.8
B	9652.8	Rockford	12859.5
C	10557.5	Albuquerque	11797.8
D	10345.8	Chattanooga	7933.5

and Albuquerque), and that sales increase with the amount of advertising. However, the advertising effects are not statistically significant (probability value 0.329). We will revisit this in Case 11 of the case study appendix and investigate whether the use of time as an additional blocking variable changes our conclusions on the significance of the treatment effects.

3.5 NOBODY ASKED US, BUT . . .

The term *analysis of variance* (ANOVA) gives no indication that the procedure is about comparing means. But as we have seen in this chapter, we test whether several means differ by comparing variances. The *F*-distribution plays the key role in the analysis, and perhaps not surprisingly, *F* stands for Fisher, the most important statistician of the 20th century. But Fisher did not invent this distribution; it was derived by George Snedecor, who named it *F* to honor Fisher.

In the randomized complete block experiment, the term *block* comes from the origins of this design in agricultural studies. Blocks were created by aggregating contiguous parcels that were homogeneous in terms of soil composition and hence fertility. Fisher described these kinds of experiments in his book *Statistical Methods for Research Workers*, which was published in 1925.

EXERCISES

Exercise 1   Consider the data from a completely randomized experiment (Table 3.2). Express the deviation of the observation from the overall mean as

$$y_{ij} - \bar{y} = (\bar{y}_i - \bar{y}) + (y_{ij} - \bar{y}_i)$$

The first component on the right-hand side compares the treatment mean to the overall mean; the second component expresses the within-sample variation. Take the square of the expression, sum the squares over both indexes i ($i = 1, 2, \ldots, k$) and j ($j = 1, 2, \ldots, n_i$), and prove the sum of squares decomposition in Section 3.2.4:

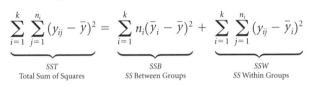

$$\underbrace{\sum_{i=1}^{k} \sum_{j=1}^{n_i} (y_{ij} - \bar{y})^2}_{\substack{SST \\ \text{Total Sum of Squares}}} = \underbrace{\sum_{i=1}^{k} n_i(\bar{y}_i - \bar{y})^2}_{\substack{SSB \\ \text{SS Between Groups}}} + \underbrace{\sum_{i=1}^{k} \sum_{j=1}^{n_i} (y_{ij} - \bar{y}_i)^2}_{\substack{SSW \\ \text{SS Within Groups}}}$$

Show that the sum of the cross products is zero; that is,

$$\sum_{i=1}^{k} \sum_{j=1}^{n_i} (\bar{y}_i - \bar{y})(y_{ij} - \bar{y}_i) = 0$$

Exercise 2   Consider the data from a randomized complete block experiment (Table 3.6). Express the deviation of the observation from the overall mean as

$$y_{ij} - \bar{y}.. = (\bar{y}_{i.} - \bar{y}..) + (\bar{y}_{.j} - \bar{y}..) + (y_{ij} - \bar{y}_{i.} - \bar{y}_{.j} + \bar{y}..)$$

The first component on the right-hand side compares the treatment mean to the overall mean; the second component compares the block mean to the overall mean; and the last component measures the error after correcting the observation for its treatment average and block average. Prove the sum of squares decomposition in Section 3.3:

$$\underbrace{\sum_{i=1}^{k} \sum_{j=1}^{b} (y_{ij} - \bar{y}..)^2}_{\substack{SST \\ \text{Total Sum of Squares}}} = \underbrace{b \sum_{i=1}^{k} (\bar{y}_{i.} - \bar{y}..)^2}_{\substack{SS \\ \text{Treatment}}} + \underbrace{k \sum_{j=1}^{b} (\bar{y}_{.j} - \bar{y}..)^2}_{\substack{SS \\ \text{Block}}} + \underbrace{\sum_{i=1}^{k} \sum_{j=1}^{b} (y_{ij} - \bar{y}_{i.} - \bar{y}_{.j} + \bar{y}..)^2}_{\substack{SS \\ \text{Error}}}$$

Exercise 3   You study the monthly amounts seventh- and eighth-grade boys and girls spend on entertainment such as movies, music CDs, and candy. Representative samples of children within the Iowa City school district were selected, and children were asked about their spending habits. The following results were obtained.

	7th-Grade Boys	8th-Grade Boys	7th-Grade Girls	8th-Grade Girls
Sample size	30	25	30	25
Mean ($)	20.1	23.2	19.6	25.0
Standard deviation ($)	6.0	5.6	5.3	7.0

(a) Test whether or not the four groups differ with respect to their mean spending amounts.

(b) Follow up on your analysis in (a) if you find differences. In particular, assess whether there are differences in the mean spending amounts of seventh- and eighth-grade boys and in the mean spending amounts of seventh- and eighth-grade girls. Test whether the yearly changes in the mean spending amounts differ between boys and girls.

Exercise 4 Your goal is to determine the effectiveness of four different TV spots.

(a) To avoid program overlap, you select four different market regions, and you assign one of the four TV spots to each region. The programs are aired for one month, and sales of the advertised product are recorded in 16 stores in each of the four markets. Store-specific sales for the previous month are also available. Discuss how you would analyze the data to learn which of the four TV spots is preferable. What additional assumption do you need to make to infer that the winning ad would also work best in future months?

(b) Assume that all your stores are in a single market, and that all four TV spots must be aired in this single market. You decide to run the spots in four consecutive months. You collect sales data on 16 stores, with each store being observed under all four TV spots. Discuss how you would analyze the data to learn which of the four TV spots is preferable. Discuss the differences to your earlier strategy in (a).

(c) Discuss the advantage and the danger of your design in (b). For example, would your analysis be affected if sales were seasonal? Discuss ways of incorporating known seasonality into your analysis. Discuss ways of blocking the experiment with respect to stores as well as months.

Exercise 5 See Exercise 11 in Chapter 2. Three different fabrics (A, B, C) are tested on a wear tester that can compare three materials in a single run. A wear tester is a mechanical device that rubs a material against an object. Our particular machine has three attachments that allow the comparison of three materials. Variability from one run to the next is expected. However, within the same run, the conditions for the three fabrics are fairly homogeneous. The assignment of the fabric to the attachment is randomized in each run. The weight losses (in milligrams) from 10 runs are given in the following table.

Test whether or not the average weight losses of the three fabrics are the same. Which fabric(s) has the lowest weight loss? What would happen to your conclusions if you ignored the run effect?

| | | | | | RUN | | | | | |
Fabric	1	2	3	4	5	6	7	8	9	10
A	36	26	31	38	28	37	22	34	25	30
B	39	27	35	42	31	39	21	37	28	34
C	40	28	34	43	30	39	22	36	27	33

Exercise 6 The female cuckoo lays her eggs into the nests of foster parents. The foster parents are usually deceived, probably because of the similarity in the sizes of the eggs. Lengths

of cuckoo eggs (in millimeters) found in the nests of hedge sparrows, robins, and wrens are shown below.

Hedge sparrow: 22.0, 23.9, 20.9, 23.8, 25.0, 24.0, 21.7, 23.8, 22.8, 23.1, 23.1, 23.5, 23.0, 23.0

Robin: 21.8, 23.0, 23.3, 22.4, 23.0, 23.0, 23.0, 22.4, 23.9, 22.3, 22.0, 22.6, 22.0, 22.1, 21.1, 23.0

Wren: 19.8, 22.1, 21.5, 20.9, 22.0, 21.0, 22.3, 21.0, 20.3, 20.9, 22.0, 20.0, 20.8, 21.2, 21.0

It is believed that the size of the egg influences the female cuckoo in her selection of the foster parent. Do the data support this hypothesis? Test whether or not the mean lengths of cuckoo eggs found in nests of the three foster-parent species are the same.

Exercise 7   The plant manager wants to investigate the productivity of three groups of workers: those with little, average, and considerable work experience. Since the productivity depends to some degree on the day-to-day variability of the available raw materials, which affects all groups in a similar fashion, the manager suspects that the comparison should be blocked with respect to day. The results (productivity, in percent) from five production days are given in the following table:

			DAY		
Experience	1	2	3	4	5
A	53	58	49	52	60
B	55	57	53	57	64
C	60	62	55	64	69

(a) Are there differences in the mean productivity among the three groups?

(b) Has the blocking made a difference?

Exercise 8   A feedlot operator wants to compare the effectiveness of three different cattle-feed supplements. He selects a random sample of 15 one-year-old heifers from his lot of more than 1,000 and divides them into three groups at random. Each group gets a different feed supplement. The weight gains over a 6-month period are shown below. One heifer in group A was lost due to an accident.

Group	Weight Gains (pounds)
A	500, 650, 530, 680
B	700, 620, 780, 830, 860
C	500, 520, 400, 580, 410

(a) Are there differences in the mean weight gains of the three feed supplements?

(b) If you could start the experiment over, would you suggest improvements that would help make the comparisons more precise? What about a blocking arrangement on initial weight?

4 | TWO-LEVEL FACTORIAL EXPERIMENTS

4.1 INTRODUCTION

In this chapter, we begin focusing on the heart of this book. In Chapter 3, we were concerned with comparing the effectiveness of several treatments, each a version of a single factor. In one example, the factor was a product display, and we tested three different displays based on a comparison of weekly store sales. Here, we extend that discussion, focusing on experiments with multiple factors.

Experimental design methods have roots in agriculture, and we use an example from that field to introduce the material we will cover. Suppose we are experimenting with ways to improve the yield of corn, and we identify three factors that seem important—type of fertilizer, variety of seed, and type of pesticide. We decide to test two fertilizer formulations, two kinds of seed, and two different pesticides. As we discussed in Chapter 1, the traditional method for testing multiple factors is to test one factor at a time. But Fisher (1935) showed that a factorial design that tests all factors simultaneously is a much better approach. Using Fisher's method, we test the $2 \times 2 \times 2 = 8$ possible combinations of fertilizers, seeds, and pesticides. We divide our experimental field into 32 equal-sized plots and randomly assign each of the eight combinations to four plots. For each of the 32 plots, we measure the number of bushels of corn produced. This factorial arrangement would allow us to compare the two fertilizers, the two kinds of seeds, and the two pesticides. It would also allow us to uncover any interactions between factors. For example, it may turn out that seed variety 1 is better than variety 2 when fertilizer 1 is used for both, but that the opposite is true when fertilizer 2 is used with both seeds.

Consider another example. An advertising agency is designing an online ad. It identifies three factors to test, with the response being the fraction of ad viewers who sign up for the advertised service. One factor is the ad copy—a traditional version or a more modern one. The second factor is the font—a traditional font or a fancier one, while the third factor is the background color—white or blue. In a factorial design, one of the eight possible ads would randomly be sent to each viewer. The question is what is important here? Is it the copy, the font, or the background color that matters? Do the factors interact in the sense that the fancier font works better with the blue background? These are the kinds of issues we will address in this chapter.

In this and the next two chapters, we discuss experiments where each factor is studied at just two levels. In Chapters 7 and 8, we consider the more general case when a factor may have more than two levels, for example, four different levels for ad copy, three background colors, and two different fonts.

4.1.1 Basic Terms

We start by defining some important terms.

The *factors* are the variables whose effects are being studied. In the advertising experiment discussed earlier, the factors are the ad copy, the font, and the background color. In an industrial experiment, the factors might be temperature, pressure, and the type of chemical catalyst. In an agricultural experiment, the factors might be type of seed, type of fertilizer, and the amount of water; whereas in a marketing experiment, the factors might be the color of the box, the price, and the dollars spent on advertising.

The *levels* are the specified values of each factor. As noted earlier, initially we will focus on 2-level designs; that is, each factor is set at one of two possible levels. For example, in the marketing experiment, the box is either red or blue.

The *response variable* is the performance measure, the dollar sales in the marketing experiment, or the number of bushels of corn in the agriculture experiment.

A *run* is a particular experiment with each factor at a specified level.

Each factor may be *continuous* or *categorical*, and as we will show in this chapter, the distinction is important. Factors such as temperature, pressure, amount of fertilizer, price, and dollars spent on advertising are continuous. Factors such as the type of ad copy, font, background color, color of the box, and catalyst are categorical.

4.2 AN EXAMPLE: REDUCING THE NUMBER OF CRACKED POTS

The following example illustrates some basic, important concepts. A company manufactures clay pots that are used to hold plants. For one of their newest products, the company has been experiencing an unacceptably high percentage of pots that crack during the manufacturing process. Company production engineers have identified three key factors they believe will affect cracking, and they decide to run an experiment to learn about the most important factor(s). The factors studied are the peak temperature in the kiln, the rate at which pots are cooled after being heated to the peak temperature, and a coefficient that describes the expansion of the clay pot. A higher peak temperature can be expected to reduce the percentage of cracks, but the higher temperature also increases operating costs. Cooling the pots at a faster rate would mean an increase in the number of pots produced per hour, but it could also increase the percentage of cracked pots. The coefficient of expansion depends on the composition of the clay. A supplier has offered the company a new clay mix that it asserts has a lower coefficient of expansion. The mix is being offered at the same price as the raw material that is currently used. A lower coefficient of expansion should decrease the incidence of cracks.

T A B L E   4 . 1

The Three Factors and Their Levels

	LEVEL	
Factors	−	+
Rate of cooling (R)	Slow	Fast
Temperature (T)	2000°F	2060°F
Coefficient of expansion (C)	Low	High

The firm wants to determine how changes in these three factors would affect the percentage of cracked pots, and the production engineers decide to experiment with each factor at two levels. The current settings are lower peak temperature, slower cooling rate, and higher coefficient of expansion. In experimental design, we use a standard notation with the low level of each factor denoted by minus (−) and the high level of each factor denoted by plus (+). Later in this chapter, we will discuss how these low and high levels would be determined in particular situations. Table 4.1 lists the three factors and their levels.

4.2.1 A Common Approach to Experimentation: Varying One Factor at a Time

A very common but inefficient approach to studying the effects of k (here $k = 3$) factors is to carry out successive experiments in which the levels of each factor are changed *one at a time*. Such experiments typically start with the current settings of the factors and begin by changing the level of the one factor that is considered the most important. The responses at the low and high settings of this one factor are compared while keeping all other factors fixed, and if there is a difference, the level at which the response is best is locked in for the next stage. The factor that is considered second most important is varied next. Again, responses at the low and high levels of this factor are compared, and the best level of this factor, if there is a difference, is locked in for all subsequent runs. This process continues until the last factor is reached.

Factor C (coefficient of expansion of the clay) was considered most critical. Applying this approach, we first set R = slow and T = 2,000, then carry out 4 runs with C = low (−) and 4 runs with C = high (+). Each run is the usual production batch of 100 pots. Suppose that the proportions of cracked pots were 5, 8, 3 and 8% for the 4 runs at the low (−) level of C, and 15, 12, 11, and 16% for the 4 runs at the high (+) level of C, resulting in averages of 6 and 13.5%, respectively. The (positive) difference of 7.5% indicates that it is better to set factor C at its low level. But it would be premature to conclude that the new clay mix with a low coefficient of expansion (factor C) decreases cracking in general. At this point, all we can say is that we observed a difference of 7.5% at a particular temperature, T = 2,000 (−), and a particular cooling rate, R = slow (−). We don't know if the same result would hold for temperature T = 2,060 or cooling rate R = fast.

Temperature was considered the second most critical factor. Next, we set the coefficient of expansion C at its low (best) level, and fix the cooling rate R at slow (−). We need to compare 4 runs with T = 2,000 and 4 runs with T = 2,060. We already have 4 runs with T = 2,000. So, we do 4 runs with T = 2,060. Suppose we found T = 2,060 to result in the better response (fewer cracked pots).

TABLE 4.2
Design Matrix for the Cracked Pots Problem

Run (standard order)	Rate of Cooling R	Temperature T	Coefficient of Expansion C
1	−	−	−
2	+	−	−
3	−	+	−
4	+	+	−
5	−	−	+
6	+	−	+
7	−	+	+
8	+	+	+

Finally, we fix temperature ($T = 2,060$) and the coefficient of expansion ($C = $ low) at their better settings, then compare 4 runs with $R = $ slow to 4 runs with $R = $ fast. We already have 4 runs with $R = $ slow, so we add four runs with $R = $ fast.

This—approach of changing one factor at a time—requires 16 runs. As a result of these 16 runs, we would only know the effect of each factor at one particular combination of settings of the other two. We would not know anything about interactions among the factors. For example, we would not know whether the effect of changing temperature from low to high depends on the level of the cooling rate. If such interactions are present, the experiment of changing one factor at a time could lead to the wrong conclusions, because it might not identify the best settings for the factors. We discuss the shortcomings of this approach in more detail in Appendix 4.1.

4.2.2 A Better Approach: Changing Factor Levels Simultaneously

Sir Ronald Fisher (1935) showed that a better approach is to vary the factors *simultaneously* and to study the response at each possible factor-level combination. With three factors at two levels each (as in our example), his factorial design requires just 8 runs, fewer than the 16 in the earlier approach of changing one factor at a time. In addition to the economy of fewer runs, the factorial design provides estimates of possible interactions and thus produces more information.

Table 4.2 shows the 8 runs of the factorial design with 3 factors at 2 levels each. We use minus (−) and plus (+) signs to represent the low and high levels of each factor.

The 8 runs are listed in the so-called standard order. For example, in run 1, all three factors are at their low levels, while in run 8, all three factors are at their high levels. The design matrix in standard order is easy to construct. We start with the first factor (in this case, R) with a minus sign and alternate the signs until we complete the column. For the second factor (T), we start with two minus signs and alternate the signs in groups of two. For the third factor (C), we start with four minus signs and alternate the signs in groups of four. This procedure gives us all $2^3 = 8$ factor-level combinations. These 8 factor-level combinations can be represented as the vertices of a cube; see Figure 4.1.

A factorial design with only two factors, A and B, has two columns, one for each factor, and 4 runs. The sequence $- + - +$ makes up the first column, and the second column

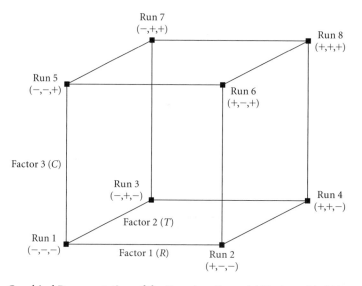

Figure 4.1 Graphical Representation of the Runs in a Factorial Design with 3 Factors, Each at 2 Levels

contains $- - + +$. The four factor-level combinations in this factorial design, arranged in standard order, are $(A = -, B = -), (A = +, B = -), (A = -, B = +),$ and $(A = +, B = +)$.

This method of generating the runs is easily extended to any number of factors. With four factors, the design consists of $2^4 = 16$ runs. For the first factor (A), we start with a minus sign and alternate signs until the levels of all 16 runs have been specified. Factor B starts with two minus signs and alternates signs in groups of two. Factor C starts with four minus signs and alternates signs in groups of four, and factor D has eight minus signs followed by eight plus signs.

This method of generating the runs in standard order is useful as it helps ensure that no combinations are missed. However, in carrying out the experiment, it is essential to perform the runs in random order. This randomization is important because there may be additional factors not included in the experiment that could influence the results. For example, there may be a day of week effect or other unknown factors that change with time. By randomizing, we ensure that the effects of these lurking or noise variables are distributed randomly across the factors. A simple way to do so is to put slips of papers into a box (numbered from 1 to 8 in the case of three factors) and draw them randomly, carrying out the runs in the order in which the slips were drawn.

4.3 THE TWO-LEVEL FACTORIAL DESIGN

In this and the next two chapters, we focus on 2-level designs. In Chapters 7 and 8, we will extend the methods developed here to cases where factors have more than two levels.

We use the notation 2^k to designate a factorial design with k factors, each having two levels. Such an experiment requires a total of 2^k distinct runs: $2^2 = 4$ runs for $k = 2$ factors, $2^3 = 8$ runs for $k = 3$ factors, $2^4 = 16$ runs for $k = 4$ factors, and so on.

TABLE 4.3
A 2^3 Factorial Design Matrix and Results for the Cracked Pots Problem

Run (standard order)	R	T	C	Percentage of Pots with Cracks
1	−	−	−	6
2	+	−	−	12
3	−	+	−	6
4	+	+	−	16
5	−	−	+	16
6	+	−	+	34
7	−	+	+	14
8	+	+	+	34

Let's return to our discussion of the ceramic pot example. Assume that a 2^3 factorial experiment has been run, resulting in the data shown in Table 4.3.

4.3.1 Calculating Main Effects

The *main effect* of a factor is defined as the change in the response variable when the level of the factor is changed from − (low) to + (high). Thus the main effect of the cooling rate R is the change in the percentage of pots that crack when the cooling rate is changed from slow to fast. It is the average percent of cracked pots at the fast cooling rate minus the average percent of cracked pots at the slow cooling rate. That is,

$$\text{cooling rate effect: } R = \frac{12 + 16 + 34 + 34}{4} - \frac{6 + 6 + 16 + 14}{4} = 24 - 10.5 = 13.5$$

In Table 4.3, notice that 12, 16, 34, and 34 are the responses when R is + (runs 2, 4, 6, and 8, respectively), while 6, 6, 16, and 14 are the responses when R is − (runs 1, 3, 5, and 7, respectively).

The main effect of temperature T is the average percent of cracked pots at the high (+) level of factor T minus the average percent at the low (−) level of T,

$$\text{temperature effect: } T = \frac{6 + 16 + 14 + 34}{4} - \frac{6 + 12 + 16 + 34}{4} = 17.5 - 17 = 0.5$$

The main effect of the coefficient of expansion C is the average percent of cracked pots at the high (+) level minus the average proportion at the low (−) level,

$$\text{expansion effect: } C = \frac{16 + 34 + 14 + 34}{4} - \frac{6 + 12 + 6 + 16}{4} = 24.5 - 10 = 14.5.$$

Note that we are using the same notation for the factors (R, T, C) and their estimated main effects. In most cases, it will be clear whether we are referring to the factor or to its estimated effect. In cases where there is the possibility of confusion, we will introduce separate notation for the estimated effects by putting parentheses around the factors, such as (R), (T), and (C).

4.3.2 Calculating 2-Factor Interactions

RC Interaction

As we will show, the effect of cooling rate is not independent of the coefficient of expansion. There is an interaction between these two factors, and we denote it by *RC*. We estimate the interaction between the cooling rate and the coefficient of expansion by comparing the effect of cooling rate (factor *R*) at the two levels of the coefficient of expansion (factor *C*).

With coefficient of expansion at $+$, the change in the percent of cracked pots when the cooling rate is changed from its low $(-)$ to high $(+)$ setting is

$$\frac{34 + 34}{2} - \frac{16 + 14}{2} = 34 - 15 = 19$$

In Table 4.3, notice that 34 and 34 are the responses when $C = +$ and $R = +$ (runs 6 and 8, respectively), and 16 and 14 are the responses when $C = +$ and $R = -$ (runs 5 and 7, respectively).

With coefficient of expansion at $-$, the change in the percent of cracked pots when the cooling rate is changed from its low $(-)$ to high $(+)$ setting is

$$\frac{12 + 16}{2} - \frac{6 + 6}{2} = 14 - 6 = 8$$

The effect of the cooling rate is much greater when the coefficient of expansion is at the $+$ level (19%) than when it is at the $-$ level (8%). If the coefficient of expansion is high, we expect more cracks when the cooling rate is increased from the slow rate to the fast rate.

By convention, the interaction between the two factors is defined as one-half of the difference between the average cooling rate effect with coefficient of expansion at $+$ and the average cooling rate effect with coefficient of expansion at $-$. Thus the interaction between factors R and C denoted by RC is given by $(19 - 8)/2 = 5.5$. Later in this chapter, we will show how to determine whether an effect is statistically significant. For now, let us assume that this interaction is statistically significant and not the result of random variation.

The square diagram in Figure 4.2 and the interaction diagram in Figure 4.3 represent convenient ways to compute and display an interaction. The square diagram lists the average response at each of the four possible combinations of settings of factors R and C. Each of the four numbers is the average of two responses. For example, when both the coefficient of expansion (C) and cooling rate (R) are at their $+$ levels (runs 6 and 8), the average response is $(34 + 34)/2 = 34$. When the coefficient of expansion C is at the low $(-)$ level, the effect of the cooling rate R is $14 - 6 = 8$, but when C is at the high $(+)$ level, the effect of the cooling rate R is $34 - 15 = 19$.

The interaction is shown graphically in the interaction diagram in Figure 4.3. Here, we connect the average response at the low and high levels of the cooling rate, and we do this separately for each level of the coefficient of expansion. Notice that the two lines have different slopes, reflecting the fact that there is an interaction. If there were no interaction, the two lines would be parallel or nearly so.

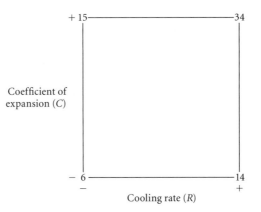

Figure 4.2   Square Diagram Showing the *CR* Interaction

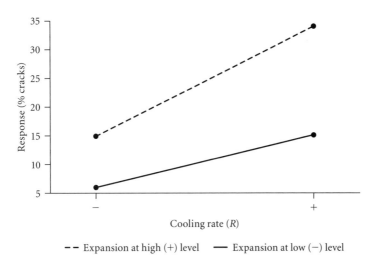

Figure 4.3   Interaction Diagram Between Coefficient of Expansion and Cooling Rate

RT Interaction

The square diagram for this interaction is shown in Figure 4.4. With temperature T at $+$, the effect of changing the cooling rate R from $-$ to $+$ is

$$\frac{16 + 34}{2} - \frac{6 + 14}{2} = 25 - 10 = 15$$

With temperature T at $-$, the effect of changing the cooling rate from $-$ to $+$ is

$$\frac{12 + 34}{2} - \frac{6 + 16}{2} = 23 - 11 = 12$$

The *RT* interaction is one-half of the difference, which is $\dfrac{15 - 12}{2} = 1.5$.

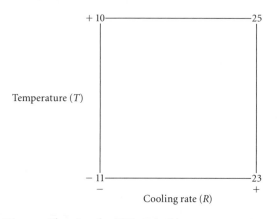

Figure 4.4 Square Diagram Showing the *RT* Interaction

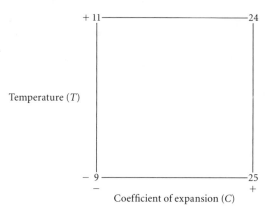

Figure 4.5 Square Diagram Showing the *TC* Interaction

TC Interaction

The square diagram for this interaction is shown in Figure 4.5. With temperature T at $+$, the effect of changing the coefficient of expansion from $-$ to $+$ is

$$\frac{14 + 34}{2} - \frac{6 + 16}{2} = 24 - 11 = 13$$

With temperature T at $-$, the effect of changing the coefficient of expansion from $-$ to $+$ is

$$\frac{16 + 34}{2} - \frac{6 + 12}{2} = 25 - 9 = 16$$

The *TC* interaction is one-half the difference, which is $\dfrac{13 - 16}{2} = -1.5$.

4.3.3 Calculating a 3-Factor Interaction

In this 3-factor example, there are three 2-factor interactions (*RC*, *RT*, *TC*) and one 3-factor interaction, denoted by *RTC*. A significant 3-factor interaction means that the

2-factor interaction between any two of the factors depends on the level of the third factor. Equivalently, it means that the effect of changing a particular factor from − to + depends on the levels of the other *two* factors. The 3-factor interaction is calculated as follows:

- Find the *RT* interaction with the coefficient of expansion *C* at +:

  With *T* = + (and *C* = +), the effect of *R* = 34 − 14 = 20.

  With *T* = − (and *C* = +), the effect of *R* = 34 − 16 = 18.

  Therefore the *RT* interaction with *C* = + is $\dfrac{20-18}{2} = 1.$

- Find the *RT* interaction with the coefficient of expansion *C* at −:

  With *T* = + (and *C* = −), the effect of *R* = 16 − 6 = 10.

  With *T* = − (and *C* = −), the effect of *R* = 12 − 6 = 6.

  Therefore the *RT* interaction with *C* = − is $\dfrac{10-6}{2} = 2.$

- The 3-factor interaction is defined as one-half the difference of these two 2-factor interactions, $RTC = \dfrac{1-2}{2} = -0.5.$ As we show later, the 3-factor interaction is not statistically significant.

To summarize, we calculated the 2-factor interaction *RT* with *C* at + and the 2-factor interaction *RT* with *C* at −, then took one-half the difference. Note that the choice of which 2-factor interaction to use in the calculation is arbitrary. We would have obtained the same result by taking half the difference of the 2-factor interaction *RC* with *T* at + and the 2-factor interaction *RC* with *T* at −, or by taking half the difference of the 2-factor interaction *TC* with *R* at + and the 2-factor interaction with *R* at −.

If there were a fourth factor (say, factor *D*), we could also calculate a 4-factor interaction. The 4-factor interaction *RTCD* is one-half of the difference between the 3-factor interaction *RTC* calculated when the fourth factor *D* is at its high (+) level and the 3-factor interaction *RTC* when *D* is at its low (−) level.

4.3.4 A Simple Method for Calculating Effects: Using Calculation Columns Obtained by Multiplying Signs

We can always estimate the effects from their definitions as we have just done, but doing so would be cumbersome. Fortunately, there is a much simpler way. Consider the expanded design matrix shown in Table 4.4. We have added four so-called calculation columns that allow us to estimate the interactions simply and directly.

The signs in the added columns (*RT, RC, TC, RTC*) were found by multiplying the signs in the design columns (*R, T, C*) row by row. For example, in the *RT* column, the sign in the first row (+) is the product of the first row of the *R* column, which is −, and the first row of the *T* column, which is also −. Similarly, observe that the sign in the seventh row of column *RTC* is −, as it is the product of − (for *R*), + (for *T*) and + (for *C*).

TABLE 4.4
Table of Signs for Calculating Effects in the 2^3 Factorial Design: Cracked Pots Example

Run	R	T	C	RT	RC	TC	RTC	Percentage of Pots with Cracks
1	−	−	−	+	+	+	−	6
2	+	−	−	−	−	+	+	12
3	−	+	−	−	+	−	+	6
4	+	+	−	+	−	−	−	16
5	−	−	+	+	−	−	+	16
6	+	−	+	−	+	−	−	34
7	−	+	+	−	−	+	−	14
8	+	+	+	+	+	+	+	34

We obtain the main effect of R by applying the signs in column R to the responses in the last column. We have

$$\text{main effect of } R = \frac{-6 + 12 - 6 + 16 - 16 + 34 - 14 + 34}{4} = \frac{54}{4} = 13.5$$

The minus and plus signs in the numerator of this expression correspond to the signs in the first column. We divide the linear combination of the responses in the numerator by the number of plus signs, which in this case is 4. This expression is equivalent to taking the average of the four responses with R at $+$ and subtracting from it the average of the four responses with R at $-$, which is how we defined and calculated the main effect previously. Similarly, for the two other main effects we have

$$\text{main effect of } T = \frac{-6 - 12 + 6 + 16 - 16 - 34 + 14 + 34}{4} = \frac{2}{4} = 0.5$$

$$\text{main effect of } C = \frac{-6 - 12 - 6 - 16 + 16 + 34 + 14 + 34}{4} = \frac{58}{4} = 14.5$$

The RT interaction is obtained by applying the signs in the RT column to the responses in the last column and dividing the result by 4, the number of plus signs in that column. We have

$$RT \text{ interaction} = \frac{6 - 12 - 6 + 16 + 16 - 34 - 14 + 34}{4} = \frac{6}{4} = 1.5$$

Similarly,

$$RC \text{ interaction} = \frac{6 - 12 + 6 - 16 - 16 + 34 - 14 + 34}{4} = \frac{22}{4} = 5.5$$

$$TC \text{ interaction} = \frac{6 + 12 - 6 - 16 - 16 - 34 + 14 + 34}{4} = \frac{-6}{4} = -1.5$$

$$RTC \text{ interaction} = \frac{-6 + 12 + 6 - 16 + 16 - 34 - 14 + 34}{4} = \frac{-2}{4} = -0.5$$

4.4 DETERMINING WHICH EFFECTS ARE SIGNIFICANT

A 2-level factorial design with k factors leads to many estimated main effects and interactions. The number of j-factor interactions is given by $\dfrac{k!}{j!(k-j)!}$, the number of combinations of k items taken j at a time. For example, consider the 2^7 factorial experiment consisting of 128 runs. There are seven main effects, $\dfrac{7!}{2!(7-2)!} = 21$ two-factor interactions, $\dfrac{7!}{3!(7-3)!} = 35$ three-factor interactions, $\dfrac{7!}{4!(7-4)!} = 35$ four-factor interactions, and so on.

Fortunately, we can expect the great majority of these effects to be negligible. Experience has shown that the Pareto principle is generally at work here, with a small number of effects constituting what the Pareto principle calls "the vital few" and the remainder comprising the "trivial many." The phrase "effects sparsity" is also used to convey the same idea. In addition, there tends to be a hierarchical ordering of effects with main effects larger in magnitude and hence more important than 2-factor interactions, 2-factor interactions larger than 3-factor interactions, and so forth. In experiments for which 4-factor and higher-order interactions can be estimated, they are almost certain to be negligible. In most cases, the 3-factor interactions will be negligible as well.

There is substantial empirical evidence of this hierarchical ordering principle based on the accumulation of experimental results in numerous settings over many years. In the case of continuous factors, there is also theoretical support. Smooth response functions can be approximated by their Taylor series expansions, with first-order terms (main effects), second-order terms (2-factor interactions), and so on, with higher-order terms corresponding to higher-order effects diminishing in magnitude.

The calculated effects are estimates that are subject to uncertainty. Repeating a particular experimental run would invariably result in a response that is somewhat different from the original one, due to experimental error. For example, in the cracked pots experiment there are numerous sources of experimental error: differences in clay composition from batch to batch, variability in actual peak kiln temperature around each of the two target settings, differences in how pots are handled by workers, and so forth. As a consequence, a second independent execution of the entire experiment would result in calculated effects that would differ from the estimates obtained before. In the light of this variability, the experimenter needs to determine which estimates are *statistically significant.* In assessing the statistical significance of an estimated effect, the question is whether the evidence is strong enough for the experimenter to conclude beyond a reasonable doubt, that the true (or mean) effect is not equal to zero.

There are four approaches to determining the statistical significance of effects in a factorial experiment, which are discussed in the next sections:

1. Replicating all or part of the design (i.e., multiple runs under the same experimental conditions),

2. Using prior information about the experimental error,

3. Assuming higher-order interactions are negligible so that their estimates represent noise (experimental error), and

4. Normal probability plots

4.4.1 Replicated Runs

Let us return to the cracked pot example. Suppose that each of the eight test conditions was run twice and that the response (percentage of cracked pots) used in calculating the effects is the average response from two runs. In this case, there are two "replications" for each experimental condition. The 16 runs were performed in random order, which is essential, and the results are shown in Table 4.5. For a particular combination of factor settings (e.g., $+ + +$), the difference in the two percentages is due to experimental error. With eight distinct combinations of factor settings (the 8 runs in standard order), we can calculate eight separate estimates of the variance of the experimental error, with each estimate having one degree of freedom. These estimates are shown in Table 4.5.

We average the eight estimates to obtain the pooled estimate

$$s_p^2 = \frac{8 + 2 + 18 + 2 + 18 + 8 + 8 + 2}{8} = 8.25$$

The pooled standard deviation $s_p = \sqrt{8.25} = 2.87$ measures the variability in the response of an individual run; it has 8 degrees of freedom (the sum of the degrees of freedom of the eight separate estimates). Imagine repeatedly carrying out a particular run, (say) run 3 ($R = -, T = +, C = -$). Because of experimental error, the outcomes would vary from run to run, with $s_p = 2.87$ estimating the variability in these responses.

Each estimated effect is the difference of two averages: the average of eight responses at the $+$ level and the average of eight responses at the $-$ level. The responses are uncertain (random variables), and hence each estimated effect is a random variable as well, with a certain unknown mean. Once the experiment is carried out, we obtain a particular value for each estimated effect. For example, we calculated that the estimated main effect of cooling rate is 13.5. If we repeated the entire experiment and recalculated the main effect of R from the 16 new runs, we would obtain another estimate for the effect of the cooling rate. Because of the variability in the 16 individual responses, this second calculated estimate would almost certainly be different from the first. If we repeated the experiment many times and averaged the estimates of R calculated each time, we would obtain something close to the long-run average or mean effect of R.

To test whether the estimate $R = 13.5$ is statistically significant, we ask the following question: If the mean effect were actually zero, how likely is it that the estimate $R = 13.5$ would occur by chance? To answer this question, we will construct a 95% confidence interval on the mean effect of cooling rate. If the confidence interval does not include 0, we will reject the hypothesis that the true mean effect of the cooling rate is 0 and conclude that the estimate 13.5 is statistically significant. For each of the estimated effects, we will follow the same procedure to determine which effects are significant.

TABLE 4.5
Results of the Cracked Pots Example with Replicated Runs

Run (standard order)	R	T	C	Percentage of Cracked Pots (individual experiments)		Average Percentage of Cracked Pots
1	−	−	−	8	4	6
2	+	−	−	11	13	12
3	−	+	−	9	3	6
4	+	+	−	17	15	16
5	−	−	+	19	13	16
6	+	−	+	36	32	34
7	−	+	+	12	16	14
8	+	+	+	33	35	34

Estimated variances:

$$s_1^2 = \frac{(8-6)^2 + (4-6)^2}{1} = 8 \qquad s_5^2 = \frac{(19-16)^2 + (13-16)^2}{1} = 18$$

$$s_2^2 = \frac{(11-12)^2 + (13-12)^2}{1} = 2 \qquad s_6^2 = \frac{(36-34)^2 + (32-34)^2}{1} = 8$$

$$s_3^2 = \frac{(9-6)^2 + (3-6)^2}{1} = 18 \qquad s_7^2 = \frac{(12-14)^2 + (16-14)^2}{1} = 8$$

$$s_4^2 = \frac{(17-16)^2 + (15-16)^2}{1} = 2 \qquad s_8^2 = \frac{(33-34)^2 + (35-34)^2}{1} = 2$$

Let σ_{effect} be the standard deviation (standard error) of an effect, and let σ_{run}^2 be the variance of the response of a single run. Each estimated effect (main effect as well as interaction) is the difference of two averages, $\bar{y}_+ - \bar{y}_-$. In the current example with 2 runs at each of the eight distinct factor-level combinations and 16 runs in total, each average is an average of the results from 8 runs. Let N be the total number of runs in the experiment, and let $n = \dfrac{N}{2}$. Recall that the variance of a sample average $\bar{y}$ from a sample of size n is given by $\sigma_{\bar{y}}^2 = \dfrac{\sigma^2}{n}$, where σ is the population standard deviation. Also, we know that the variance of the difference of two independent random variables is the sum of the variances. Hence we find

$$var(\text{effect}) = var(\bar{y}_+ - \bar{y}_-) = var(\bar{y}_+) + var(\bar{y}_-)$$

$$= \frac{\sigma_{\text{run}}^2}{n} + \frac{\sigma_{\text{run}}^2}{n} = \frac{2\sigma_{\text{run}}^2}{n}$$

$$= \frac{4}{N}\sigma_{\text{run}}^2$$

Replacing σ_{run}^2 by its estimate s_p^2 leads to

$$\text{estimated } var(\text{effect}) = s_{\text{effect}}^2 = \frac{4}{N}s_p^2$$

TABLE 4.6

Confidence Intervals for Main Effects and Interactions: Cracked Pots Example

	Estimate	95% Confidence Interval
MAIN EFFECTS		
Cooling rate R	**13.5**	$13.5 \pm 2.306 \times 1.44 = 13.5 \pm 3.3$
Temperature T	0.5	$0.5 \pm 2.306 \times 1.44 = 0.5 \pm 3.3$
Coefficient of expansion C	**14.5**	$14.5 \pm 2.306 \times 1.44 = 14.5 \pm 3.3$
TWO-FACTOR INTERACTIONS		
RT	1.5	$1.5 \pm 2.306 \times 1.44 = 1.5 \pm 3.3$
RC	**5.5**	$5.5 \pm 2.306 \times 1.44 = 5.5 \pm 3.3$
TC	-1.5	$-1.5 \pm 2.306 \times 1.44 = -1.5 \pm 3.3$
THREE-FACTOR INTERACTION		
RCT	-0.5	$-0.5 \pm 2.306 \times 1.44 = -0.5 \pm 3.3$

NOTE: Significant effects are shown in boldface.

In our example, the total number of runs $N = 16$ and $s_p^2 = 8.25$. Therefore, the estimated variance of an effect is $s_{\text{effect}}^2 = \dfrac{4}{16}(8.25) = 2.0625$, and the standard error of an effect is $s_{\text{effect}} = \sqrt{2.0625} = 1.44$.

We use the standard error of an effect to construct a 95% confidence interval for the mean effect. In our example we use the 97.5th percentile (0.025 tail probability) of the t-distribution with 8 degrees of freedom (2.306), because the variance s_p^2 was pooled from eight separate variances. Suppose in this example, there were three replications at each combination of settings for a total of 24 runs. Each of the eight separate variance estimates would have 2 degrees of freedom, and the pooled estimate would have 16 degrees of freedom. The confidence intervals for the seven effects are given in Table 4.6. Significant effects are shown in boldface. They are the main effects of cooling rate R and coefficient of expansion C, and the RC interaction.

Instead of constructing confidence intervals, equivalently, we can compare the t-ratios (estimated effects divided by their standard error) with ± 2.306. Effects larger than 2.306 in absolute value are considered significant.

Interpretation of Results

The estimated main effect of temperature T is very small and not statistically significant. Also, the factor T does not interact with either of the other factors. Since increasing the temperature of the kiln would be more costly, it is clear from these results that it would be best to keep the current kiln temperature. The main effects of cooling rate R and coefficient of expansion C, and the interaction between these two factors are statistically significant. The main effect of a factor should be looked at by itself only if the factor does not have a statistically significant interaction with another factor. Because of the RC interaction, we should and will examine the two factors jointly.

From the square diagram in Figure 4.2 we observe the nature of the interaction. With coefficient of expansion C at $+$ (the current material), increasing the cooling rate from $-$ to $+$ increases the percentage of cracked pots from 15% to 34%. With coefficient of expansion at $-$ (the new material), increasing the cooling rate from $-$ to $+$ increases the percentage of cracked pots from 6% to 14%. The lowest percentage of cracked pots, 6%, occurs with both coefficient of expansion C and cooling rate R at their low $(-)$ levels. Since the cost of the new material with the lower coefficient of expansion is the same as the cost of the current material, it is clear from the experiment that the new material is preferred. The question of the best cooling rate would require additional analysis. The issue is whether the cost savings from lowering the proportion of broken pots from 14% to 6% is greater than the cost of the decrease in productivity that would result from operating at the current (slower) cooling rate compared to the faster rate.

4.4.2 Prior Information about σ^2_{run}, the Variance of the Experimental Error

Sometimes, based on previous extensive experiments, the experimenter can assume that the variance of the response from a single run, σ^2_{run}, is known. In this case, we substitute this value into the equation for the standard error shown in the previous subsection. Also, we replace the percentiles of the t-distribution (t-values) by the percentiles of the normal distribution (z-values).

4.4.3 Assuming Higher-Order Interactions Are Negligible

We continue with the cracked pots example. Suppose after analyzing the results of the 2^3 factorial experiment, management decides to add an additional factor to the test. One of the production workers suggests that cracking may be due to movement of the pots as they travel on the conveyor and are handled by workers. She suggests using a rubberized carrier instead of the current metal carrier. We add that factor, labeled D, to the design, with the $-$ level being the current metal carrier and the $+$ level being the new rubberized carrier. The design matrix with the 16 runs in standard order, the results of each run (only a single run was taken at each experimental condition), and the estimated effects are shown in Table 4.7.

You may want to check the value of the 4-factor interaction. Obtain the calculation column $RTCD$ by multiplying the $+$ and $-$ signs of the four design columns, apply the signs of this calculation column to the observations, and divide the result by the number of plus signs, which is 8. You will find that $RTCD = 1.500$.

There are four 3-factor interactions and one 4-factor interaction. If we assume that each of these interactions is negligible (i.e., the true (mean) effects are zero) we can use these five estimated effects to obtain an estimate of the standard error of an effect. We can think of these five values as a sample from a distribution with mean 0 and unknown, but common standard deviation (the standard error of an effect). To estimate the variance, we take each value minus 0, square the result, sum the 5 squared deviations, and divide by 5. Since we assume that the mean is 0, we do not use the sample average in the calculation of the variance estimate, and we divide by 5, the number of observations, rather than $5 - 1$. Our estimate has 5 degrees of freedom.

$$\text{estimated } var(\text{effect}) = s^2_{\text{effect}}$$

$$= \frac{(-1.5 - 0)^2 + (-1.75 - 0)^2 + (0.25 - 0)^2 + (-0.75 - 0)^2 + (1.5 - 0)^2}{5}$$

$$= \frac{8.1875}{5} = 1.6375$$

TABLE 4.7
*Design Matrix, Results, and Estimated Effects for
Cracked Pots Problem with Four Factors*

Run	R	T	C	D	Percentage of Cracked Pots
1	−	−	−	−	14
2	+	−	−	−	16
3	−	+	−	−	8
4	+	+	−	−	22
5	−	−	+	−	19
6	+	−	+	−	37
7	−	+	+	−	20
8	+	+	+	−	38
9	−	−	−	+	1
10	+	−	−	+	8
11	−	+	−	+	4
12	+	+	−	+	10
13	−	−	+	+	12
14	+	−	+	+	30
15	−	+	+	+	13
16	+	+	+	+	30

Estimated effects:

	Effect
Average	17.625
R	**12.500**
T	1.000
C	**14.500**
D	**−8.250**
RT	1.250
RC	**5.250**
RD	−0.500
TC	−0.250
TD	0.500
CD	1.000
RTC	−1.500
RTD	−1.750
RCD	0.250
TCD	−0.750
RTCD	1.500

NOTE: Significant effects are shown in boldface.

$$\text{standard error}(\text{effect}) = s_{\text{effect}} = \sqrt{1.6375} = 1.28$$

The 97.5th percentile of a t-distribution with 5 degrees of freedom is given by 2.571. Hence, the 95% confidence interval for each effect is obtained by adding and subtracting $(2.571) \times (1.28) = 3.29$ from the estimate. Any effect with absolute value greater than 3.29 is statistically significant. We find that R, C, D, and RC are significant, and they are shown in boldface in Table 4.7. The rubberized carrier (D) reduces the percentage of cracked pots by 8.25%.

4.4.4 Normal Probability Plot

How can we determine which effects are significant if there are no replications? In that case, a simple graphical procedure called a normal probability plot will be useful. Assume that the response does not depend on any of the factors, and that observations vary around a constant level. Then all main and interaction effects, which are linear combinations of the responses, should vary around zero. Furthermore, because of the central limit effect, their distribution should be approximately normal, because estimated effects average the observations (with half of the weights at -1 and half of the weights at $+1$). A dot plot or a histogram of the effects is useful as such plots can highlight big effects that do not follow a normal distribution around zero. However, with seven estimated effects (in the factorial with $k = 3$ factors) or 15 estimated effects (in the factorial with $k = 4$), the construction of a histogram and a check of whether it is bell-shaped around zero are not very reliable, given that there are just too few observations. A normal probability plot of the effects, on the other hand, provides a useful tool.

Let us denote the m estimated effects by $f_1, f_2, \ldots, f_m$. In general, there will be $m = 2^k - 1$ effects. For example, for $k = 3$, there will be seven estimated effects. The procedure is the following. First, we order the effects from smallest to largest. Next, as described below, we plot the observed (empirical) cumulative probabilities associated with the estimated effects against the estimated effects. The x-axis represents the effects, and the y-axis represents the cumulative probabilities. The scale of the y-axis is constructed in such a way that if the data points follow a normal distribution, the cumulative probabilities will plot as a straight line. For effects that are from a normal distribution with mean zero, the plot of the effects should approximate a straight line with the line passing through the point $(x = 0, y = 0.5)$. Significant effects much different from 0 will fall away from this line. Effects that are unusually small or large and fail to follow the straight line pattern are judged to be significant. Statistical software packages can readily construct such normal probability plots.

To illustrate the procedure, consider the 2^3 design in Table 4.4, from which we estimate the seven effects shown in Table 4.6. We order the seven effects from small to large: $TC = -1.5, RCT = -0.5, T = 0.5, RT = 1.5, RC = 5.5, R = 13.5, C = 14.5$. The smallest among the seven effects ($TC = -1.5$) represents a cumulative probability between 0 and 1/7 and is assigned a cumulative probability (y-value) at the midpoint of that interval, which is 0.0714. The second smallest among the seven effects ($RCT = -0.5$) represents a cumulative probability between 1/7 and 2/7 and is assigned a y-value at the midpoint of that interval, which is 0.2143. The third smallest effect is assigned a cumulative probability at the midpoint of the interval 2/7 to 3/7, and so forth. In general, with m effects, the ith smallest effect is plotted at a cumulative probability of $(i - 0.5)/m$.

Now let us return to the cracked pots example in which four factors were tested, the original three factors plus factor D, the type of carrier. Table 4.7 lists the data and the $m = 15$ effects that can be estimated from this unreplicated experiment. Which effects are significant? As noted above, a simple dot plot of the 15 effects (there are 15 effects in a 2^4 design) could highlight big effects that are far from zero, but a better approach is to construct a normal probability plot. Figure 4.6 shows the normal probability plot created by Minitab. Note that Minitab uses a slightly different definition of empirical percentiles; the ith largest effect is plotted at a cumulative probability (y-value) of $(i - 0.3)/(m + 0.4)$, instead of at $(i - 0.5)/m$. However, this change is of little practical significance. Observe that the scale

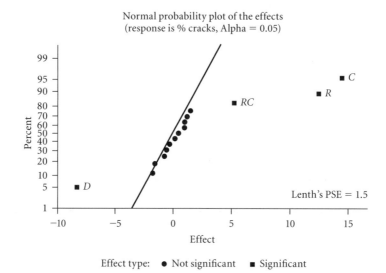

Figure 4.6 Normal Plot of Effects in the Cracked Pots Problem: 16 Runs, with 4 Factors at 2 Levels Each

on the y-axis (cumulative probability, in percent) in Figure 4.6 is not linear. This is where the normal distribution comes into play. With m effects, the ith smallest effect is associated with a cumulative probability of $(i - 0.3)/(m + 0.4)$. The percentile of order $100(i - 0.3)/(m + 0.4)$ implied by the standard normal distribution is called the *normal score* or *z-score* of the ith smallest effect. A normal probability plot is a plot of normal scores against the estimated effects. For illustration, the normal score of the smallest effect $D = -8.25$ with cumulative probability 0.0455 is given by -1.690 (the 4.55th percentile from the standard normal). The normal score of the second smallest among the 15 effects ($RTD = -1.75$) with cumulative probability 0.1104 is -1.224, and so on. The label on the y-axis denotes the cumulative probabilities multiplied by 100. Minitab makes it easy to distinguish between insignificant effects (with mean 0) and significant effects by fitting a straight line to the middle portion of the graph. It also adds information on Lenth's (1989) PSE, a method that we explain in Appendix 4.3. The effects R, C, D, and RC are significant, which is consistent with the results we obtained by assuming that 3- and 4-factor interactions are zero.

4.5 CASE STUDY: A DIRECT MAIL CREDIT CARD OFFER

The financial industry—including insurance, investment, credit card, and banking firms—was among the first to use experimental design techniques for marketing testing. The project described here is from a leading Fortune 500 financial products and services firm. The company name and proprietary details have been removed, but the test strategy, designs, results, and insights are accurate.

The focus of this particular experiment was on increasing the response rate: the number of people who respond to a credit card offer. The marketing team decided to study the ef-

TABLE 4.8
Description of the Four Factors

Factor	(−) Control	(+) New Idea
A Annual fee	Current	Lower
B Account-opening fee	No	Yes
C Initial interest rate	Current	Lower
D Long-term interest rate	Low	High

fects of interest rates and fees, using the four factors shown in Table 4.8. These factors are the annual fee, a fee for opening an account, the initial interest rate, and the long-term interest rate. The company wanted to test the effects of lowering the annual fee and initiating an account-opening fee. Although the account-opening fee was likely to reduce the response, one manager thought the fee would give an impression of exclusivity that would mitigate the magnitude of the response decline. The team also wanted to test the effect of a small increase in the long-term interest rate. At the same time, they wanted to test the effect of two different initial interest rates, both lower than the long-term rate.

To study interactions along with all main effects, the consultant recommended a 2^4 design. The marketing team used columns A–D of the test matrix in Table 4.9 to create the 16 mail packages. Many advertisements had to be sent to targeted customers as the average response rate to such mail ads is only in the 2–3% range. Each of the 16 different mailings was sent to 7,500 customers, requiring a total of 120,000 mailings. The numbers of orders received and the order rates are given in Table 4.9. In total, 2,837 orders were received, resulting in an overall order rate of $100(2,837/120,000) = 2.364\%$.

The $+/-$ combinations in the 11 interaction (product) columns are used solely for the statistical analysis of the results. The 15 main effects and interactions (the six 2-factor interactions, the four 3-factor interactions, and one 4-factor interaction) are obtained by calculating linear combinations of the response rates using the weights (± 1) in the design and interaction columns, and dividing the results by 8. For example, the main effect of factor A is given by

$$(A) = [-2.45 + 3.36 - 2.16 + 2.29 - 2.49 \cdots - 2.04 + 2.03]/8 = 0.4075.$$

The (ABC) interaction is obtained by first forming the product column ABC (which is given by $-, +, +, -, +, -, -, +, -, +, +, -, +, -, -, +$) and calculating

$$(ABC) = [-2.45 + 3.36 + 2.16 - 2.29 + 2.49 \cdots - 2.04 + 2.03]/8 = -0.0525.$$

We have put parentheses around the effects to distinguish them from the factors and calculation columns. The estimated effects are shown in Table 4.10. Significant effects exceeding two standard errors are indicated in boldface. The calculation of standard errors is explained in Section 4.6.4.

Figure 4.7 graphs the effects in the order of their absolute magnitudes. The broken line indicates estimates that are larger than two standard errors. For simplicity, we use the factor 2 to approximate the 97.5th percentile of the standard normal distribution (1.96).

T A B L E 4.9

Results of the 2⁴ Factorial Experiment

Test Cell	A ANNUAL FEE	B ACCOUNT-OPENING FEE	C INITIAL INTEREST RATE	D LONG-TERM INTEREST RATE	AB	AC	AD	BC	BD	CD	ABC	ABD	ACD	BCD	ABCD	Orders	Response Rate
								INTERACTIONS									
1	−	−	−	−	+	+	+	+	+	+	−	−	−	−	+	184	2.45%
2	+	−	−	−	−	−	−	+	+	+	+	+	+	−	−	252	3.36%
3	−	+	−	−	−	+	+	−	−	+	+	+	−	+	−	162	2.16%
4	+	+	−	−	+	−	−	−	−	+	−	−	+	+	+	172	2.29%
5	−	−	+	−	+	−	+	−	+	−	+	−	+	+	−	187	2.49%
6	+	−	+	−	−	+	−	−	+	−	−	+	−	+	+	254	3.39%
7	−	+	+	−	−	−	+	+	−	−	−	+	+	−	+	174	2.32%
8	+	+	+	−	+	+	−	+	−	−	+	−	−	−	−	183	2.44%
9	−	−	−	+	+	+	−	+	−	−	−	+	+	+	+	138	1.84%
10	+	−	−	+	−	−	+	+	−	−	+	−	−	+	+	168	2.24%
11	−	+	−	+	−	+	−	−	+	−	+	−	+	−	+	127	1.69%
12	+	+	−	+	+	−	+	−	+	−	−	+	−	−	−	140	1.87%
13	−	−	+	+	+	−	−	−	−	+	+	+	−	−	+	172	2.29%
14	+	−	+	+	−	+	+	−	−	+	−	−	+	−	−	219	2.92%
15	−	+	+	+	−	−	−	+	+	+	−	−	−	+	−	153	2.04%
16	+	+	+	+	+	+	+	+	+	+	+	+	+	+	+	152	2.03%

TABLE 4.10
Estimated Effects

	Effect (%)	
Constant	2.3642	
(A)	**0.405**	
(B)	**−0.518**	
(C)	**0.252**	
(D)	**−0.498**	
(AB)	**−0.302**	
(AC)	0.002	
(AD)	−0.108	
(BC)	−0.048	
(BD)	0.102	
(CD)	**0.158**	(close to two standard errors, 0.175)
(ABC)	−0.052	
(ABD)	0.088	
(ACD)	0.008	
(BCD)	−0.108	
(ABCD)	−0.052	

NOTE: Estimates that exceed 2 standard errors,
±(2)(0.087) = ±0.175, are shown in boldface.

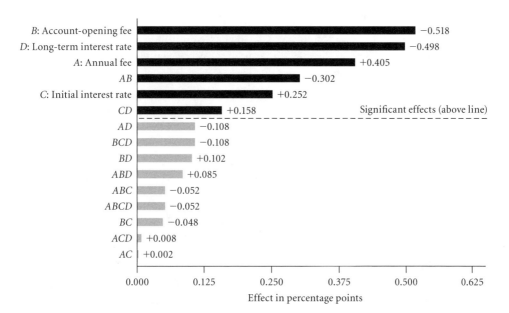

Figure 4.7   Estimated Effects Ordered by Their Absolute Magnitudes

As shown in Figure 4.7, all four main effects and one (and perhaps a second) interaction (the *AB* and the *CD* interactions) are significant. Note that the *CD* interaction is just slightly smaller than 2 times the standard error.

B−: No account-opening fee. One manager thought that charging an initial fee would give the impression of exclusivity. However, this fee had the largest negative effect, reducing the response rate by 0.518 percentage points.

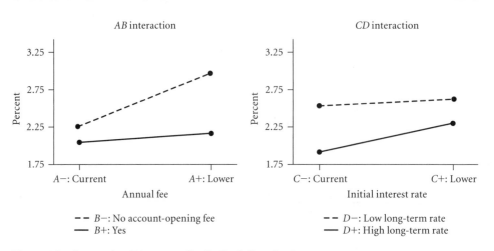

Figure 4.8 Interaction Diagrams: Credit Card Case Study

D−: Low long-term interest rate. An attempt at increasing the interest rate showed that the long-term interest rate had to stay low. Raising the interest rate reduced the response on average by 0.498 percentage points.

A+: Lower annual fee. The annual fee was not charged until the end of the first year, but the fee was stated in the mailing. As with the other charges, a lower fee was better, increasing the response by 0.405 percentage points.

C+: Lower initial interest rate. Results showed that reducing the initial interest rate increased the response by 0.252 percentage points.

The main effects are quite strong. However, the significant interactions (*AB* and *CD*) imply that one needs to look at the effects of *A* and *B*, and *C* and *D* jointly. The interaction diagrams in Figure 4.8 show the nature of the interactions. The *AB* interaction supports both main effects, but provides additional important insights. With an account opening fee (*B+*), the lower annual fee results in only a small increase in response from 2.05% to 2.16%; but with no account-opening fee (*B−*), a lower annual fee results in a very large increase in response from 2.27% to 2.98%. The estimated response of 2.98% is much higher for the combination *A+B−*, with both the lower annual fee and no account-opening fee. The *AB* interaction expresses the fact that *A+* and *B−* together develop synergies that increase the response rate beyond what can be expected by each of the two factors separately. This is extremely valuable information. Using its financial models, the company found that the increase in response resulting from no account-opening fee and a lower annual fee (*A+B−*) was much greater than the loss in revenue that would result from reducing these fees.

The *CD* interaction shows that when the long-term rate is low (*D−*), the effect of a lower initial rate is small and not statistically significant (a change in the response from 2.57% to 2.66%). It is clear that offering the lower initial rate would not be profitable if the lower long-term rate were also offered. However, when the long-term rate is high (*D+*), the lower initial rate has a large impact, with the response changing from 1.91% to 2.32%. In contrast to the main effects that suggest both interest rates should be low, these results, followed by additional analysis using the company's financial models, showed that a lower long-term rate coupled with the current (higher) initial rate would be the most profitable.

Overall, these interactions give important insight into the true relationship among the factors and help to better quantify their effects on profitability. By combining all main effects and significant interactions into one model, the marketing team could analyze different combinations and estimate response rates and profits more accurately.

4.6 ADDITIONAL ISSUES IN DESIGNING AND ANALYZING FACTORIAL EXPERIMENTS

In this section we discuss some additional important issues related to factorial designs. We begin with a discussion of how to set factor levels such as the plus and minus values for kiln temperature. Next, we show how a factorial design can be represented as a regression model, and how this model can be used to predict the response as a function of the factor settings. We then define and discuss an important mathematical property, *orthogonality*, which 2-level factorial designs have, and which leads to the independent estimation of effects. Also in this section, we consider the special characteristics of experiments such as direct mail, where the response variable is the fraction of people who respond. We show how to determine the needed sample size and we explain how to assess which effects are statistically significant. Two-level factorial designs are linear models that will be inadequate if the relationship between the response variable and one or more factors is nonlinear. We end the section by showing how to check for curvature by adding runs to the original design.

4.6.1 Choosing the Levels for Each Factor

It is standard notation to use plus ($+$) and minus ($-$) to denote the high and low levels of the factors. The units, of course, depend on the particular situation at hand. For a continuous factor such as the price of a relatively inexpensive product like a hamburger, the low and high levels may represent prices that are 10 cents lower/higher than the usual price. For an expensive product such as a car, the low and high levels may be prices that are $1,000 lower/higher than normal. For advertising, the low/high levels may represent advertising expenditures that are 10% lower and 10% higher than the standard. Usually it takes careful thought to specify the levels of continuous factors. The levels should not be too close, because then not much of a change in response would be expected, and its magnitude might be overwhelmed by the inherent variability. On the other hand, the levels should not be too different, either, because then the effect might no longer be linear over the studied range. Also, note that in 2-level experiments it is not possible to detect nonlinearity, since with just two levels only a straight line can be fit to the responses. In Section 4.6.5, we discuss how to use additional runs to check for nonlinearity.

In general, these considerations do not apply to unordered categorical factors such as type of ad copy, font, color, and catalyst, because levels in-between the categories are not meaningful. Nevertheless, it is important to select categories that are sufficiently *bold*, in the sense that they will lead to differences in the response.

4.6.2 Regression Equation Implied by Factorial Experiments

Consider again the results of the cracked pot problem with four factors. This experiment was not replicated, and the significant effects were determined either from the normal probability plot or by assuming that the four 3-factor interactions and the one 4-factor interaction were negligible. Using either approach, we found that the significant effects were R, C, D, and RC. Table 4.7 shows that the estimated effects are $R = 12.5, C = 14.5, D = -8.25$, and $RC = 5.25$.

These estimated effects are the same as twice the estimates in the equivalent regression model that regresses the observations on the design and interaction columns in which $+1$ and -1 replace $+$ and $-$. For those with some background in linear algebra, Appendixes 4.4 and 4.5 at the end of this chapter continue this discussion in more detail.

In the estimated regression equation we include only terms for the significant effects. It is given by

$$\hat{y} = 17.625 + 6.25x_R + 7.25x_C - 4.125x_D + 2.625x_Rx_C$$

x_R, x_C, x_D represent the levels of the design factors, and they are -1 if the factor is at its low level, and $+1$ if the factor is at its high level. They correspond to the columns in Table 4.7, with -1 representing $-$, and $+1$ representing $+$. The product x_Rx_C represents the calculation column that is obtained by multiplying the entries in the two design columns. The intercept in this equation (17.625) is the average of all observations. The coefficient for each of the other terms is one-half of the corresponding effect. For example, 6.25 is 12.5/2, 7.25 is 14.5/2, and so on. The coefficients in the regression equation represent the slope or the change in the response per unit change of the factor. Because each estimated effect represents a 2-unit change (from -1 to $+1$), we divide each effect by 2 to obtain the corresponding regression coefficient.

The regression equation can be used to obtain the predicted response at any combination of the levels of the significant factors. The lowest response (lowest proportion of cracked pots) occurs for $R = -1$, $C = -1$, and $D = +1$. The predicted response is

$$\hat{y} = 17.625 + 6.25(-1) + 7.25(-1) - 4.125(+1) + 2.625(-1)(-1) = 2.625$$

The predicted response at the current settings of the factors $(R = -1, C = +1, D = -1)$ is

$$\hat{y} = 17.625 + 6.25(-1) + 7.25(+1) - 4.125(-1) + 2.625(-1)(+1) = 20.125$$

The regression approach can be used to obtain estimates of the effects and their standard errors directly. In the case of a replicated full factorial design, the replications allow for an error term in the regression model. Statistical software such as Minitab or JMP can be used to carry out the regression, and the output from these programs provides an estimate of the error variance σ^2_{run}, the regression coefficients, and their standard errors. In the case in which there are no replications, a simpler regression model without terms for 3-factor and higher-order interactions can be specified. This results in a pooling of the omitted interac-

TABLE 4.11
The 2^3 Design with Its Calculation Columns

Run	A	B	C	AB	AC	BC	ABC	Response
1	−1	−1	−1	+1	+1	+1	−1	y_1
2	+1	−1	−1	−1	−1	+1	+1	y_2
3	−1	+1	−1	−1	+1	−1	+1	y_3
4	+1	+1	−1	+1	−1	−1	−1	y_4
5	−1	−1	+1	+1	−1	−1	+1	y_5
6	+1	−1	+1	−1	+1	−1	−1	y_6
7	−1	+1	+1	−1	−1	+1	−1	y_7
8	+1	+1	+1	+1	+1	+1	+1	y_8

tions and leads to the standard error of the estimates. This is equivalent to our approach in Section 4.4.3 of assuming that higher-order interactions are negligible.

The regression approach has some advantages. It is more flexible for analyzing experiments that include missing observations. Also, the regression approach is needed when factors are continuous with more than two levels, and if one wishes to model the functional relationship between the response and the factors. Imagine an experiment that assesses the relationship between the sales of a magazine and its cover price. Experiments at four different price levels—$1.0, $1.5, $2.0, and $3.0 — may have been conducted, and one may wish to model the functional relationship between sales and price, determine whether the sales-price relationship is linear or quadratic, and find the price at which sales are maximized. For that, one needs regression.

4.6.3 Orthogonality

Definition. A design is orthogonal if for any two design factors, each factor-level combination has the same number of runs.

The 2^k factorial design is an *orthogonal design*. In the 2^k factorial design each pair of design factors is studied at four possible combinations, and at each of these combinations, 2^{k-2} runs are carried out. Consider the 2^3 design for factors A, B, and C shown in Table 4.11. The four level-combinations of factors A and B, for example, are $(-1, -1), (+1, -1), (-1, +1),$ and $(+1, +1)$, and 2 runs are conducted at each combination. The same is true for the other two pairs: factors A and C, and factors B and C.

Now, ignore the response column, and consider any *two* columns (design, as well as calculation columns) in the matrix in Table 4.11. Multiply the entries in each row of the two selected columns, and sum the products. It will give zero for *any* pair of columns. For illustration, take the product of columns C and ABC; you obtain the sum $+1 - 1 - 1 + 1 + 1 - 1 - 1 + 1 = 0$. This characteristic is a property of orthogonal 2-level designs.

Because of this orthogonal design structure, effects are estimated independently—for example, the main effect of A does not depend on the main effect of B because the corresponding columns are uncorrelated. It is easy to see why this is the case. Whenever A is at +1, B is equally likely to be at +1 or −1. Similarly, whenever A is at −1, B is equally likely to be at +1 or −1. As a result, any change in one effect is canceled out in the estimate of any other effect. For example, suppose whenever A is +1 the response is increased by some amount Δ. In calculating the main effect of B, the amount Δ will be added for each of the

2 runs when $A = +1$ and $B = +1$ (runs 4 and 8) and subtracted for each of the 2 runs when $A = +1$ and $B = -1$ (runs 2 and 6).

4.6.4 Determining the Significance of the Effects When the Response Is a Proportion

In many service applications, especially in marketing, each run of the experiment results in a *proportion* of subjects who have responded positively (yes) to a particular stimulus. For example, in a typical direct-mail experiment, the response variable is the proportion of the recipients who accept the offer. These proportions become the averages $\bar{y}$ in Section 4.4.1 that go into the calculation of the main effects and interactions.

Suppose N represents the total number of people in the test. The number of people will be divided equally and randomly among the runs (factor-level combinations) of the test. For each effect, $N/2$ people will see the $+$ level, and $N/2$ people will see the $-$ level. For example, in a particular direct mailing, half the people would receive a green envelope, while half the people would receive a red envelope.

The effect (main effect of a factor or an interaction) is the difference, $\bar{p}_+ - \bar{p}_-$, where $\bar{p}_+$ is the proportion of positive responses among the $N/2$ subjects exposed to the $+$ level, and $\bar{p}_-$ is the proportion of positive responses among the $N/2$ subjects exposed to the $-$ level. For each effect, the question is whether the difference in these two proportions is statistically significant. This problem is equivalent to taking two independent, equal-sized random samples and testing whether the observed difference in the proportions of successes is statistically significant. Formally, we are testing the null hypothesis $H_0: \pi_1 = \pi_2$ or $\pi_1 - \pi_2 = 0$ against the alternative hypothesis $H_1: \pi_1 - \pi_2 \neq 0$, where π_1 is the true proportion of successes for subjects exposed to the $+$ level, and π_2 is the true proportion of successes for subjects exposed to the $-$ level. Under the null hypothesis that the true (population) proportions are equal, the estimated standard error of the difference of the two sample proportions is given by

$$standard\ error(\text{effect}) = \sqrt{\frac{\bar{p}(1-\bar{p})}{N/2} + \frac{\bar{p}(1-\bar{p})}{N/2}} = \sqrt{4\frac{\bar{p}(1-\bar{p})}{N}}$$

where $\bar{p}$ is the overall success proportion, pooled over all runs. The total sample size in the case study in Section 4.5 was $N = 120,000$, and the overall success rate (over all runs and samples) was $\bar{p} = 2,387/120,000 = 0.02364$. Hence, the estimated standard error of an effect is

$$standard\ error(\text{effect}) = \sqrt{4\frac{(0.02364)(0.97636)}{120,000}} = 0.000877, \text{ or } 0.0877\%$$

This is the standard error in Table 4.10 that was used to assess the significance of the estimated effects. We reject the null hypothesis that the two proportions are equal and conclude that an estimated effect is statistically significant (at the 5% level) if the estimated effect $(\bar{p}_+ - \bar{p}_-)$ has absolute value greater than (1.96) *standard error*(effect). For simplicity, we substitute 2 for 1.96 and, in Table 4.10, identify an effect as significant if its estimate has absolute value greater than $(2)(0.0877) = 0.175\%$.

4.6.5 Determining the Required Sample Size
When the Response Is a Proportion

In the case study of Section 4.5, the total sample size was 120,000. A fundamental question in problems of this type is how large a sample size is needed? Statistical packages, including Minitab and JMP, provide useful software for making this determination. Appendix 2.1 discusses the theory behind their calculations. Suppose that, based on prior experience in mailings, the financial services company described in the case study estimated that the overall response rate would be about 0.025 or 2.5%. Further, suppose the firm decided that a change in response of 0.25% was economically meaningful. Hence, the firm wanted to be able to detect such a change (either an increase from 0.025 to 0.0275, or a decrease from 0.025 to 0.0225) with high probability.

Here we illustrate Minitab's power and sample size routines. Minitab includes a function for determining the sample size in the comparison of two proportions ("Stat > Power and Sample Size > 2 Proportions"). In the case study, the total sample size was 120,000 with 7,500 people receiving the package mailing defined by each of the 16 runs. Each effect is the difference in two sample proportions $(\bar{p}_+ - \bar{p}_-)$, with 60,000 people exposed to the + level and 60,000 people exposed to the − level. This means that in estimating each effect we are comparing two independent samples of size 60,000 each. We enter 60,000 for the sample size, 0.025 for proportion 1, and "not equal" in the options for the alternative hypothesis. This setup tests the null hypothesis H_0: $\pi_1 = \pi_2 = 0.025$ or $\pi_1 - \pi_2 = 0$ against the alternative hypothesis H_1: $\pi_1 - \pi_2 \neq 0$. We use a 0.05 significance level when testing the null hypothesis that the two (population) proportions equal 0.025, and that there is no difference between the proportions.

The power of a test is the probability of rejecting the null hypothesis if the alternative hypothesis is true; it is 1 minus the probability of a Type II error. We want the probability of rejecting the null hypothesis to be large if there are economically meaningful differences in the response rates at the plus and minus levels. In this case, we want to be able to detect a difference of 0.25%. To find the power of the test, we first enter 0.0225 for proportion 2, which corresponds to $\pi_1 - \pi_2 = 0.0025$ (or 0.25%). The resulting power is 0.812. We then repeat the procedure by entering 0.0275 for proportion 2, which corresponds to $\pi_1 - \pi_2 = -0.0025$. The resulting power is 0.773. Thus, a total sample size of 120,000 (for each effect, 60,000 people are exposed to the + level, and 60,000 people are exposed to the − level) results in a power of about 80%. Hence, we are 80% likely to detect economically meaningful changes.

This particular Minitab function is very useful. Here, we have specified the sample size and used Minitab to calculate the power of the test. Alternatively, the user can specify the desired power, and the program will provide the required sample size. Similar functions are included in other programs such as JMP.

4.6.6 Checking for Curvature in the Response

One advantage of 2-level factorial designs is that they are very efficient in terms of the number of runs that need to be carried out. A disadvantage of these designs is that in the case of continuous factors, the effects are assumed to be linear, and there is no way to check whether this assumption is reasonable without adding runs to the design. To illustrate, consider a single continuous factor with two levels. With two response averages—one at the low and one at the high level—we can fit a straight line perfectly, but we cannot check whether the linear model is appropriate.

If we want to fit models with linear and quadratic effects, we need at least three factor levels. However, often this is costly in terms of the number of required runs. At the initial stage, where usually one starts with many factors that may or may not have an effect on the response, this is not a practical approach. Factorials with factors at three or more levels may be appropriate at a later stage, after the experimenter has reduced the number of factors to a smaller set.

At the initial screening stage, only a simple check for nonlinearity is needed. This can be achieved by adding to the factorial design one or more runs at the center point. The *center point* of a 2-level factorial experiment sets each factor equal to the average of its low and high levels. In coded units, it is the run with $x_1 = x_2 = \cdots = x_k = 0$. A center point is appropriate for experiments with continuous factors, but not for categorical factors where an in-between level has no meaning.

Assume that we collect n_c independent replications at the center point and obtain their average $\bar{y}_c$ and standard deviation s_c. The standard error of the average of the n_c observations at the center point is given by $s_c/\sqrt{n_c}$.

Next, consider the average $\bar{y}$ of the measurements at the 2^k factor-level combinations. If the response function is linear, then this average will also be an estimate of the level (mean) at the center point. However, this is not the case if the response function is nonlinear.

The difference between the two averages $\bar{y}$ and $\bar{y}_c$ is a measure of the curvature in the response function. A large nonzero difference points to a nonlinear relationship. The standard error of the difference is needed to assess the statistical significance. It is reasonable to assume that the variability of individual responses at the factorial design points is similar to the variability at the center point. Hence, the standard error of $\bar{y}$, the average of the observations at the 2^k factor-level combinations, is given by $s_c/\sqrt{2^k}$. Because of the independence of the two averages, the standard error of their difference is given by

$$\text{standard error } (\bar{y} - \bar{y}_c) = s_c\sqrt{\frac{1}{2^k} + \frac{1}{n_c}}$$

Comment. Observe that the calculation of the standard error of the difference requires information on the variability of individual responses. Often, as is done above, the standard deviation is obtained from independent replications at the center point. Sometimes, as shown in the following example, the standard deviation is obtained from replications at all design points.

Example Case 2 (Magazine Price Test) is used as an illustration. You should refer to the case study appendix for a detailed discussion of the experiment and the resulting data. A 2^3 factorial experiment with three continuous factors—cover price ($3.99 and $5.99), subscription price ($1 and $3), and number of newsstand copies (1/3 less than current, and

TABLE 4.12
Results of a Magazine Price Test

A Cover Price	B Subscription Price	C Copies on Newsstand	Percent Change in Sales
−	−	−	−2.31
+	−	−	−5.54
−	+	−	−1.62
+	+	−	−3.10
−	−	+	18.30
+	−	+	1.41
−	+	+	22.60
+	+	+	−0.73
0	0	0	2.08

1/3 more than current)—was carried out with the objective of assessing the impact of these factors on magazine sales. A center point, with a \$4.99 cover price, a \$2 subscription price, and the currently used number of newsstand copies was also considered. Each of the nine combinations was run over a 5-week period, and the resulting average weekly percent changes in sales are shown in Table 4.12. The week-to-week variation was used as a measure of experimental error. Variances among weekly percent changes, calculated for each of the nine runs, were averaged, resulting in an estimated standard deviation of the change in weekly sales. This standard deviation was calculated to be 5%.

Main effects and interactions are estimated in the case, and the main effects plots shown there illustrate mostly linear relationships between sales and the three studied factors. The average of the responses at the eight factorial points is $\bar{y} = 3.63$, and its distance to the response at the center point is $3.63 - 2.08 = 1.55$. The percent changes listed in Table 4.12 are averages of $n = 5$ weekly observations with standard deviation 5, and hence their standard deviation is given by $5/\sqrt{n} = 2.236$. This becomes the estimate s_c from the earlier discussion. With a single response at the center point ($n_c = 1$), the standard error of this difference is

$$standard\ error\ (\bar{y} - \bar{y}_c) = 2.236\sqrt{\frac{1}{8} + 1} = 2.37$$

The standard error exceeds the observed difference $3.63 - 2.08 = 1.55$, and thus there is no evidence of curvature in the response. If the observed difference were large (say, 2 times the standard error), that would be evidence that the linear model is inadequate. In that case, an experiment with each factor at 3 levels (we discuss these designs in Chapter 7) would be needed.

4.7 NOBODY ASKED US, BUT . . .

As discussed in this chapter, experience has shown that there is a hierarchical ordering of effects with main effects larger than 2-factor interactions, 2-factor interactions larger than 3-factor interactions, and so forth. We noted that for continuous factors the Taylor series expansion of a continuous response function provides theoretical support for this finding. But this is only true for continuous factors with smooth response functions. With categorical factors, there is no such theoretical justification. In some cases, for example, it may be true that a 3-factor interaction is just as large (or even larger) than main effects and 2-factor interactions. To illustrate how this might occur, consider plant growth as the response (a continuous variable), and the three categorical factors: water (no/yes), fertilizer (no/yes), and temperature (0 degrees/25 degrees). Only one of the eight factor-level combinations (water, fertilizer, and temperature of 25 degrees) leads to plant growth, resulting in a large 3-factor interaction among water, fertilizer, and temperature. This and similar examples do not mean that hierarchical ordering does not apply to qualitative factors at all. Experience has shown that in general it does. It just means that when the factors under investigation are qualitative (red background/blue background, headline 1/headline 2, and so forth) some caution is needed before one automatically assumes that 3-factor (and higher-order) interactions will be negligible.

In Section 4.4.1, we described how replicated runs are used to estimate the variance of the experimental error (variance of a run) and to find the standard error of an effect to determine which effects are statistically significant. It is essential that the repeated runs at a particular combination of factor settings be *genuine independent replications*. For example, in the cracked pots example a production batch consisted of 100 pots. Simply following one batch with a second would not constitute a genuine independent replication. The variation in the percentage of broken pots between these two batches is very likely to underestimate the experimental error. A true replicate requires that each setup procedure in the process be done independently before each run. This would mean carrying out the steps needed to set the peak temperature in the kiln, setting the conveyor speeds that determine the cooling rate, and choosing a batch of clay material from the appropriate supply (low or high coefficient of expansion) in a fashion that reflects the variability that exists in this raw material.

In general, a common mistake in manufacturing processes is to take repeated measurements from the same run and treat each as a replicate. But this only captures measurement error, which is typically only a small part of the total experimental error.

Randomization is important in experimentation and no less so when replicates are included in the experiment. Carrying out 2 runs in succession at the same factor settings would likely lead to underestimating the experimental error because these responses would be more alike than if the order of the 2 runs were determined randomly.

In the cracked pot example, the 8-run factorial design was entirely replicated, resulting in a total of 16 runs. For a factorial design with 5 factors or even 4, replicating all of the runs might be uneconomical. For example, in a 4-factor, 16-run factorial design, the experimenter might randomly choose (say) 8 of the 16 runs to replicate. In that case, the calculation of effects and their standard errors would have to be done using regression because only half of the 16 experimental conditions would be run twice. Also, there would be 8 degrees of freedom associated with the standard error of an effect compared to the 16, if all 16 runs were repeated. The resulting confidence intervals for the effects would be wider, because a *t*-value from a distribution with 8 degrees of freedom would be larger.

Designed experiments are part of a scientific learning process. The goals of this process are to confirm or refute prior knowledge and to suggest new hypotheses for future study. Clearly, it is important that this experimental approach be efficient and lead to the right answers. We showed that factorial designs where factors are varied simultaneously are more efficient and provide more information than experiments that vary each factor one at a time.

INEFFICIENCY OF APPROACHES THAT

CHANGE ONE FACTOR AT A TIME

All too often the effects of k factors are studied by carrying out successive experiments in which the levels of each factor are changed *one at a time*. Such experiments start with the standard settings of the k factors, then change the levels of the one factor that is considered the most influential. The responses at the low and high settings of this one factor are compared while keeping all other factors fixed, and the level at which the response is best is locked in for the next stage. The factor that is considered second most important is varied next. Again, responses at the low and high levels of this factor are compared, and the best level of this factor gets locked in for all subsequent runs. Then on to the third factor, and so on, until the last factor is reached.

Compared to the factorial (multifactor) experiments where the levels of all factors are changed *together*, the approach of changing one factor at a time is inefficient for several reasons:

- It requires more runs to achieve the same precision for the effects estimates.
- It may miss the optimum altogether.
- It cannot estimate interactions.
- It does not provide general conclusions about factor effects, given that the estimates depend on specific levels of the remaining factors.

We illustrate these shortcomings in the context of the 2^2 factorial design. In the 2^2 factorial without replications, we conduct a total of 4 runs. The main effects of both factors are estimated by comparing two observations at the low and the high levels of each factor. Assume that the approach of changing one factor at a time begins with $(x_1 = +, x_2 = +)$ and varies the levels of factor 1 first. To obtain the same precision for the estimates, one must start with four observations, 2 runs each at $(x_1 = -, x_2 = +)$ and $(x_1 = +, x_2 = +)$. Locking in the best level for factor 1 (assume that it is $x_1 = -$), one proceeds to the next comparison where one varies the levels of factor 2 and studies the response at $(x_1 = -, x_2 = -)$ and $(x_1 = -, x_2 = +)$. The 2 runs at $(x_1 = -, x_2 = +)$ have already been obtained in the first step; but 2 more runs at $(x_1 = -, x_2 = -)$ are required. This leads to a total of 6 runs, as compared to the 4 runs in the 2^2 factorial design. This shows that the approach of changing one factor at a time requires more runs to obtain estimates with the same precision.

Also note that the approach of changing one factor at a time may miss the optimum. Consider the situation with the following four factor-level combinations and their responses that are supposed to be maximized:

$$\begin{bmatrix} (x_1 = -, x_2 = +):y = 80 & (x_1 = +, x_2 = +):y = 70 \\ (x_1 = -, x_2 = -):y = 90 & (x_1 = +, x_2 = -):y = 110 \end{bmatrix}$$

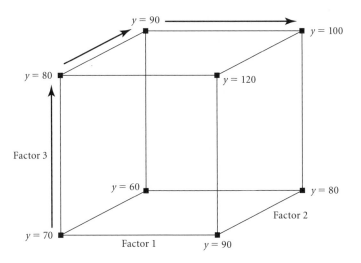

Figure 4.A.1 Illustration of the Approach of Changing One Factor at a Time in the Factorial Experiment

Starting with $(x_1 = +, x_2 = +)$, the approach of changing one factor at a time would set the first factor at its low level (80 is larger than 70). Locking in the low level of factor 1 and varying factor 2, one would select the low level of factor 2 (90 is larger than 80). However, the combination $(x_1 = -, x_2 = -)$ with $y = 90$ has not located the overall maximum $y = 110$ at $(x_1 = +, x_2 = -)$. The reason why the approach of changing one factor at a time fails is because of the interaction between factors 1 and 2.

It is not possible to estimate an interaction with the data generated under the approach of changing one factor at a time. We cannot compare the main effects of a factor at 2 levels of the other factor if we don't have data at all four factor-level combinations.

Furthermore, the main effects estimates from the approach of changing one factor at a time cannot be generalized because they are main effects at specific levels of the other factors. Since we are uncertain whether there is an interaction, we cannot generalize these effects to other levels of the factors.

The same points can be made in the context of a 2^3 factorial design, which can be visualized as the vertices of a cube; see Figure 4.A.1. Four runs at each of the low and the high settings of each factor are used to estimate main effects. The approach that changes one factor at a time does not consider all 8 level combinations, but only the 4 that are outlined in Figure 4.A.1. Again, we learn the following: (1) The approach of changing one factor at a time is inefficient in terms of the number of runs. For the same precision, we need 8 runs for establishing the main effect of the first factor and the best level of the first factor; we need 4 more to establish the level of the second factor; and 4 more for the third factor. Hence, we need a total of 16 runs to estimate the main effects with the same precision that is achieved by the factorial design. (2) It is not possible to estimate interaction terms. (3) In the presence of interaction, the procedure of changing one factor at a time can miss the optimum. The objective in Figure 4.A.1 is to find the maximum. Here, we start by varying factor 3 first, with factor 2 second. The method misses the optimum $y = 120$ at $(x_1 = +, x_2 = -, x_3 = +)$. Note that the correct optimum is reached if we vary factor 1 first. However, the appropriate order is not known. (4) Main effects at the studied settings cannot be extended to other settings.

How does this generalize to comparisons with k factors. For the same precision, the approach of changing one factor at a time needs a total of $2^k + 2^{k-1}(k - 1) = 2^k(1 + (k - 1)/2)$ runs. The number of runs is increased by a factor of $(1 + (k - 1)/2)$. For $k = 4$, this factor amounts to 2.5. Moreover, there is no guarantee that this procedure will find the optimum.

MULTIPLE COMPARISONS

The procedure for determining which effects are significant involves multiple comparisons of many effects. In the 2^4 factorial experiment, for example, we assess the significance of 15 effects. In $m = 15$ comparisons, it would not be unreasonable to see one effect outside the critical value (margin of error), $\pm(1.96)$ *standard error*(effect), just by chance even though there are no active factors. To guard against the error of declaring a factor significant incorrectly, we can apply multiple comparison procedures that increase the critical value. A simultaneous 5% margin of error is obtained by replacing the 97.5th percentile (0.025 upper tail probability) with the percentile of order $(1 + 0.95^{1/m})/2$. This simultaneous margin of error uses the fact that estimates of the effects are independent. For example, for $m = 7$ comparisons, the percentile of order $(1 + 0.95^{1/7})/2 = 0.9963$ from the standard normal distribution is 2.68. This is larger than 1.96, the factor used in the critical value without a multiple-comparison adjustment.

We can apply the multiple comparisons procedure to the results of the replicated 3-factor cracked pots example. In Table 4.6, the confidence intervals and significance tests for the effects are based on a *t*-value of 2.306, which corresponds to 97.5th percentile of a *t*-distribution with 8 degrees of freedom. Applying the multiple comparison method with $m = 7$, the appropriate *t*-value is 3.56, the 99.63th percentile of the *t*-distribution with 8 degrees of freedom. The confidence interval for each effect becomes wider; Estimated effect $\pm(3.56)(1.44)$, or Estimated effect ±5.13. Any effect with absolute value greater than 5.13 is statistically significant. The main effects are still significant, but the significance of the *RC* interaction (estimated effect of 5.5) becomes borderline.

Adjustments for multiple comparisons guard against the error of judging too many factors as important. One could argue against the use of such adjustments on the ground that it is usually not a serious mistake to consider an insignificant effect as significant. Most experimentation is a sequential activity, and not a one-shot affair. Including borderline significant factors at a subsequent stage certainly involves more work as extra factors need to be carried along. However, one will learn at the next stage that such factors are not needed, and not much harm is done by not ruling them out immediately. On the other hand, disposing of a factor too quickly may pose a more serious risk.

LENTH'S APPROACH FOR DETERMINING STATISTICAL

SIGNIFICANCE IN UNREPLICATED FACTORIAL DESIGNS

Lenth, in his paper "Quick and Easy Analysis of Unreplicated Factorials," discusses another useful strategy for assessing the significance of effects in unreplicated experiments. His procedure is based on the following simple formula for the standard error of an estimated effect. If none of the factors are active, the standard deviation of the m estimated effects $f_1, f_2, \ldots, f_m$ serves as the standard error of the estimated effects. However, if some effects are active, this estimate is too large, as it not only incorporates random variability but also the effects of active factors. Hence, one needs to omit from the calculation of the standard deviation the estimates of all active factors. The normal probability plot discussed previously does this informally when determining the best-fitting straight line from just the estimates in the linear portion of the middle part of the graph, not from the estimates on the extreme left and right side that do not appear to fit the line through the middle.

Lenth (1989) uses the fact that the median of the absolute values of the estimated non-active effects, suitably normalized, provides an estimate of the standard deviation, and he calculates

$$s = (1.5)\text{Median}(|f_1|, |f_2|, \ldots, |f_m|)$$

The factor 1.5 in the normalization arises from the relationship between the standard deviation and the median of the absolute value of a mean zero normal random variable. In the next step, Lenth (1989) omits from this calculation all estimates with absolute values larger than $2.5s$, and he calculates a revised standard deviation

$$\text{PSE} = (1.5)\underset{|f_i|<2.5s}{\text{Median}}(|f_1|, |f_2|, \ldots, |f_m|)$$

He calls this the *pseudo standard error* (PSE) and uses it in the calculation of the confidence intervals for the effects. The 95% confidence interval for an effect, *estimated effect*$\pm(t)(\text{PSE})$, uses the 97.5th percentile of a t-distribution with $m/3$ degrees of freedom. For a standard error that is estimated with reasonable confidence and that comes from many observations, one would use the 97.5th percentile of the standard normal distribution, or simply $t = 2$. However, in the unreplicated situation the PSE comes from very few observations, and Lenth (1989) found through simulations that the t-distribution with $m/3$ degrees of freedom works best. For $m = 7$ effects, $t = 3.76$; for $m = 15$, $t = 2.57$, and for $m = 31$, $t = 2.22$. Lenth recommends displaying the estimated effects on a bar chart, and adding to this chart the margin of error, $\pm(t)(\text{PSE})$. If an estimate exceeds these limits, then it is likely that this particular factor is active (i.e., significant). We should point out that Lenth suggests even larger margins of errors by incorporating simultaneous adjustments for the multiple comparisons (see Appendix 4.2 for a discussion of multiple comparisons). Minitab displays Lenth's PSE in the context of its normal probability plot (see Figure 4.6).

A BRIEF PRIMER ON REGRESSION

Note that regression software is readily available and all that is needed in practice is an understanding of how to interpret the program output. While a detailed knowledge of regression is not necessary for applying the design approach put forward in this chapter, it will help you understand certain issues in Chapters 7 and 8. Also, if you have had prior exposure to regression, the material in the following two appendixes will give you a brief concise summary of the main results.

Consider the simple linear regression model $y = \beta_0 + \beta_1 x + \varepsilon$. Ignoring the noise ε, this model represents a straight line if plotted on an x-y graph. The noise component introduces random scatter around the model line $\beta_0 + \beta_1 x$. Assume that the noise component has mean zero and variance σ^2.

Assume that there are n pairs of observations $(x_1, y_1), (x_2, y_2), \ldots, (x_n, y_n)$, which are graphed on a scatter plot. Figure 2.7 in Section 2.3 is an example of such a plot. The objective is to determine the line that fits the data best. Least squares estimation selects the estimates for β_0 and β_1, which we denote by $\hat{\beta}_0$ and $\hat{\beta}_1$, by minimizing the sum of the squared vertical distances $\sum_{i=1}^{n} [y_i - (\beta_0 + \beta_1 x_i)]^2$. The estimates can be calculated quite easily. Expressions for the estimates can be written down in vector/matrix format,

$$\hat{\boldsymbol{\beta}} = \begin{bmatrix} \hat{\beta}_0 \\ \hat{\beta}_1 \end{bmatrix} = (X'X)^{-1}X'\mathbf{y} \tag{4A.1}$$

The $n \times 1$ column vector $\mathbf{y}$ consists of the response (y) observations. The $n \times 2$ matrix X consists of two $n \times 1$ columns: a vector of ones, denoted by $\mathbf{1}$, and the vector $\mathbf{x}$ containing the values of the regressor (x) variable. That is,

$$\mathbf{y} = \begin{bmatrix} y_1 \\ y_2 \\ \cdot \\ \cdot \\ \cdot \\ y_n \end{bmatrix}; \quad X = \begin{bmatrix} \mathbf{1} & \mathbf{x} \end{bmatrix} = \begin{bmatrix} 1 & x_1 \\ 1 & x_2 \\ \cdot & \cdot \\ \cdot & \cdot \\ \cdot & \cdot \\ 1 & x_n \end{bmatrix}$$

The matrix $X' = \begin{bmatrix} 1 & 1 & \cdot & \cdot & \cdot & 1 \\ x_1 & x_2 & \cdot & \cdot & \cdot & x_n \end{bmatrix}$ is the transpose of the matrix X, $(X'X)$ is the matrix product of X' and X, and $(X'X)^{-1}$ is the inverse of $(X'X)$. The matrix expression for the estimates in equation (4A.1) is very convenient as it also works for more general models.

The fitted values from the regression fit, $\hat{y}_i = \hat{\beta}_0 + \hat{\beta}_1 x_i$, are obtained by replacing the regression coefficients in the model equation by their estimates. The residuals are the differences between the observations and the fitted values, $y_i - \hat{y}_i = y_i - (\hat{\beta}_0 + \hat{\beta}_1 x_i)$. An estimate of the variance σ^2 is obtained from

$$s^2 = \frac{\sum_{i=1}^{n} [y_i - (\hat{\beta}_0 + \hat{\beta}_1 x_i)]^2}{n - p} \tag{4A.2}$$

where $p = 2$ is the number of regression coefficients; here, there are two coefficients, β_0 and β_1.

Another important result specifies the covariance matrix of the regression estimates $\hat{\boldsymbol{\beta}}$. It can be shown that

$$V(\hat{\boldsymbol{\beta}}) = \begin{bmatrix} V(\hat{\beta}_0) & \mathrm{Cov}(\hat{\beta}_0, \hat{\beta}_1) \\ \mathrm{Cov}(\hat{\beta}_0, \hat{\beta}_1) & V(\hat{\beta}_1) \end{bmatrix} = s^2 (X'X)^{-1} \tag{4A.3}$$

The variances of the regression estimates are in the diagonal of this matrix. Their square roots, the standard errors of the estimates, are used to construct confidence intervals for the regression coefficients. A diagonal covariance matrix implies that the estimates are uncorrelated.

A nice feature of these matrix expressions is the fact that they work for general models. Consider the general linear regression model with k regressors (factors); that is,

$$y = \beta_0 + \beta_1 x_1 + \beta_2 x_2 + \cdots + \beta_k x_k + \varepsilon$$

Define the observations of the ith case (the ith experimental run) as y_i (for the response), and $x_{i1}, x_{i2}, \ldots, x_{ik}$ (for the studied values of the k factors). Note that the first subscript in this double subscript notation is the index for the run $(1, 2, \ldots, n)$, and the second subscript is the index for the factor $(1, 2, \ldots, k)$. The $n \times (k + 1)$ matrix X is given by

$$X = \begin{bmatrix} \mathbf{1} & \mathbf{x}_1 & \mathbf{x}_2 & \cdots & \mathbf{x}_k \end{bmatrix} = \begin{bmatrix} 1 & x_{11} & x_{12} & \cdots & x_{1k} \\ 1 & x_{21} & x_{22} & \cdots & x_{2k} \\ \cdot & \cdot & \cdot & \cdots & \cdot \\ \cdot & \cdot & \cdot & \cdots & \cdot \\ \cdot & \cdot & \cdot & \cdots & \cdot \\ 1 & x_{n-1,1} & x_{n-1,2} & \cdots & x_{n-1,k} \\ 1 & x_{n1} & x_{n2} & \cdots & x_{nk} \end{bmatrix}$$

The equations for the least squares estimates in equation (A4.1) (now there are $p = k + 1$ estimates, $\hat{\beta}_0, \hat{\beta}_1, \hat{\beta}_2, \ldots, \hat{\beta}_k$) and for the covariance matrix in equation (A4.3) (which is now a $k + 1 \times k + 1$ matrix, with variances in the diagonal) carry over to the general case. The computational aspects (in particular, taking the inverse $(X'X)^{-1}$) are more difficult, but nothing that a computer program cannot handle.

The fitted values from the regression fit, $\hat{y}_i = \hat{\beta}_0 + \hat{\beta}_1 x_{i1} + \cdots + \hat{\beta}_k x_{ik}$, are obtained by replacing the parameters in the model equation by their estimates. The residuals are the differences between the observations and the fitted values, $y_i - \hat{y}_i = y_i - (\hat{\beta}_0 + \hat{\beta}_1 x_{i1} + \cdots + \hat{\beta}_k x_{ik})$. An estimate of the variance σ^2 is obtained from

$$s^2 = \frac{\sum_{i=1}^{n} [y_i - (\hat{\beta}_0 + \hat{\beta}_1 x_{i1} + \cdots + \hat{\beta}_k x_{ik})]^2}{n - k - 1}$$

The numerator $SSE = \sum_{i=1}^{n}[y_i - (\hat{\beta}_0 + \hat{\beta}_1 x_{i1} + \cdots + \hat{\beta}_k x_{ik})]^2$ is referred to as the error sum of squares; it measures the variability that is not explained by the model. The sum of the squared distances of the observations from their sample mean, $SST = \sum_{i=1}^{n}[y_i - \bar{y}]^2$, is called the total sum of squares. It expresses the variability in the observations, without any adjustment for the explanatory variables. The difference between the total sum of squares and the error sum of squares expresses the sum of squares that is explained by the regression. It is called the regression sum of squares,

$$SSR = SSR(\mathbf{x}_1, \ldots, \mathbf{x}_k) = SST - SSE$$
$$= \sum_{i=1}^{n}[y_i - \bar{y}]^2 - \sum_{i=1}^{n}[y_i - (\hat{\beta}_0 + \hat{\beta}_1 x_{i1} + \cdots + \hat{\beta}_k x_{ik})]^2$$

The coefficient of determination, $R^2 = SSR/SST$, measures the proportion of the variation that is explained by the regression model.

Computer software, such as Minitab, JMP, and even Excel, provides detailed regression output including the estimates and their standard errors, the sums of squares, s^2 and R^2, and the fitted values and residuals. All a user has to do is enter the data into a worksheet and specify the column of the response and the columns containing the regressor variables.

Special Case. Consider how these matrix results specialize for the linear regression model with a single regressor $(k = 1)$. Multiplying the transpose $X' = \begin{bmatrix} 1 & 1 & \cdot & \cdot & \cdot & 1 \\ x_1 & x_2 & \cdot & \cdot & \cdot & x_n \end{bmatrix}$

with the matrix $X = \begin{bmatrix} 1 & x_1 \\ 1 & x_2 \\ \cdot & \cdot \\ \cdot & \cdot \\ \cdot & \cdot \\ 1 & x_n \end{bmatrix}$, leads to the product $(X'X) = \begin{bmatrix} n & \sum x_i \\ \sum x_i & \sum (x_i)^2 \end{bmatrix}$. Furthermore, $X'\mathbf{y} = \begin{bmatrix} \sum y_i \\ \sum x_i y_i \end{bmatrix}$. All sums go from $i = 1$ through $i = n$. The inverse of $(X'X)$ is

$$(X'X)^{-1} = \frac{1}{n\sum (x_i)^2 - [\sum x_i]^2} \begin{bmatrix} \sum (x_i)^2 & -\sum x_i \\ -\sum x_i & n \end{bmatrix} = \frac{1}{n\sum (x_i - \bar{x})^2} \begin{bmatrix} \sum (x_i)^2 & -\sum x_i \\ -\sum x_i & n \end{bmatrix}$$

Multiplying the inverse $(X'X)^{-1}$ with $X'\mathbf{y}$ leads—after some algebra—to the least squares estimates

$$\hat{\beta}_1 = \frac{\sum (x_i - \bar{x})y_i}{\sum (x_i - \bar{x})^2} \quad \text{and} \quad \hat{\beta}_0 = \bar{y} - \hat{\beta}_1 \bar{x}$$

Substituting the inverse $(X'X)^{-1}$ shown above into the expression $V(\hat{\boldsymbol{\beta}}) = s^2(X'X)^{-1}$, leads—again after some algebra—to the variances of the least squares estimates

$$V(\hat{\beta}_1) = \frac{s^2}{\sum (x_i - \bar{x})^2} \quad \text{and} \quad V(\hat{\beta}_0) = \frac{s^2 \sum x_i^2/n}{\sum (x_i - \bar{x})^2} = s^2 \left[\frac{1}{n} + \frac{\bar{x}^2}{\sum (x_i - \bar{x})^2} \right]$$

REGRESSION APPROACH APPLIED TO THE ANALYSIS OF

2-LEVEL FACTORIAL EXPERIMENTS, AND THE FORTUNATE

CONSEQUENCES OF ORTHOGONALITY

In Section 4.3 we defined and calculated main and interaction effects, and we showed that they are linear combinations of the responses, with weights coming from the design vectors and the calculation columns (obtained by multiplying elements of the design vectors). An alternative way of obtaining main and interaction effects is to write down a *regression model* for the response and to obtain the estimates of the regression coefficients. Denote the vector of responses as $\mathbf{y}$, the k design vectors consisting of ± 1 values as $\mathbf{x}_1, \mathbf{x}_2, \ldots, \mathbf{x}_k$, and the calculation columns as $\mathbf{x}_{12}, \mathbf{x}_{13}, \ldots, \mathbf{x}_{k-1,k}$ (each a product of two design columns), $\mathbf{x}_{123}, \ldots$ (each a product of three design columns), $\ldots$, all the way to $\mathbf{x}_{12\ldots k}$ (the product of all k design columns). Including the column of ones, $\mathbf{x}_0 = \mathbf{1}$, there are 2^k columns and each column is of length 2^k.

Examples of these vectors are given in Tables 4.4 and 4.9. There we list the vector of responses, as well as the design and calculation columns. The only difference is that the columns are denoted by factor labels ($R, T, \ldots, RTC$ in Table 4.4; and $A, B, \ldots, ABCD$ in Table 4.9), instead of $\mathbf{x}_1, \mathbf{x}_2, \ldots, \mathbf{x}_{12\ldots k}$. Also, Tables 4.4 and 4.9 do not list the column of ones.

The regression model can be written as

$$\mathbf{y} = \beta_0 \mathbf{1} + \beta_1 \mathbf{x}_1 + \beta_2 \mathbf{x}_2 + \cdots + \beta_k \mathbf{x}_k + \beta_{12} \mathbf{x}_{12} + \beta_{13} \mathbf{x}_{13} + \cdots + \beta_{k-1,k} \mathbf{x}_{k-1,k}$$
$$+ \beta_{123} \mathbf{x}_{123} + \cdots + \beta_{1234} \mathbf{x}_{1234} + \cdots + \beta_{123\ldots k} \mathbf{x}_{123\ldots k} \qquad (A4.4)$$

We regress the vector of responses $\mathbf{y}$ on 2^k regressor vectors $\mathbf{1}, \mathbf{x}_1, \mathbf{x}_2, \ldots, \mathbf{x}_k, \mathbf{x}_{12}, \mathbf{x}_{13}, \ldots,$ $\mathbf{x}_{k-1,k}, \mathbf{x}_{123}, \ldots, \mathbf{x}_{1234}, \ldots, \mathbf{x}_{123\ldots k}$. There is no error term in this regression as this is a fully saturated model, with the same number of coefficients as number of observations. Of course, not all effects need to be included. For example, the model in Section 4.6.2 considered only main effects of R, C, and D, and the 2-factor interaction RC. Also, one may be interested in just main effects and 2-factor interactions. In this case, one would regress $\mathbf{y}$ on $1 + k + ((k - 1)k/2)$ vectors $\mathbf{1}, \mathbf{x}_1, \mathbf{x}_2, \ldots, \mathbf{x}_k, \mathbf{x}_{12}, \mathbf{x}_{13}, \ldots, \mathbf{x}_{k-1,k}$. The noise component in this model would reflect interactions of order three and higher.

Computer software can be employed to carry out the regression. The responses and the design vectors are entered as columns into a spreadsheet; calculation columns are formed by multiplying various subsets of the design vectors; and the regression command is executed.

The vector of regression coefficients $\boldsymbol{\beta}$ consists of elements β_0 (the constant), $\beta_1, \beta_2, \ldots, \beta_k$ (main effects), $\beta_{12}, \beta_{13}, \ldots, \beta_{k-1,k}$ (2-factor interactions), $\beta_{123}, \ldots$ (3-factor interactions), $\beta_{1234} \ldots$ (4-factor interactions), $\ldots$, and $\beta_{12\ldots k}$ (k-factor interaction). X is the $2^k \times 2^k$ matrix containing the regressor vectors. Regression theory in Appendix 4.4 shows that the least squares estimate of the regression coefficients $\boldsymbol{\beta}$ is given by

$$\hat{\boldsymbol{\beta}} = (X'X)^{-1}X'\mathbf{y}$$

where X' is the transpose of the matrix X, and $(X'X)^{-1}$ is the inverse of the $2^k \times 2^k$ matrix $X'X$.

The orthogonality of the factorial design (see the discussion in Section 4.6.3) implies a *diagonal $X'X$ matrix.* You can check this with the X matrix in Table 4.11 that results from the 2^3 design. The multiplication of the transpose X' with the matrix X leads to an 8×8 diagonal matrix with 8 in the diagonal.

For the general 2^k factorial design, the entries in the diagonal of $X'X$ are all equal to 2^k, and all off-diagonal elements of $X'X$ are zero. The diagonal structure of $X'X$ implies that $(X'X)^{-1}$ is diagonal with diagonal elements $1/2^k$. Hence the estimate of an element of $\boldsymbol{\beta}$ is given by

$$\hat{\beta} = \frac{1}{2^k} \sum_{i=1}^{2^k} c_i y_i = \frac{1}{2} \left(\frac{1}{2^{k-1}} \sum_{i=1}^{2^k} c_i y_i \right) = \frac{1}{2}(\text{effect})$$

where the weights c_i are the elements in the corresponding design/calculation column. Apart from the different normalization, these estimates coincide with our previous definition of main and interaction effects in Section 4.3. The only difference is the factor $\frac{1}{2}$. The definition of effects in Section 4.3 looks at the difference in the average responses at the high and low settings of a factor. The regression estimates cut this into half; the coefficients in the regression equation represent the slope or the change in the response per unit change of the factor.

The orthogonality of the design has several fortunate consequences as far as estimation is concerned.

1. The estimates of the effects are uncorrelated. The diagonal $X'X$ matrix implies a diagonal covariance matrix $V(\hat{\boldsymbol{\beta}})$.

2. The estimates do not change when we omit factors from the model. Let us illustrate this with data from the factorial experiment that includes 3 factors and the 8 runs in Table 4.4. Let us ignore the third factor and consider the regression model with just factors 1 and 2,

$$\mathbf{y} = \beta_0 \mathbf{1} + \beta_1 \mathbf{x}_1 + \beta_2 \mathbf{x}_2 + \beta_{12} \mathbf{x}_{12} + \boldsymbol{\varepsilon}$$

In this case, the X matrix has fewer columns (only four as compared to the eight if factor 3 is included). The matrix $X'X$ is still diagonal, although of smaller dimension (4×4), and its diagonal elements are still 8. The inverse $(X'X)^{-1}$ is diagonal with diagonal elements $1/8$, and the estimates in $\hat{\boldsymbol{\beta}} = (X'X)^{-1}X'\mathbf{y}$, consisting of $\hat{\beta}_0, \hat{\beta}_1, \hat{\beta}_2, \hat{\beta}_{12}$, are the same as the estimates in the model that includes all eight regressors.

3. In orthogonal designs, the joint (combined) regression sum of squares of the effects can be partitioned into the sum of the regression sums of squares of the individual effects. That is,

$$SSR(\mathbf{x}_0, \mathbf{x}_1, \ldots, \mathbf{x}_k, \mathbf{x}_{12}, \mathbf{x}_{13}, \ldots, \mathbf{x}_{k-1,k}, \ldots, \mathbf{x}_{12\ldots k})$$

$$= SSR(\mathbf{x}_0) + SSR(\mathbf{x}_1) + \cdots + SSR(\mathbf{x}_k) + SSR(\mathbf{x}_{12}) + SSR(\mathbf{x}_{13})$$

$$+ \cdots + SSR(\mathbf{x}_{k-1,k}) + \cdots + SSR(\mathbf{x}_{12\ldots k})$$

The regression sums of squares are additive. The regression sums of squares from regressing the response vector $\mathbf{y}$ on each single column $\mathbf{x}$ of the design matrix separately can be added to obtain the regression sum of squares of the complete model. This decomposition does not work for nonorthogonal designs with nondiagonal $X'X$ matrices.

EXERCISES

Exercise 1   Consider Case 1 (Eagle Brands) from the case study appendix.

(a) Assume that there is one factor that increases average store sales by $100. You want to be 80% confident that a 5% significance test can detect such a large increase. Determine the sample size. Use computer software such as Minitab or JMP to check your calculations.

What if you wanted to be 70% confident to detect a change as large as $80?

Hint: Use the approach outlined in Appendix 2.1. However, note that here the effect compares two *averages*, instead of two proportions. Assume that the standard deviation of individual sales is given by σ. The variance σ^2 replaces $\pi(1 - \pi)$, the variance of the 0/1 random variable in Appendix 2.1. This substitution leads to the expression for the required sample size that is shown here:

$$n \cong \frac{2\sigma^2[z_{1-\alpha} + z_{1-\beta}]^2}{\delta^2}$$

Also note that n is the sample size of each of the two groups. The sample size of the factorial experiment is obtained by multiplying the above expression by 2.

(b) Now that you know the sample size, discuss the advantages of a multifactor experiment over the approach of changing one factor at a time.

(c) Eagle Brands wants to learn about the effects of six factors. A full 2^6 factorial in 64 runs could be considered. Discuss the advantages and disadvantages of such a design.

(d) Discuss the protocol that you would use to carry out the experiment.

(e) Discuss whether one should analyze absolute or relative (proportional) changes in sales.

Exercise 2   Consider Case 2 (Magazine Price Test) from the case study appendix.

(a) Consider Sales. Estimate the main effects (A, B, C) and the interaction effects (AB, AC, BC, ABC), and construct a normal probability plot. Assess the significance of the estimated effects. Note that Minitab will calculate Lenth's (1989) PSE (see Appendix 4.3) and help you with the assessment.

Note: This amounts to assessing the significance of effects from an unreplicated design. You will notice that A, C and AC are large and significant.

(b) Obtain the coefficients in the regression model of sales on main effects and interactions and convince yourself that the coefficients are one-half of the estimated effects. Run the regression twice: Once with the eight factorial responses and once with all 9 runs, including the response at the center point. You will notice that all coefficients in these two regressions, except the intercept, are the same. Explain these findings.

Hint: Use the regression formulation in Appendix 4.4. The intercept is the average response; hence, the intercept in the regression that includes the center point is 8/9 (average response from factorial runs) + 1/9 (response at center point).

(c) Consider subscriptions. Estimate the main effects (A, B, C) and the interaction effects (AB, AC, BC, ABC), and construct a normal probability plot. Assess the significance of the estimated effects.

Note: You will notice that A, B, and AB are large and significant.

(d) Recreate the main and interaction plots that are given in this case.

(e) Averages in the table are calculated from five weekly percent changes. The case also provides an estimate of the standard deviation of weekly percent changes: 5% for sales changes, and 15% for subscription changes. Use these estimates to obtain standard errors of the estimated effects for both sales and subscriptions (see Section 4.4.2). Check whether these standard errors change the conclusions you reached in (a) and (b). Discuss the assumptions that one makes when using week-to-week changes to estimate the variability.

(f) Use the standard deviations of weekly percent changes to test for curvature in both sales and subscriptions (see Section 4.6.6).

Exercise 3   A 2^2 factorial experiment with two independent replications at each of the four design points was conducted.

Factor 1	Factor 2	Responses
−1	−1	9, 12
1	−1	26, 22
−1	1	15, 19
1	1	32, 27

(a) Estimate the main effects and the interaction.

(b) Use the independent replications to obtain a standard error of the effects and assess the significance of the effects.

(c) Test the hypothesis that the two main effects are the same.

Hint: Use the fact that this design is orthogonal and that the estimates are statistically independent. This implies that $var(\text{effect } 1 - \text{effect } 2) = var(\text{effect } 1) + var(\text{effect } 2)$

Exercise 4   Consider three categorical factors at two levels each. Assume that only one of the eight experimental conditions has an effect on the response (response is 10 at $(+, +, +)$), while the seven others have no effect (response is zero). Analyze the data. Estimate main and interaction effects. Discuss our comment in Section 4.7 that effect sparcity and effect hierarchy, which are useful design principles for experiments that involve continuous factors, may have less applicability for categorical factors.

Exercise 5 Montgomery (1996, p. 543) used a 2^4 factorial experiment in developing a nitride etch process on a single-wafer plasma etcher. The etching process uses C_2F_6 (perfluoroethane) as the reactant gas. Four factors can be varied: the gas flow, the power applied to the cathode, the pressure in the reaction chamber, and the gap between the anode and the cathode. The response variable is the etch rate for silicon nitride (in angstroms per minute). Each factor is varied at a high- and a low-level setting. The objective is to find the factor-level settings that maximize the etch rate. The levels for gap (factor A) are 0.8 and 1.2 cm; the levels for pressure (factor B) are 450 and 550 mTorr; the levels for the C_2F_6 flow (factor C) are 125 and 200 sccm (standard cc/minute); the levels for power (factor D) are 275 and 325 watts. For further background on the etching process and details of the experiment, you can consult the original source for this exercise, Yin and Jillie (1987).

Run	A (gap)	B (pressure)	C (flow)	D (power)	Response (etch rate)
1	-1	-1	-1	-1	550
2	1	-1	-1	-1	669
3	-1	1	-1	-1	604
4	1	1	-1	-1	650
5	-1	-1	1	-1	633
6	1	-1	1	-1	642
7	-1	1	1	-1	601
8	1	1	1	-1	635
9	-1	-1	-1	1	1,037
10	1	-1	-1	1	749
11	-1	1	-1	1	1,052
12	1	1	-1	1	868
13	-1	-1	1	1	1,075
14	1	-1	1	1	860
15	-1	1	1	1	1,063
16	1	1	1	1	729

Analyze the results of the 2^4 factorial experiment. Find the important main effects and interactions. Assess their significance by using normal probability plots and/or Lenth's (1989) PSE approach. How would you select the factor level settings so that you achieve high etch rates?

Exercise 6 Meredith Corporation, the publisher of *Ladies' Home Journal* magazine, sends more than a million letters each year to potential subscribers hoping to secure as many subscriptions as possible. The marketing team looks for the right mix of promotional materials, and it experiments constantly with various aspects of the brochure, order card, enclosed testimonials, and offers. The June 2005 campaign, for example, tested different versions of the front page of the brochure, and different messages on the front and the back side of the order card.

> *Front side of brochure.* One version (level -1) shows a radiant-looking Kelly Ripa (the star of the ABC show *Live with Regis and Kelly*), while the other version (level $+1$) features Dr. Phil (known from his nationally syndicated TV show and publications on life strategies and relationships).

Front side of the order card. Level 1 (−1) highlights the message "Double our Best Of-fer," while level 2 (+1) draws attention to the message "We never had a bigger sale."

Back side of the order card. Level 1 (−1) emphasizes "Two extra years free," while level 2 (+1) features magazine covers of previous issues.

The results (number of letters sent and the number of orders that were received) are shown below.

Order Card Front	Order Card Back	Brochure	Letters Sent	Orders	Proportion
−1	−1	−1	15,042	573	0.0380933
1	−1	−1	15,042	644	0.0428135
−1	1	−1	15,042	563	0.0374285
1	1	−1	15,042	616	0.0409520
−1	−1	1	15,042	564	0.0374950
1	−1	1	15,042	550	0.0365643
−1	1	1	15,042	575	0.0382263
1	1	1	15,042	553	0.0367637

Analyze the data. Estimate main and interaction effects. Display the effects graphically through main effects and interaction plots.

Assess the significance of the effects, using the approach discussed in Section 4.6.4.

Summarize your conclusions.

Exercise 7 This exercise applies the general regression results in Appendix 4.4 to the special model without intercept, $y = \beta x + \varepsilon$, that relates a response vector $\mathbf{y}$ to a single regressor vector $\mathbf{x}$. Show that

(a) $\hat{\beta} = [\sum x_i y_i]/[\sum x_i^2]$

(b) $SSR(\mathbf{x}) = [\sum x_i y_i]^2/[\sum x_i^2]$

> *Hint:* The matrix X in Appendix 4.4 is the $n \times 1$ vector $\mathbf{x}$. The total sum of squares in a model without an intercept is given by $\sum y_i^2$. The error sum of squares is given by $\sum(y_i - \hat{\beta}x_i)^2$. The regression sum of squares is the difference, $SSR(\mathbf{x}) = \sum y_i^2 - \sum(y_i - \hat{\beta}x_i)^2$. Result (b) follows after substituting result (a) into this equation.

Exercise 8 This exercise was inspired by a real problem described in the article, "Strategic Testing Stops Leaky Litter Cartons in Their Tracks" (*Packaging Digest*, August 2001). The exercise resembles what the actual company did, but the data are not real.

The makers of "Cats Love It" cat litter are facing a serious problem. Retail customers are reporting that cartons of the firm's premium brand cat litter are leaking the product onto store shelves. The company realizes that while cat lovers are used to cleaning stray sprays of litter tracked through the house, they are not willing to put up with cartons that leak on the way home.

Management has determined that the problem is with the carton-sealing process. Cartons are filled and sealed on a production line run by 20 workers. The company decides to perform a 3-factor factorial experiment. A run consists of filling and sealing 200 cartons.

The factors to be tested and levels of each are shown below. Factor A is line speed with the minus level at 22 cartons per minute and the plus level at 30 cartons per minute. Factor B is the pressure applied by the gluing machine, with the minus level being lower pressure and the plus level being higher pressure. Factor C is the amount of glue used, with the plus level being the current amount and the minus level being 40% less glue.

The design matrix and the estimated effects are shown below. The response is the proportion of cartons that leak.

	LEVEL	
	$-$	$+$
A: Line speed	Slow	Fast
B: Glue pressure	Lower	Higher
C: Amount of glue	Less	More

Run	A	B	C	Response
1	$-$	$-$	$-$	8
2	$+$	$-$	$-$	45
3	$-$	$+$	$-$	47
4	$+$	$+$	$-$	10
5	$-$	$-$	$+$	8
6	$+$	$-$	$+$	40
7	$-$	$+$	$+$	41
8	$+$	$+$	$+$	8

Estimation results:

$$\text{Average} = 25.875$$

$$A = \cdots \qquad B = 1.25 \qquad C = 3.25$$

$$AB = -34.75 \qquad AC = \cdots \qquad BC = -0.75 \qquad ABC = 2.25$$

(a) What is the estimated main effect of factor A? What is the estimated AC interaction?

(b) Suppose each response is the average of 2 replicated runs (note that the numbers have been rounded). Suppose the pooled estimate of the variance of the response of an individual run is equal to 16. Based on 95% confidence intervals for the effects, which effects are significant?

(c) Based on the results of this experiment, what levels would you recommend for each factor?

(d) What is the regression prediction equation, and what is the predicted response (proportion of leaking cartons) if your recommended settings are used?

Exercise 9   A 2^3 factorial experiment is to be conducted. The variance of the response of an individual run is known to be equal to 4 from previous experiments. Suppose that we want the width of a 95% confidence interval for the mean of an effect to be 1.8 or smaller. How many runs need to be made for each test condition, and how many runs are needed in total? Assume we have the same number of runs for each test condition.

TWO-LEVEL FRACTIONAL FACTORIAL DESIGNS

5.1 INTRODUCTION

In Chapter 4, we discussed 2-level factorial designs. These designs are very useful when there are relatively few factors. But as k, the number of factors, increases, the required number of runs in a 2^k factorial design grows rapidly, with each additional factor doubling the number of runs. With 4 factors, there are $2^4 = 16$ runs; with 5 factors, $2^5 = 32$ runs; with 6 factors, $2^6 = 64$ runs; and so forth. With 10 factors, there would be $2^{10} = 1,024$ runs! Obviously, an experiment with that many runs would be out of the question. If full factorial designs were the only choice for the experimenter, experimental design tools would have limited value. But as we will see in this chapter, fractional designs in which the experimenter performs only a fraction of the number of runs required in a full factorial design offer an extremely powerful approach to experimentation.

5.2 SPILLING THE BEANS: A FRACTIONAL DESIGN
FOR 5 FACTORS IN 16 RUNS

A company supplies freshly roasted coffee to restaurants and gourmet food stores. In a recent blind taste test, the company's Kenya AA coffee was judged inferior to the same variety of coffee produced by a competitor. In light of this disappointing outcome, the firm's chief coffee roaster, with the help of a statistical consultant, decides to conduct an experiment aimed at improving the taste of the Kenya AA. (This example is based on an actual study. For simplicity, some minor details have been changed, but the essential elements of the real experiment, including the conclusions, have not been altered.)

The chief roaster has identified 5 factors likely to be important. The factors are the initial temperature of the roasting machine when the green (unroasted) beans are put into it (factor 1); the temperature of the flame (factor 2), which determines how quickly the beans are roasted; the color of the beans when they are removed from the roaster (factor 3); the supplier of the green beans (factor 4); and the roasting machine (factor 5). Two small (5-pound) roasting machines are used in the experiment. The operating ranges for flame temperature are the same for both machines.

TABLE 5.1
Factors and Levels in the Coffee Experiment

	Factor	LEVEL	
		−	+
1	Initial temperature	Lower	Higher
2	Flame temperature	Lower	Higher
3	Color	Lighter	Darker
4	Supplier	Current	New
5	Machine	Current	New

The 5 factors and levels are shown in Table 5.1. For initial temperature, based on experience and the operating ranges recommended by equipment makers, the levels are set an equal distance above and below the normal setting. For flame temperature, the minus level is the minimum temperature required to roast the beans, while the plus level is the highest temperature in the operating range. In practice, the chief roaster varies the color of roasted beans depending on the variety of the coffee, but favors lighter roasting, recognizing that coffee roasted too dark will have a bitter and burnt taste. For the color factor, the minus level corresponds to his normal color for Kenya AA, while the plus level is considerably darker. The two suppliers for the test are the existing one and a well-regarded competitor. The last factor is the roasting machine. The company has a small (5-pound capacity) roasting machine that it uses to test new sources of green beans. The chief roaster wants to evaluate another small machine made by a different manufacturer and sees this experiment as an opportunity to do so.

Suppose the company is willing and able to do only 16 runs rather than the 32 runs of a full 2^5 factorial design. What is the best design for doing so, and what is lost by carrying out only 16 runs?

5.2.1 Constructing the Design Matrix

We build the 16-run design in Table 5.2. We begin by writing down the full factorial design for 4 factors, 1, 2, 3, and 4, which consists of 16 runs. The four columns are shaded to highlight the familiar pattern of pluses and minuses that make up the 2^4 factorial design. For now, ignore the shaded column for factor 5. The remaining columns represent the 11 interactions in the 2^4 design. There are six 2-factor interactions, four 3-factor interactions, and one 4-factor interaction. The signs for these columns, also referred to as the calculation columns in Chapter 4, were found by multiplying the signs of the design columns. For example, the signs for column 123 were found by multiplying the signs of columns 1, 2, and 3. As we discussed in Chapter 4, the 15 columns are pairwise orthogonal and consequently each of the corresponding 15 effects in the 2^4 factorial design is estimated independently of every other effect.

Factor 5 completes the design. We made the signs in column 5 identical to the signs for the 1234 interaction. We will explain this choice in a moment. But first, suppose instead we had made the signs in column 5 identical to the signs in column 1. This would mean that whenever factor 5 was at its minus level, factor 1 would also be minus, and whenever factor 5

TABLE 5.2

Constructing a Fractional Factorial Design for 5 Factors in 16 Runs

	FACTOR					CALCULATION COLUMNS FOR THE 2^4 DESIGN IN FACTORS 1, 2, 3, AND 4										
Run	1	2	3	4	5	12	13	14	23	24	34	123	124	134	234	1234
1	−	−	−	−	+	+	+	+	+	+	+	−	−	−	−	+
2	+	−	−	−	−	−	−	−	+	+	+	+	+	+	−	−
3	−	+	−	−	−	−	+	+	−	−	+	+	+	−	+	−
4	+	+	−	−	+	+	−	−	−	−	+	−	−	+	+	+
5	−	−	+	−	−	+	−	+	−	+	−	+	−	+	+	−
6	+	−	+	−	+	−	+	−	−	+	−	−	+	−	+	+
7	−	+	+	−	+	−	−	+	+	−	−	−	+	+	−	+
8	+	+	+	−	−	+	+	−	+	−	−	+	−	−	−	−
9	−	−	−	+	−	+	+	−	+	−	−	−	+	+	+	−
10	+	−	−	+	+	−	−	+	+	−	−	+	−	+	+	+
11	−	+	−	+	+	−	+	−	−	+	−	+	−	+	−	+
12	+	+	−	+	−	+	−	+	−	+	−	−	+	−	−	−
13	−	−	+	+	+	+	−	−	−	−	+	+	+	−	−	+
14	+	−	+	+	−	−	+	+	−	−	+	−	−	+	−	−
15	−	+	+	+	−	−	−	−	+	+	+	−	−	−	+	−
16	+	+	+	+	+	+	+	+	+	+	+	+	+	+	+	+

was at its plus level, factor 1 would be plus as well. In this case, the average of the responses when 5 (=1) is at the plus level ($\bar{y}_+$) minus the average of the responses when 5 (=1) is at the minus level ($\bar{y}_-$) is actually an estimate of the main effect of 5 *plus* the main effect of 1. With this arrangement, the two main effects are said to be *confounded*, and it is not possible to separate them. The main effect of factor 5 and the main effect of factor 1 are called *aliases* of each other. The calculated effect ($\bar{y}_+ - \bar{y}_-$) might be due to the main effect of factor 5, the main effect of factor 1, or some combination of the two main effects. Confounding two main effects in this way would be a poor choice since main effects tend to be the largest and most important effects.

The best choice is to confound the main effect of factor 5 with an effect that is least likely to be important, which is the 4-factor interaction 1234. Therefore we set 5 = 1234, confounding the main effect of 5 with the 1234 interaction. Taking the average of the responses when 5 (=1234) is at the plus level ($\bar{y}_+$) minus the average of the responses when 5 (=1234) is at the minus level ($\bar{y}_-$), estimates the main effect of factor 5 *plus* the 4-factor interaction 1234. Since 4-factor interactions are almost certain to be negligible, we are left with an estimate of the main effect of factor 5.

Effects Are Confounded in Pairs. By setting 5 = 1234, we not only confound these two effects, but all other effects become confounded in pairs as well. For example, consider in Table 5.2 the column of signs representing the 12 interaction. Writing this column as a row to save space, we have

$$12 = + - - + + - - + + - - + + - - +$$

Now multiply the signs for factors 3, 4, and 5 to obtain a column representing the 345 interaction. It is

$$345 = + - - + + - - + + - - + + - - +$$

The two columns are identical; the effects 12 and 345 are confounded. Taking the average of the responses when the signs in column 12 (=345) are plus ($\bar{y}_+$) minus the average of the responses when the signs in column 12 (=345) are minus ($\bar{y}_-$) results in an estimate that is the *sum* of the 2-factor interaction 12 and the 3-factor interaction 345.

The entire confounding pattern can be found in the same way, by multiplying columns of signs for every interaction and identifying pairs of columns that are identical. However, this brute force approach is not necessary; in Section 5.2.3 we will present a much simpler method for determining which columns have identical signs.

5.2.2 The Design Matrix, Confounding Pattern, and Results of the Coffee Experiment

Table 5.3 shows the 16-run design that is used in the coffee experiment. The runs are performed in random order and a sample of roasted beans is taken from each run and brewed using the same coffee maker. A blind taste test is carried out, with each sample rated on a scale from 1 (lowest quality) to 10 (highest quality). The last column in Table 5.3 shows the quality ratings of the brewed coffee that resulted from the various roasts.

The lower part of the table shows the 15 effects that are independently estimated and the confounding patterns that arise from this design. Each main effect is confounded with a 4-factor interaction, and each 2-factor interaction is confounded with a 3-factor interaction. In showing the confounding pattern, we introduce some new notation.

Notice that the column of signs associated with each estimate is identical to one of 15 columns in the 2^4 factorial design. To calculate each effect, we apply the signs in its column to the observations and divide the result by the number of plus signs, 8. Since each is a linear function of the observations and compares two averages (response averages at the plus and the minus levels of that column), we refer to the estimated effect as a *linear contrast*. We use the letter l to denote the estimate (l for linear), and we use the column label as a subscript to identify the column that is involved. For example, the estimate l_5 applies the signs in column 5 to the responses, obtains the sum, and divides the sum by 8. The estimated effect

$$l_5 = \frac{+3 - 7 - 6 + 5 - 10 + 7 + 8 - 9 - 5 + 3 + 3 - 4 + 7 - 9 - 10 + 8}{8}$$

$$= \frac{3 + 5 + 7 + 8 + 3 + 3 + 7 + 8}{8} - \frac{7 + 6 + 10 + 9 + 5 + 4 + 9 + 10}{8}$$

$$= 5.5 - 7.5 = -2.0$$

is a difference (contrast) between the two averages at the plus and minus levels of column 5. We use the notation $l_5 \rightarrow 5 + 1234$ to show that l_5 estimates the main effect of factor 5 plus the 1234 interaction. The arrow means "estimates." Similarly, to calculate l_{12}, for example, we apply the signs in column 12 to the responses, sum them, and divide by 8. This contrast estimates 12 + 345, and we indicate this by writing $l_{12} \rightarrow 12 + 345$.

In this design, if we assume that 3-factor and 4-factor interactions are negligible, which is very likely, we are left with clear estimates of all main effects and 2-factor interactions.

TABLE 5.3

Design Matrix, Estimated Effects, and Confounding Patterns in the Coffee Experiment

Run	FACTOR					Response Rating
	1	2	3	4	5	
1	−	−	−	−	+	3
2	+	−	−	−	−	7
3	−	+	−	−	−	6
4	+	+	−	−	+	5
5	−	−	+	−	−	10
6	+	−	+	−	+	7
7	−	+	+	−	+	8
8	+	+	+	−	−	9
9	−	−	−	+	−	5
10	+	−	−	+	+	3
11	−	+	−	+	+	3
12	+	+	−	+	−	4
13	−	−	+	+	+	7
14	+	−	+	+	−	9
15	−	+	+	+	−	10
16	+	+	+	+	+	8

Effects that may be estimated and their confounding pattern:

$l_0 = 6.5 \rightarrow$ average

$l_1 = 0 \rightarrow 1 + 2345$ $l_{12} = -0.25 \rightarrow 12 + 345$ $l_{24} = 0 \rightarrow 24 + 135$

$l_2 = 0.25 \rightarrow 2 + 1345$ $l_{13} = -0.50 \rightarrow 13 + 245$ $l_{25} = 0.75 \rightarrow 25 + 134$

$\mathbf{l_3 = 4.00 \rightarrow 3 + 1245}$ $l_{14} = -0.25 \rightarrow 14 + 235$ $l_{34} = 0.75 \rightarrow 34 + 125$

$l_4 = 0.75 \rightarrow 4 + 1235$ $l_{15} = 0.50 \rightarrow 15 + 234$ $l_{35} = 0 \rightarrow 35 + 124$

$\mathbf{l_5 = -2.00 \rightarrow 5 + 1234}$ $l_{23} = 0.25 \rightarrow 23 + 145$ $l_{45} = 0.25 \rightarrow 45 + 123$

NOTE: The signs in the factor 5 column are identical to the signs in the 1234 calculation column. Significant effects are shown in boldface.

And this is accomplished with only 16 runs, compared to the $2^5 = 32$ runs that would be required in a full factorial design.

To determine which estimated effects are significant, we examine the normal probability plot of the estimated effects in Figure 5.1 that was generated with the Minitab software. Two effects are significant: the main effect of factor 3 (color), and the main effect of factor 5 (machine). A change in the color of the roasted beans from lighter to darker increases the taste rating by 4 points on average, while a change to the new machine decreases the rating by 2 points.

The average of the 16 responses is 6.5. Given this average and our estimates of the two significant effects, the implied regression prediction equation is $\hat{y} = 6.5 + 2x_3 - 1x_5$. At the best settings, when factor 3 is at + (darker color) and factor 5 is at − (current machine), the predicted taste rating is $\hat{y} = 6.5 + 2(+1) - 1(-1) = 9.5$.

The conclusions from this experiment are clear: Keep the current machine and, most important, roast the coffee to the darker color. The chief roaster realized immediately that he had been roasting the Kenya AA too light, and from that point on, he began roasting it to the darker color.

This story began when a competitor's coffee was judged superior in a blind taste test. All concerned parties at the roasting company were extremely pleased when, in the next

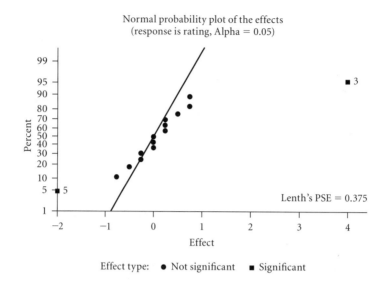

Figure 5.1   Normal Probability Plot of the Estimated Effects in the Coffee Experiment

blind taste test against a fresh batch of the competitor's coffee, the chief roaster's experimentally designed Kenya AA was judged best.

5.2.3 Finding the Confounding Pattern: Generator and Defining Relation

The 16-run fractional design for the coffee experiment was found by first writing out the design matrix for a 4-factor factorial design and then setting column 5 equal to column 1234. The relation $5 = 1234$ is called the *generator* of the design. The generator is used to find the complete confounding pattern for the design. Before we describe this procedure, we introduce several basic and important rules for multiplying columns of signs.

1. The capital letter I denotes a column of all plus signs. Multiplying any column (of plus and minus signs) by itself results in I. Multiplying a plus entry by itself yields a plus, and multiplying a minus entry by itself yields a plus, also.

2. Multiplying a column by I leaves the column unchanged. This is analogous to multiplication by 1 in ordinary arithmetic.

3. When multiplying columns together, the order of multiplication does not matter. For example, $2123 = 2213 = 2132$.

Now we proceed to find the confounding pattern for the 5-factor 16-run design with generator $5 = 1234$. Multiplying both sides of the generator by column 5, we obtain

$$5 \times 5 = 1234 \times 5$$

$$I = 12345$$

$I = 12345$ is called the *defining relation* of the design.

We use the defining relation to find the confounding pattern among the 15 independent effect estimates (linear combinations of the responses) that can be calculated in this design. For example, to find what is confounded with 1 (the main effect of factor 1), we multiply both sides of the defining relation by column 1:

$$1(I) = 112345$$

$$1 = 2345$$

because $1(I) = 1$ and $(1)(1) = I$. Thus the signs in column 1 and the 2345 interaction column are identical, and the main effect of factor 1 is confounded with the 2345 interaction. To check that this is correct, multiply the signs of columns 2, 3, 4, and 5 and show that they are identical to the signs in column 1.

Similarly, multiplying both sides of the defining relation by (column) 34 results in

$$34(I) = 3412345$$

$$34 = 125$$

The right-hand side is 125 because $34 \times 34 = I$, and $I \times 125 = 125$. The 2-factor interaction 34 is confounded with the 3-factor interaction 125. The average of the responses when 34 ($=125$) is plus ($\bar{y}_+$) minus the average of the responses when 34 ($=125$) is minus ($\bar{y}_-$) estimates the sum of the 34 and the 125 interactions.

The 5-factor design is called a half-fraction of the full 2^5 factorial design. It consists of half of the 32 runs that are required for a full factorial design with 5 factors. We use the notation 2^{5-1} (2 to the power of 5 minus 1) to denote this design. The 2 denotes that there are 2 levels for each factor, the 5 indicates there are 5 factors, and the 1 in the exponent tells us that it is a half fraction involving one generator. The notation also expresses the fact that there are 16 runs ($2^{5-1} = 2^4 = 16$).

The Importance of Maintaining Orthogonality. Setting $5 = 1234$ is the best choice in this case, but we could have set the signs of column 5 equal to the signs of any one of the 15 orthogonal columns in the 2^4 factorial design. By choosing one of these 15 columns, we maintain the important property of orthogonality. Each of the 15 effect estimates in the fractional design uses one of the 15 columns in the 2^4 design, which means that they are independently estimated. But now each estimated effect is actually the sum of two effects, either a main effect and a 4-factor interaction or a 2-factor interaction and a 3-factor interaction. The price we pay for reducing the number of runs from 32 to 16 is the introduction of confounding. But in this case, since 3- and 4-factor interactions are likely to be negligible, we have lost very little.

5.3 CRACKED POTS REDUX

Let us return to the cracked pots example in Chapter 4. In that problem we examined 4 factors in a 2^4 factorial design. The factors were cooling rate, temperature of the kiln, coefficient of expansion, and carrier (metal or rubberized). The results of that experiment are

TABLE 5.4

The Cracked Pots Problem: Design Matrix, Estimated Effects, and Confounding Pattern for a 4-Factor Experiment in 8 Runs

	FACTOR				INTERACTION COLUMNS FOR THE CALCULATION OF REMAINING EFFECTS			RESPONSE
Run	1 Cooling Rate	2 Temperature of Kiln	3 Coefficient Expansion	4 = 123 Carrier	12	13	14 (=23)	Percentage of Cracked Pots
1	−	−	−	−	+	+	+	15
2	+	−	−	+	−	−	+	8
3	−	+	−	+	−	+	−	3
4	+	+	−	−	+	−	−	18
5	−	−	+	+	+	−	−	12
6	+	−	+	−	−	+	−	34
7	−	+	+	−	−	−	+	21
8	+	+	+	+	+	+	+	27

Effects that may be estimated and their confounding pattern:
$l_0 = 17.25 \rightarrow$ average
$\mathbf{l_1 = 9.0 \rightarrow 1 + 234}$ $l_{12} = 1.5 \rightarrow 12 + 34$
$l_2 = 0.0 \rightarrow 2 + 134$ $\mathbf{l_{13} = 5.0 \rightarrow 13 + 24}$
$\mathbf{l_3 = 12.5 \rightarrow 3 + 124}$ $l_{14} = 1.0 \rightarrow 14 + 23$
$\mathbf{l_4 = -9.5 \rightarrow 4 + 123}$

NOTE: Significant effects are shown in boldface.

shown in Table 4.7 of Chapter 4. We found very large main effects for cooling rate and coefficient of expansion and a significant interaction between these two factors. We also found a significant main effect for carrier.

5.3.1 Testing 4 Factors in 8 Runs

Suppose the company had decided to do only 8 runs rather than the 16 that were carried out. What design would have been best? We label the factors 1, 2, 3, and 4 for cooling rate, temperature, coefficient of expansion, and carrier, respectively. The design matrix for an 8-run experiment is shown in Table 5.4. Using an approach that is analogous to the one we used in the coffee experiment, we start with the first three columns of the 8-run 2^3 design. The generator 4 = 123 sets the signs for factor 4 equal to the signs for the 123 interaction column (the highest order interaction in the 2^3 design). The defining relation is $I = 1234$. From the defining relation, we can easily find the confounding pattern using the method described above. For example, 2 = 134 (the main effect of 2 is confounded with the 134 interaction), while 13 = 24 (the 2-factor interaction 13 is confounded with the 2-factor interaction 24). You can multiply signs to confirm that these are correct. The table also shows the interaction columns in this design that are needed to calculate three of the effects. In this design, each main effect is confounded with a 3-factor interaction, and each 2-factor interaction is confounded with another 2-factor interaction. Compared to the fractional design in the coffee experiment (5 factors in 16 runs), the confounding in this design is worse. Here if we assume that 3-factor interactions are negligible, we have clear estimates of each main effect. But now each 2-factor interaction is confounded with another 2-factor interaction.

5.3.2 Results of the Experiment

Assume the experiment was carried out with the results of each run (% cracked pots) shown in the last column of the table. We calculate each effect estimate from its column of signs. For example we have

$$l_{13} = \frac{15 - 8 + 3 - 18 - 12 + 34 - 21 + 27}{8}$$

$$= \frac{15 + 3 + 34 + 27}{4} - \frac{8 + 18 + 12 + 21}{4}$$

$$= 19.75 - 14.75 = 5$$

Suppose that based on previous experiments, the company is confident that the variance of the response of a run is 8.5. In Section 4.4.1 of Chapter 4, we showed that the estimated variance of an effect $s_{\text{effect}}^2 = (4/N)s_p^2$, where N is the total number of runs, which in this case is 8. Thus we have that $s_{\text{effect}}^2 = (4/8)(8.5) = 4.25$ and $s_{\text{effect}} = \sqrt{4.25} = 2.06$. A 95% confidence interval for the mean of each effect is given by the estimated effect $\pm 2.06(1.96)$. Effects larger than $2.06(1.96) = 4.04$ are statistically significant. There are four significant effects: $(1 + 234)$, $(3 + 124)$, $(4 + 123)$, and $(13 + 24)$. Assuming that 3-factor interactions are zero, we would conclude that the first three significant effects are estimates of the main effect of 1 (cooling rate), the main effect of 3 (coefficient of expansion), and the main effect of 4 (carrier), respectively. The three estimates have values that are close to what we found in the $2^4 = 16$ runs of experiment of Chapter 4.

In the full factorial experiment, there was no confounding of course, and we found that there was a significant interaction between cooling rate and coefficient of expansion, here labeled as the 13 interaction. But in this 8-run design, there is some uncertainty. The 13 interaction is confounded with the 24 interaction (an interaction between kiln temperature and carrier). If this 8-run experiment had been run rather than the 16-run full factorial, would we have been able to identify with confidence the interaction between cooling rate and coefficient of expansion (the 13 interaction as we have labeled it here)? It is hard to say for sure. The fact that the two main effects, cooling rate and coefficient of expansion, are large would lead us to believe that the significant effect $(13 + 24)$ is due to the 13 interaction but an interaction between kiln temperature and carrier (factors 2 and 4) is also conceivable. We have gained by cutting the required number of runs in half, but we have introduced confounding and thus some uncertainty in the interpretation of the results.

5.4 DESIGN RESOLUTION

We have discussed two fractional factorial designs one with 5 factors and another with 4. The full factorial 2^5 design requires 32 runs, and there is no confounding among the estimated effects. The half fraction 2^{5-1} design is more economical with only 16 runs, but it induces confounding: Main effects are confounded with 4-factor interactions, while 2-factor interactions are confounded with 3-factor interactions. The 2^{4-1} design is a half-fraction of the $2^4 = 16$-run full factorial design. Compared to the 2^{5-1} design, its confounding pattern is worse. For this 8-run design for 4 factors, main effects are confounded with 3-factor interactions, and 2-factor interactions are confounded with other 2-factor interactions.

The *resolution R* of a fractional design is an index (usually written as a roman numeral) that expresses the degree of confounding.

1. A design of resolution $R = $ III confounds main effects with 2-factor interactions. (We will see examples of these designs shortly).

2. A design of resolution $R = $ IV confounds main effects with 3-factor interactions, and 2-factor interactions with other 2-factor interactions.

3. A design of resolution V confounds main effects with 4-factor interactions, and 2-factor interactions with 3-factor interactions.

The 16-run design for 5 factors with generator $5 = 1234$ has resolution V. It confounds main effects with 4-factor interactions, and 2-factor interactions with 3-factor interactions. We denote this design by 2_V^{5-1}. It consists of $2^{5-1} = 2^4 = 16$ runs. The "1" in the exponent expresses the fact that there is a single generator. The subscript "V" denotes its resolution. The 8-run design for 4 factors with generator $4 = 123$ has resolution IV. It confounds main effects with 3-factor interactions, and each 2-factor interaction with another 2-factor interaction. We denote it by 2_{IV}^{4-1}.

In constructing our 2^{5-1} design, we could have used *any* interaction or main effect column to accommodate the 5th factor. We chose the generator $5 = 1234$, which yields the half-fraction with the highest possible resolution. (Similarly, in the 2^{4-1} design we set $4 = 123$). Suppose we had set $5 = 123$ instead. Then the defining relation would be $I = 1235$, and the design would be resolution IV. In general, with k factors, the generator $k = 123 \ldots (k-1)$ produces a half-fraction with highest possible resolution. For example, with $k = 3$, the best generator would be $3 = 12$.

The resolution of a design can be determined directly from its defining relation. Each term in the defining relation to the right of I is called a "word." For example, for the 2^{5-1} design with generator $5 = 1234$, the defining relation $I = 12345$ consists of the single word 12345. For the 2^{4-1} design with generator $4 = 123$, the defining relation $I = 1234$ consists of the single word 1234. Shortly, we will see examples of designs with defining relations that consist of more than one word. The resolution of a design is the length of the shortest word in the defining relation. In the first case ($I = 12345$), the length of the single word is 5 (it consists of 5 numbers), and the design is resolution V. In the second case ($I = 1234$), the length of the (shortest) word is 4 (it consists of 4 numbers), and the design is resolution IV.

In choosing a design, the experimenter must consider both design resolution and cost. Higher-resolution designs have more attractive confounding patterns but require more runs and are therefore more costly. Resolution V designs are especially useful because we can obtain unambiguous estimates of main effects and 2-factor interactions if we assume that interactions of order three and higher are negligible. In instances where the costs associated with each run are relatively small, these designs are particularly attractive, because the experimenter is confident beforehand of achieving unambiguous results.

5.5 FRACTIONAL DESIGNS IN 8 RUNS

Table 5.5 shows fractional designs for 4, 5, 6, and 7 factors. All 8-run fractional designs are constructed from the same building block—the design matrix and the calculation (interaction) columns of the 2^3 factorial design. In each case, the design matrix starts with columns A, B, and C (the darker shaded area in Table 5.5) and is completed by associating each additional factor in the design with a column of signs from the lighter shaded area containing the four interaction columns. The column for each generator is set equal to one of the four interaction columns (AB, AC, BC, or ABC).

The lower part of the table has one row for each design. The first column identifies the design and its resolution, the second column shows the generator(s), and the last seven columns show the seven effects that are independently estimated in the design.

5.5.1 Five Factors in 8 Runs: The 2^{5-2} Design

The 2_{III}^{5-2} resolution III design (main effects are confounded with 2-factor interactions) has two generators: $D = AB$ and $E = AC$ (the "2" in the exponent denotes this). From these two generators, we construct the defining relation. Multiplying both sides of the first generator by D, and both sides of the second generator by E, results in the equation

$$I = ABD = ACE$$

But since $ABD = I$ and $ACE = I$, it is also true that their product equals I; that is $I = ABD \times ACE = BCDE$. Hence the defining relation is given by

$$I = ABD = ACE = BCDE$$

It consists of three words. The length of the shortest word is three (letters), which shows that this is a resolution III design.

To find the confounding pattern for this or *any* other 8-run fractional design, those in the table or designs that use other generators, we always follow the same procedure. We multiply the defining relation by each of the seven columns A, B, C, AB, AC, BC, ABC that make up the design matrix and the calculation columns of the building block, the 2^3 factorial design. For example, multiplying by A, we have

$$AI = AABD = AACE = ABCDE$$
$$A = BD = CE = ABCDE$$

Multiplying by BC, we have

$$BCI = BCABD = BCACE = BCBCDE$$
$$BC = ACD = ABE = DE$$

Multiplying by ABC, we have

$$ABCI = ABCABD = ABCACE = ABCBCDE$$
$$ABC = CD = BE = ADE$$

TABLE 5.5
8-Run Fractional Factorial Designs: Generators, Confounding Patterns, and Resolution

Resolution IV design:

4 factors: 2_{IV}^{4-1} $D = ABC$

Resolution III designs:

5 factors: 2_{III}^{5-2} $D = AB, E = AC$
6 factors: 2_{III}^{6-3} $D = AB, E = AC, F = BC$
7 factors: 2_{III}^{7-4} $D = AB, E = AC, F = BC, G = ABC$

Run	A	B	C	AB	AC	BC	ABC
1	−	−	−	+	+	+	−
2	+	−	−	−	−	+	+
3	−	+	−	−	+	−	+
4	+	+	−	+	−	−	−
5	−	−	+	+	−	−	+
6	+	−	+	−	+	−	−
7	−	+	+	−	−	+	−
8	+	+	+	+	+	+	+

	A	B	C	AB	AC	BC	ABC
2_{IV}^{4-1}	A	B	C	AB + CD	AC + BD	BC + AD	D
2_{III}^{5-2}	A + BD + CE	B + AD	C + AE	D + AB	E + AC	BC + DE	BE + CD
2_{III}^{6-3}	A + BD + CE	B + AD + CF	C + AE + BF	D + AB + EF	E + AC + DF	F + BC + DE	AF + BE + CD
2_{III}^{7-4}	A + BD + CE + FG	B + AD + CF + EG	C + AE + BF + DG	D + AB + CG + EF	E + AC + BG + DF	F + AG + BC + DE	G + AF + BE + CD

2_{IV}^{4-1} $D = ABC$
2_{III}^{5-2} $D = AB$, $E = AC$
2_{III}^{6-3} $D = AB$, $E = AC$, $F = BC$
2_{III}^{7-4} $D = AB$, $E = AC$, $F = BC$, $G = ABC$

NOTE: The darker-shaded area represents the runs of the 2^3 factorial building block design. The lighter-shaded area represents the calculation columns that are available for generating the levels of additional factors. The expressions below the columns of plus and minus signs specify the confounding patterns of the estimated effects. For example, in the 2_{III}^{6-3} design, the linear contrast l_B estimates $B + AD + CF$. Interactions of order 3 or higher are assumed to be zero throughout the table.

TABLE 5.6

A Fractional Design for 5 Factors in 8 Runs with Generators $D = AB$ and $E = AC$, and Its Confounding Pattern

	FACTORS					
Run	A	B	C	$D = AB$	$E = AC$	Response
1	−	−	−	+	+	y_1
2	+	−	−	−	−	y_2
3	−	+	−	−	+	y_3
4	+	+	−	+	−	y_4
5	−	−	+	+	−	y_5
6	+	−	+	−	+	y_6
7	−	+	+	−	−	y_7
8	+	+	+	+	+	y_8

Effects that may be estimated and their confounding pattern (3-factor and higher-order interactions are assumed to be zero):

$l_0 = \bar{y} \rightarrow$ average

$l_A \rightarrow A + BD + CE$ $l_C \rightarrow C + AE$ $l_{BC} \rightarrow BC + DE$

$l_B \rightarrow B + AD$ $l_{AB} = l_D \rightarrow D + AB$ $l_{ABC} = l_{CD} \rightarrow CD + BE$

    $l_{AC} = l_E \rightarrow E + AC$

The confounding pattern for each design (assuming that 3-factor and higher-order interactions are zero) is given in Table 5.5, with each estimated effect shown with its aliases. For the 2_{III}^{5-2} design, the seven estimated effects are shown as $(A + BD + CE)$, $(B + AD)$, $(C + AE)$, $(D + AB)$, $(E + AC)$, $(BC + DE)$, and $(BE + CD)$.

The design matrix, estimated effects and confounding pattern for this 5-factor design in 8 runs is shown in Table 5.6. Each of the seven contrasts uses one of the columns of signs in the 2^3 building block. We emphasize that important fact in labeling the contrasts.

5.5.2 Six Factors in 8 Runs: The 2^{6-3} Design

The 6-factor 2_{III}^{6-3} resolution III design has three generators: $D = AB$, $E = AC$, and $F = BC$. For more than two generators, as in this case, the procedure for finding the defining relation is analogous to the procedure used when there are two generators. We multiply both sides of the first generator by D, both sides of the second by E, and both sides of the third by F. The result is that each right-hand side is equal to I, forming the first three terms (or words) in the defining relation,

$$I = ABD = ACE = BCF$$

However, there are more terms in the defining relation as all products of these three terms, $(ABD)(ACE)$, $(ABD)(BCF)$, $(ACE)(BCF)$, and $(ABD)(ACE)(BCF)$ are I. The complete defining relation consists of seven words,

$$I = ABD = ACE = BCF = BCDE = ACDF = ABEF = DEF$$

The length of the shortest word is three; hence this is a resolution III design. The confounding pattern shown in Table 5.5 is found by multiplying the defining relation by each of the seven columns A, B, C, AB, AC, BC, ABC that make up the design matrix and the calculation columns of the 2^3 factorial building block design. For example

$$BC = ACD = ABE = F = DE = ABDF = ACEF = BCDEF$$

and

$$ABC = CD = BE = AF = ADE = BDF = CEF = ABCDEF$$

5.5.3 Seven Factors in 8 Runs: The 2^{7-4} Design

The final 8-run design shown in the last row of Table 5.5 is the 7-factor 2_{III}^{7-4} resolution III design. Its four generators use all four available interaction columns (AB, AC, BC, and ABC) of the 2^3 factorial building block design. Because of this, the design is said to be *saturated*. We will discuss this design in detail in Section 5.7.

5.6 FRACTIONAL DESIGNS IN 16 RUNS

Table 5.7 shows a set of 16-run fractional factorial designs with the number of factors ranging from 5 to 15. The table is constructed in the same fashion as Table 5.5. In that table for 8-run designs, the building block was the 8-run 2^3 factorial design with its four interaction columns. Here, the building block is the 16-run 2^4 factorial design and its 11 interaction columns. For each 16-run fractional design, the design matrix starts with columns A, B, C, and D (the darker shaded area in Table 5.7) and is completed by associating each additional factor in the design with a column of signs from the lighter shaded area containing the 11 interaction columns.

The confounding patterns shown in Table 5.7 ignore 3-factor and higher-order interactions. Table 5.7 shows the generators for each of these designs, but to save space, we have omitted the confounding patterns for designs with 10 through 14 factors. Software such as Minitab and JMP provide the generators and the confounding patterns for all of the designs in Tables 5.5 and 5.7 automatically. The user simply enters the number of factors and the number of runs. In Section 8.3 of Chapter 8, we discuss the capabilities of these software programs in more detail.

5.6.1 Seven Factors in 16 Runs: The 2^{7-3} Design

Table 5.7 specifies the generators of this design as $E = ABC$, $F = BCD$, and $G = ACD$. The defining relation is given by

$$I = ABCE = BCDF = ACDG = ADEF = BDEG = ABFG = CEFG$$

The last four words are obtained by forming all products of the first three words of the defining relation. The length of the shortest word is four; hence, this is a resolution IV design. The 15 estimated effects and their confounding pattern are shown in the lower part of Table 5.7. To find the confounding pattern for this or *any* 16-run fractional design, we always follow the same procedure. We multiply its defining relation by each of the 15 effect columns A, B, C, D, AB, AC, AD, BC, BD, CD, ABC, ABD, ACD, BCD, and $ABCD$ that make up the design matrix and the calculation columns of the 2^4 factorial design. For example, multiplying by *ABD* we have

$$ABD = CDE = ACF = BCG = BEF = AEG = DFG = ABCDEFG$$

The contrast l_{ABD} estimates *ABD* plus the sum of six other 3-factor interactions plus a 7-factor interaction. However, none of these interactions are visible in Table 5.7 because aliases of order 3 and higher are not shown. Since 3-factor interactions are usually negligible, this contrast is an estimate of the noise.

The design matrix, estimated effects and confounding pattern for this design are shown in Table 5.8. We have labeled each contrast to emphasize the fact that each effect uses one of the 15 columns of signs of the 2^4 building block. By using this approach and listing the effects in the order of the 15 columns of the building block, we also provide a systematic way to identify each of the 15 effects.

5.6.2 Testing 15 Factors in 16 Runs: The 2^{15-11} Saturated Fractional Factorial Design

This is a very efficient design in terms of the number of runs as 15 effects are estimated from just 16 runs. Table 5.7 shows that all 11 interaction (calculation) columns of the 2^4 full factorial building block design are assigned to the additional factors (*E* through *P*). Because all interaction columns are used to define the generators, it is a saturated design. There are 11 generators, and the defining relation consists of many words (in fact, $2^{11} - 1$ words). This is a resolution III design, and as shown in the last row of Table 5.7, each main effect is confounded (aliased) with a string of seven 2-factor interactions. This design is useful in screening experiments where it is reasonable to ignore 2-factor interactions in the initial stage of an investigation. In these cases the experimenter wants to test many factors in relatively few runs to identify a few important ones for further study. Also, as we discuss in the next section, it is an easy matter to add 16 runs to this design to create a 32-run design of resolution IV.

5.7 RESOLVING AMBIGUITIES IN FRACTIONAL FACTORIAL DESIGNS

Experimentation often proceeds sequentially, starting with an initial experiment followed by an additional set of runs to resolve the ambiguities that arise from the confounding of effects. Frequently, a series of experiments would begin with a resolution III design, which is quite economical in terms of the number of runs required. For example, we saw earlier that 7 factors could be studied in 8 runs, and 15 factors in 16 runs. But resolution III designs confound main effects and 2-factor interactions, and at the conclusion of the experiment, the decision maker would not know whether a significant effect is the result of a main effect or one or several 2-factor interactions among other factors. If experiments were only one-shot affairs, these resolution III designs would have less value. But as we show in this section, it is possible and desirable to augment an initial experiment with additional runs that can clarify open questions.

TABLE 5.7
16-Run Fractional Factorial Designs: Generators, Confounding Patterns, and Resolution

Resolution V design:

| 5 factors: | 2_V^{5-1} | $E = ABCD$ |

Resolution IV designs:

6 factors:	2_{IV}^{6-2}	$E = ABC, F = BCD$
7 factors:	2_{IV}^{7-3}	$E = ABC, F = BCD, G = ACD$
8 factors:	2_{IV}^{8-4}	$E = ABC, F = BCD, G = ACD, H = ABD$

Resolution III designs:

9 factors:	2_{III}^{9-5}	$E = ABC, F = BCD, G = ACD, H = ABD, J = ABCD$
10 factors:	2_{III}^{10-6}	$E = ABC, F = BCD, G = ACD, H = ABD, J = ABCD, K = AB$
11 factors:	2_{III}^{11-7}	$E = ABC, F = BCD, G = ACD, H = ABD, J = ABCD, K = AB, L = AC$
12 factors:	2_{III}^{12-8}	$E = ABC, F = BCD, G = ACD, H = ABD, J = ABCD, K = AB, L = AC, M = AD$
13 factors:	2_{III}^{13-9}	$E = ABC, F = BCD, G = ACD, H = ABD, J = ABCD, K = AB, L = AC, M = AD, N = BC$
14 factors:	2_{III}^{14-10}	$E = ABC, F = BCD, G = ACD, H = ABD, J = ABCD, K = AB, L = AC, M = AD, N = BC, O = BD$
15 factors:	2_{III}^{15-11}	$E = ABC, F = BCD, G = ACD, H = ABD, J = ABCD, K = AB, L = AC, M = AD, N = BC, O = BD, P = CD$

Run	A	B	C	D	AB	AC	AD	BC	BD	CD	ABC	ABD	ACD	BCD	ABCD
1	−	−	−	−	+	+	+	+	+	+	−	−	−	−	+
2	+	−	−	−	−	−	−	+	+	+	+	+	+	−	−
3	−	+	−	−	−	+	+	−	−	+	+	+	−	+	−
4	+	+	−	−	+	−	−	−	−	+	−	−	+	+	+
5	−	−	+	−	+	−	+	−	+	−	+	−	+	+	−
6	+	−	+	−	−	+	−	−	+	−	−	+	−	+	+
7	−	+	+	−	−	−	+	+	−	−	−	+	+	−	+
8	+	+	+	−	+	+	−	+	−	−	+	−	−	−	−
9	−	−	−	+	+	+	−	+	−	−	−	+	+	+	−
10	+	−	−	+	−	−	+	+	−	−	+	−	−	+	+
11	−	+	−	+	−	+	−	−	+	−	+	−	+	−	+
12	+	+	−	+	+	−	+	−	+	−	−	+	−	−	−
13	−	−	+	+	+	−	−	−	−	+	+	+	−	−	+
14	+	−	+	+	−	+	+	−	−	+	−	−	+	−	−
15	−	+	+	+	−	−	−	+	+	+	−	−	−	+	−
16	+	+	+	+	+	+	+	+	+	+	+	+	+	+	+

Design (generators)	A	B	C	D	AB	AC	AD	BC	BD	CD	DE	CE	BE	AE	E
2_V^{5-1} $E=ABCD$	A	B	C	D	AB	AC	AD	BC	BD	CD	DE	CE	BE	AE	E
2_{IV}^{6-2} $E=ABC$ $F=BCD$	A	B	C	D	AB+CE	AC+BE	AD+EF	BC+AE+DF	BD+CF	CD+BF	E	3–f int	3–f int	F	AF+DE
2_{IV}^{7-3} $E=ABC$ $F=BCD$ $G=ACD$	A	B	C	D	AB+CE+FG	AC+BE+DG	AD+CG+EF	BC+AE+DF	BD+CF+EG	CD+AG+BF	E	3–f int	G	F	AF+BG+DE
2_{IV}^{8-4} $E=ABC$ $F=BCD$ $G=ACD$ $H=ABD$	A	B	C	D	AB+CE+DH+FG	AC+BE+DG+FH	AD+BH+CG+EF	BC+AE+DF+GH	BD+AH+CF+EG	CD+AG+BF+EH	E	H	G	F	AF+BG+CH+DE
2_{III}^{9-5} $E=ABC$ $F=BCD$ $G=ACD$ $H=ABD$ $J=ABCD$	A+FJ	B+GJ	C+HJ	D+EJ	AB+CE+DH+FG	AC+BE+DG+FH	AD+BH+CG+EF	BC+AE+DF+GH	BD+AH+CF+EG	CD+AG+BF+EH	E+DJ	H+CJ	G+BJ	F+AJ	J+AF+BG+CH+DE
⋯															
2_{III}^{15-11} $E=ABC$ $F=BCD$ $G=ACD$ $H=ABD$ $J=ABCD$ $K=AB$ $L=AC$ $M=AD$ $N=BC$ $O=BD$ $P=CD$	A+ BK CL DM EN FJ GP HO	B+ AK CN DO EL FP GJ HM	C+ AL BN DP EK FO GM HJ	D+ AM BO CP EJ FN GL HK	K+ AB CE DH FG JP LN MO	L+ AC BE DG FH JO KN MP	M+ AD BH CG EF JN KO LP	N+ AE BC DF GH JM KL OP	O+ AH BD CF EG JL KM NP	P+ AG BF CD EH JK LM NO	E+ AN BL CK DJ FM GO HP	H+ AO BM CJ DK EP FL GN	G+ AP BJ CM DL EO FK HN	F+ AJ BP CO DN EM GK HL	J+ AF BG CH DE KP LO MN

NOTE: The expressions below the columns of plus and minus signs from the previous page specify the confounding patterns of the estimated effects. I, which denotes the column of plus signs, is not used as a factor. Interactions of order three or higher are assumed zero.

TABLE 5.8

A 2_{IV}^{7-3} Design with Generators E = ABC, F = BCD, and G = ACD

Run	FACTORS						
	A	B	C	D	E	F	G
1	−	−	−	−	−	−	−
2	+	−	−	−	+	−	+
3	−	+	−	−	+	+	−
4	+	+	−	−	−	+	+
5	−	−	+	−	+	+	+
6	+	−	+	−	−	+	−
7	−	+	+	−	−	−	+
8	+	+	+	−	+	−	−
9	−	−	−	+	−	+	+
10	+	−	−	+	+	+	−
11	−	+	−	+	+	−	+
12	+	+	−	+	−	−	−
13	−	−	+	+	+	−	−
14	+	−	+	+	−	−	+
15	−	+	+	+	−	+	−
16	+	+	+	+	+	+	+

Effects that may be estimated and their confounding pattern (3-factor and higher-order interactions are assumed to be zero):

$l_0 = \bar{y} \rightarrow$ average

$l_A \rightarrow A$ $l_{AB} \rightarrow AB + CE + FG$ $l_{ABC} = l_E \rightarrow E$

$l_B \rightarrow B$ $l_{AC} \rightarrow AC + BE + DG$ $l_{ABD} \rightarrow 3$ factor interactions

$l_C \rightarrow C$ $l_{AD} \rightarrow AD + CG + EF$ $l_{ACD} = l_G \rightarrow G$

$l_D \rightarrow D$ $l_{BC} \rightarrow BC + AE + DF$ $l_{BCD} = l_F \rightarrow F$

$l_{BD} \rightarrow BD + CF + EG$ $l_{ABCD} = l_{AF} \rightarrow AF + BG + DE$

$l_{CD} \rightarrow CD + AG + BF$

5.7.1 An Example: Improving Online Learning

A university that specializes in online learning wants to improve the effectiveness of its 8-week experimental design course. To accomplish this goal, it seems natural to conduct an experiment to test some key factors. Professor John Pesky, head of the statistics group, identifies 7 factors to test at 2 levels each, as shown in Table 5.9. The effectiveness of each change will be measured by final exam scores. (Note that there have been some major experiments examining the relationship between class size and learning. But except for these studies, in searching the literature we found few examples of statistical experiments in education and even fewer that studied more than one factor. We believe there are many opportunities to use experimental design methods to improve the effectiveness of education. The example in this section is not an actual one, but it demonstrates the kind of experiments that could be performed.)

The Factors and Levels. Factor A is the textbook, and Pesky wants to compare a new book to the one he has been using. He also wants to see if additional readings would improve performance (factor B). Factor C is the amount of homework, and the two levels are the current 5 hours per week (which students have complained about) and a less demanding 3 hours per week. Factor D's two levels compare a new software package to the existing one, while factor E is the number of lectures. Currently there are 4 one-hour video lectures per

TABLE 5.9
Factors and Levels in the Online Learning Example

	LEVEL	
Factor	−	+
A Textbook	Current	New
B Readings	No	Yes
C Homework	5 Hours	3 Hours
D Software	Current	New
E Sessions	4 per week	3 per week
F Review	No	Yes
G Lecture notes	No	Yes

week, but Professor Pesky is under pressure from the administration to cut back the number to reduce costs. His superior Dean Takahashi believes that three sessions per week would be just as effective. Finally, the last two factors will test two other changes to the course, adding an online review session for the final (factor F) and the addition of a set of lecture notes (factor G).

The Design. The professor decides to use the 8-run 2_{III}^{7-4} saturated design shown in Table 5.10. The generators are $D = AB$, $E = AC$, $F = BC$, $G = ABC$, and the defining relation consists of 15 words.

$$I = ABD = ACE = BCF = ABCG = AFG = BEG = CDG = DEF$$

$$= ABEF = ACDF = ADEG = BCDE = BDFG = CEFG = ABCDEFG$$

The first four words in the defining relation correspond to the four generators. The other words were determined by multiplying these four words, taking two at a time (six combinations), three at a time (four combinations), and all four together. The confounding pattern (ignoring 3-factor and higher-order interactions) is shown in the last row of Table 5.5. This is a resolution III design with main effects confounded with 2-factor interactions.

Each of the 8 runs defines the characteristics of a section, and 20 students are randomly assigned to each of the eight sections. At the end of the course, each student takes a final exam. The response variable shown in Table 5.10 is the average score for each section of 20 students,

Which Effects Are Significant? The sample variance calculated from the test scores of the same section provides an estimate of the variability of an individual test score. The variances can be averaged across the eight sections to obtain an even better estimate of the variability, resulting in the pooled estimate s_p^2 with $(8)(19) = 157$ degrees of freedom. Section 4.4.1 of Chapter 4 showed that the estimated variance of an effect is given by $s_{effect}^2 = (4/N)s_p^2$, where $N = (8)(20) = 160$ is the total number of students in the experiment.

Professor Pesky performs this calculation and finds that $s_p^2 = 106.4$. Hence the variance of an effect $s_{effect}^2 = (4/N)s_p^2 = (4/160)(106.4) = 2.66$, and the standard error of an effect $s_{effect} = \sqrt{2.66} = 1.63$. A 95% confidence interval for the mean of an effect is the estimated effect $\pm (1.96)(1.63)$; an estimated effect is statistically significant if its absolute value is greater than $(1.96)(1.63) = 3.19$. Two estimated effects exceed this threshold: $(A + BD + CE + FG)$ and $(E + AC + BG + DF)$.

TABLE 5.10
Online Learning Example: Results of the Initial 2_{III}^{7-4} Experiment

Run				FACTOR				Response Average Score
	A	B	C	$D = AB$	$E = AC$	$F = BC$	$G = ABC$	
1	−	−	−	+	+	+	−	63.6
2	+	−	−	−	−	+	+	76.8
3	−	+	−	−	+	−	+	60.3
4	+	+	−	+	−	−	−	80.3
5	−	−	+	+	−	−	+	67.2
6	+	−	+	−	+	−	−	71.3
7	−	+	+	−	−	+	−	68.3
8	+	+	+	+	+	+	+	74.3

Effects (rounded to one significant figure) that may be estimated and their confounding patterns (3-factor and higher-order interactions are assumed to be zero):
$l_A = 10.8 \rightarrow A + BD + CE + FG$
$l_B = 1.1 \rightarrow B + AD + CF + EG$
$l_C = 0 \rightarrow C + AE + BF + DG$
$l_D = 2.2 \rightarrow D + AB + CG + EF$
$l_E = -5.8 \rightarrow E + AC + BG + DF$
$l_F = 1.0 \rightarrow F + AG + BC + DE$
$l_G = -1.2 \rightarrow G + AF + BE + CD$

Suppose we (and Professor Pesky) assume that in $(A + BD + CE + FG)$ and $(E + AC + BG + DF)$, the six 2-factor interactions are negligible. Then we would have estimates of the two main effects: A and E. With this interpretation the best levels are + for factor A (textbook), and − for factor E (sessions per week). Changing to the new book increases the average score by almost 11 points, while reducing the number of sessions per week from the current four sessions to three sessions *decreases* the average score by nearly 6 points. A graduate student leaks the news to Dean Takahashi, who is not pleased to hear it.

But what if the 2-factor interactions are not negligible? Then the observed estimates might be due to one or more 2-factor interactions rather than the main effects. Because of these uncertainties, Professor Pesky decides to do a second set of 8 runs designed to clarify the initial results.

The experimental design course is about to be offered again. Pesky creates eight sections by randomly assigning 20 students to each run in this second design, and at the end of the course, he calculates the final exam averages for each section.

A Second 8-Run Experiment Switching the Signs of Column A. Table 5.11 shows the original design matrix (runs 1–8) followed by the design matrix for the second experiment (runs 9–16). This second set of 8 runs was constructed from the original design matrix by switching the signs in column A while leaving the other columns unchanged.

Reversing the signs in column A means that the signs for each interaction column involving A are reversed as well. For example, as Table 5.10 shows, in the original design the signs of columns B, AD, CF, and EG are identical, and l_B (the average of the responses when B is at the plus level minus the average of the responses when B is at the minus level) estimates $(B + AD + CF + EG)$. In the second design, the signs for columns B, CF, and EG are still identical, but the signs for column AD are now reversed. As a result, the average of the responses when B is at the plus level minus the average of the responses when B is at the minus level, which we denote by l_B^f (with superscript f for "follow-up"), estimates

TABLE 5.11

The 2_{III}^{7-4} Design (Runs 1–8) Joined by the Design That Switches the Signs of Factor A:
Online Learning Example

| | | FACTORS | | | | | | | Response Average |
	Run	A	B	C	D	E	F	G	Score
2_{III}^{7-4}	1	−	−	−	+	+	+	−	63.6
	2	+	−	−	−	−	+	+	76.8
	3	−	+	−	−	+	−	+	60.3
	4	+	+	−	+	−	−	−	80.3
	5	−	−	+	+	−	−	+	67.2
	6	+	−	+	−	+	−	−	71.3
	7	−	+	+	−	−	+	−	68.3
	8	+	+	+	+	+	+	+	74.3
2_{III}^{7-4} with	9	+	−	−	+	+	+	−	79.1
signs of factor A	10	−	−	−	−	−	+	+	58.7
switched	11	+	+	−	−	+	−	+	77.1
	12	−	+	−	+	−	−	−	63.1
	13	+	−	+	+	−	−	+	72.7
	14	−	−	+	−	+	−	−	70.2
	15	+	+	+	−	−	+	−	69.4
	16	−	+	+	+	+	+	+	65.4

Estimated effects: Original 2_{III}^{7-4}
$l_A = 10.8 \to A + BD + CE + FG$
$l_B = 1.1 \to B + AD + CF + EG$
$l_C = 0 \to C + AE + BF + DG$
$l_D = 2.2 \to D + AB + CG + EF$
$l_E = -5.8 \to E + AC + BG + DF$
$l_F = 1.0 \to F + AG + BC + DE$
$l_G = -1.2 \to G + AF + BE + CD$

Estimated effects: Follow-up (signs of A switched)
$l_A^f = 10.2 \to A - BD - CE - FG$
$l_B^f = -1.4 \to B - AD + CF + EG$
$l_C^f = -0.1 \to C - AE + BF + DG$
$l_D^f = 1.2 \to D - AB + CG + EF$
$l_E^f = 7.0 \to E - AC + BG + DF$
$l_F^f = -2.6 \to F - AG + BC + DE$
$l_G^f = -2.0 \to G - AF + BE + CD$

Combining the estimated effects:
$(1/2)(l_A + l_A^f) = 10.5 \to A$
$(1/2)(l_B + l_B^f) = -0.2 \to B + CF + EG$
$(1/2)(l_C + l_C^f) = 0 \to C + BF + DG$
$(1/2)(l_D + l_D^f) = 1.7 \to D + CG + EF$
$(1/2)(l_E + l_E^f) = 0.6 \to E + BG + DF$
$(1/2)(l_F + l_F^f) = -0.8 \to F + BC + DE$
$(1/2)(l_G + l_G^f) = -1.6 \to G + BE + CD$

$(1/2)(l_A - l_A^f) = 0.3 \to BD + CE + FG$
$(1/2)(l_B - l_B^f) = 1.3 \to AD$
$(1/2)(l_C - l_C^f) = 0.1 \to AE$
$(1/2)(l_D - l_D^f) = 0.5 \to AB$
$(1/2)(l_E - l_E^f) = -6.4 \to AC$
$(1/2)(l_F - l_F^f) = 1.8 \to AG$
$(1/2)(l_G - l_G^f) = 0.4 \to AF$

$(B - AD + CF + EG)$. As shown in the table, each of the other five contrasts (l_C^f through l_G^f) that include a 2-factor interaction involving A are changed in the same manner.

Now, consider column A. In the original design the signs of columns A, BD, CE, and FG are identical, and l_A estimates $(A + BD + CE + FG)$. In the second design, the signs for columns BD, CE, and FG are still identical. But now in every run, the signs in these three 2-factor interaction columns are the opposite of the sign in column A. As a result, l_A^f (the average of the responses when A is plus minus the average of the responses when A is minus) estimates $(A - BD - CE - FG)$.

Combining the Estimates from the Two 8-Run Experiments. We use two simple algebraic operations (addition and subtraction) to combine the two sets of estimated effects and to reveal the confounding pattern for the entire 16-run experiment. Consider l_A and l_A^f. From Table 5.11

$$l_A \rightarrow A + BD + CE + FG$$

$$l_A^f \rightarrow A - BD - CE - FG$$

Hence

$$\frac{l_A + l_A^f}{2} \rightarrow \frac{A + BD + CE + FG + A - BD - CE - FG}{2} = A$$

$$\frac{l_A - l_A^f}{2} \rightarrow \frac{A + BD + CE + FG - A + BD + CE + FG}{2} = BD + CE + FG$$

These two operations separate A from $(BD + CE + FG)$. In our example $(1/2)(l_A + l_A^f) = (10.8 + 10.2)/2 = 10.5$ is an estimate of the main effect of A, and $(1/2)(l_A - l_A^f) = (10.8 - 10.2)/2 = 0.3$ is an estimate of $(BD + CE + FG)$.

In Table 5.11, we perform these two operations repeatedly to combine the estimates from the two 8-run experiments. The result is not only an estimate of the main effect of A that is no longer confounded, but clear estimates of all 2-factor interactions involving A as well. The two largest effects are $A = 10.5$ and $AC = -6.4$. We obtained these estimates by combining estimates from the two 8-run experiments. Equivalently, we can obtain the estimates directly from the combined 16-run design. For example, to estimate AC, we determine the signs for the AC column by multiplying the signs in columns A and C and then apply these signs to the responses. We have

$$l_{AC} = \frac{63.6 + 60.3 + 71.3 + 74.3 + 58.7 + 63.1 + 72.7 + 69.4}{8}$$

$$- \frac{76.8 + 80.3 + 67.2 + 68.3 + 79.1 + 77.1 + 70.2 + 65.4}{8}$$

$$= 66.675 - 73.05 = -6.4$$

The interaction diagram in Figure 5.2 shows the nature of the interaction between A (textbook) and C (homework). Notice that the new textbook is always better than the old one. If the new textbook is used, more homework is better, but with the old textbook, more homework is worse. One explanation is that with the current textbook, assigning more homework increases frustration and hurts final exam performance. The new book with more homework appears to be the best option. The other effects are not significant. In particular, class notes, additional readings, and a review class for the final apparently offer no benefits.

There is one more very important finding. A main effects interpretation of the original experiment identified the main effect of E (number of sessions) as significant, with four sessions per week being better than three sessions per week. But that conclusion was wrong. As we determined after the second experiment, it was the AC interaction, not the main effect of E, that was responsible for the statistically significant estimate of $(E + AC + BG + DF)$. In the combined results of Table 5.11, the estimate of $(E + BG + DF)$ is very small (0.6). As Dean Takahashi will be happy to learn, the results show that there is actually no difference in student performance if three rather than four lectures are given each week.

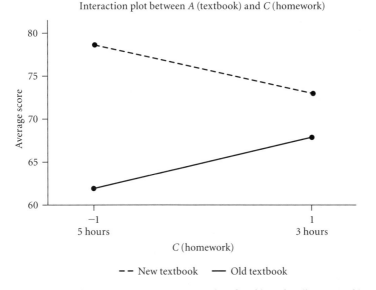

Figure 5.2 shows: Interaction plot between A (textbook) and C (homework)

Figure 5.2 Interaction Diagram Between Factors A (textbook) and C (homework) Using Results of All 16 Runs

5.7.2 Foldover: Switching the Signs of Every Column

Suppose in the previous example in carrying out 8 more runs, we switched the signs in every column. The original 2_{III}^{7-4} design is shown in the top panel of Table 5.12 (runs 1–8) with the second design (runs 9–16) shown below it. We refer to the second design as a (complete) foldover. What is the confounding pattern for the foldover design, and what is the confounding pattern for the combined 16-run design?

In the original design, each main effect is confounded with three 2-factor interactions. For example, as shown in Table 5.10, l_A estimates $A + BD + CE + FG$. In the foldover design, switching the signs of *all* columns leaves the signs of every 2-factor interaction column unchanged, and now the signs in column A are the opposite of those in the three interaction columns. As a result l_A^f estimates $A - BD - CE - FG$. Using the same procedure as in Table 5.11, we combine the two estimates: $(1/2)(l_A + l_A^f) \rightarrow A$ and $(1/2) \times (l_A - l_A^f) \rightarrow BD + CE + FG$. The main effect of A is no longer confounded with the three 2-factor interactions. Combining each of the estimates in this way, the result is that we now have clear (unconfounded) estimates of all main effects, while 2-factor interactions remain confounded as before.

The original 8-run design was resolution III, while the 16-run combined design is resolution IV. The generators for the combined design are $E = BCD$, $F = ACD$, and $G = ABC$. The defining relation for this design consists of the following seven words:

$$I = BCDE = ACDF = ABCG = ABEF = ADEG = BDFG = CEFG$$

Notice that these are 7 of the 15 words in the defining relation for the original 2_{III}^{7-4} design, which we showed at the beginning of this section. They remain in the defining relation for the combined design, whereas the other eight words in the original defining relation drop

TABLE 5.12

The 2_{III}^{7-4} Design (Runs 1–8) Joined by the Design That Switches the Signs of All Columns (Runs 9–16)

	Run	A	B	C	D	E	F	G
					FACTORS			
2_{III}^{7-4}	1	−	−	−	+	+	+	−
	2	+	−	−	−	−	+	+
	3	−	+	−	−	+	−	+
	4	+	+	−	+	−	−	−
	5	−	−	+	+	−	−	+
	6	+	−	+	−	+	−	−
	7	−	+	+	−	−	+	−
	8	+	+	+	+	+	+	+
2_{III}^{7-4} with signs of all factors switched	9	+	+	+	−	−	−	+
	10	−	+	+	+	+	−	−
	11	+	−	+	+	−	+	−
	12	−	−	+	−	+	+	+
	13	+	+	−	−	+	+	−
	14	−	+	−	+	−	+	+
	15	+	−	−	+	+	−	+
	16	−	−	−	−	−	−	−

Estimated effects: Original 2_{III}^{7-4}

$l_A \rightarrow A + BD + CE + FG$
$l_B \rightarrow B + AD + CF + EG$
$l_C \rightarrow C + AE + BF + DG$
$l_D \rightarrow D + AB + CG + EF$
$l_E \rightarrow E + AC + BG + DF$
$l_F \rightarrow F + AG + BC + DE$
$l_G \rightarrow G + AF + BE + CD$

Estimated effects: Follow-up (foldover of all signs)

$l_A^f \rightarrow A - BD - CE - FG$
$l_B^f \rightarrow B - AD - CF - EG$
$l_C^f \rightarrow C - AE - BF - DG$
$l_D^f \rightarrow D - AB - CG - EF$
$l_E^f \rightarrow E - AC - BG - DF$
$l_F^f \rightarrow F - AG - BC - DE$
$l_G^f \rightarrow G - AF - BE - CD$

Combining the estimated effects

$(1/2)(l_A + l_A^f) \rightarrow A$ $(1/2)(l_A - l_A^f) \rightarrow BD + CE + FG$
$(1/2)(l_B + l_B^f) \rightarrow B$ $(1/2)(l_B - l_B^f) \rightarrow AD + CF + EG$
$(1/2)(l_C + l_C^f) \rightarrow C$ $(1/2)(l_C - l_C^f) \rightarrow AE + BF + DG$
$(1/2)(l_D + l_D^f) \rightarrow D$ $(1/2)(l_D - l_D^f) \rightarrow AB + CG + EF$
$(1/2)(l_E + l_E^f) \rightarrow E$ $(1/2)(l_E - l_E^f) \rightarrow AC + BG + DF$
$(1/2)(l_F + l_F^f) \rightarrow F$ $(1/2)(l_F - l_F^f) \rightarrow AG + BC + DE$
$(1/2)(l_G + l_G^f) \rightarrow G$ $(1/2)(l_G - l_G^f) \rightarrow AF + BE + CD$

out. Why? The seven words that stay in all have four letters (an even number), while the eight words that drop out such as *ABD* and *ACE* have an odd number of letters (seven have three letters each while one, *ABCDEFG*, has seven letters). Switching the signs of all columns leaves the signs of the four letter words unchanged, and they remain in the defining relation of the foldover design.

But in the defining relation of the foldover design the eight words with an odd number of letters appear with their signs changed. For example, $I = -ABD$. In the combined design *ABD* is no longer equal to I ($ABD = I$ in runs 1–8, but $ABD = -I$ in runs 9–16). As a result, *ABD* and the other words with an odd number of letters drop out of the defining relation for the combined design.

From the defining relation for the combined 16-run design we can find the entire confounding pattern in the usual way. For example, multiplying the defining relation by *AD* and ignoring interactions of order 3 or higher, we have that $AD = CF = EG$, which is identical to what we found by combining the two 8-run designs as shown in Table 5.12. In estimating the effects for the combined design, we can combine the estimates from the two 8-run designs as shown in Table 5.12. But the simplest approach is to estimate the effects directly from the combined 16-run design. For example, for AD $(= CF = EG)$ we multiply the signs in column A and D to determine the signs for the AD column. Then we calculate l_{AD}, the difference between the average response when AD is at the plus level and the average response when AD is at the minus level. This is an estimate of $AD + CF + EG$.

Suppose we had followed this complete foldover procedure in the online learning example of the last section. What would the result have been? In the original 8 runs, there were two significant estimates $l_A = 10.8 \rightarrow A + BD + CE + FG$ and $l_E = -5.8 \rightarrow E + AC + BG + DF$. Switching the signs of column A gave us clear estimates of the main effect of A and the AC interaction which were the two significant effects.

A complete foldover would have given us a clear estimate of the main effect of A. In addition, it would have separated E from $(AC + BG + DF)$. The estimate of the main effect of E would have been small, while the estimate of $(AC + BG + DF)$ would have been similar to -5.8. Given that the main effect of A was large, we might have guessed that the AC interaction was responsible for this significant estimate, but we would not have been certain. In this case, switching the sign of column A turned out to be a better choice.

Implementing the Sequential Approach. We have discussed two distinct methods for resolving the ambiguities in a resolution III design: switching the signs of one column and switching the signs of all columns (foldover). A foldover increases the resolution of the design from resolution III to resolution IV. The decision regarding which method to use should be made *after* the results of the initial experiment have been obtained. If the initial experiment shows that a number of estimated effects are significant, a foldover is probably the preferred choice. On the other hand, if the initial experiment shows that only one estimate (as in our online learning example) or perhaps two are significant, switching the signs of the factor associated with the largest estimate is likely to be the best choice because this will unconfound all the effects involving that seemingly important factor.

Decisions about the appropriate follow-up design should be made after the results of the initial experiments have been analyzed. For example, it does not make much sense to decide on a foldover before the results are known, since in that case the experimenter could have implemented a 16-run resolution IV design in the first place.

A complete foldover does not make sense if the initial experiment is already a resolution IV design since in this case the main effects are already isolated. Switching the column signs for one factor, however, may be useful as it will unconfound all interactions involving that factor.

5.8 A 2_{IV}^{8-4} DESIGN: IMPROVING E-MAIL ADVERTISING

With consistent growth and solid profit from their stores, an office supplies company that we call *ABC* decided to expand an e-mail program that directs potential small-business customers to its Web site. After brainstorming ideas and trimming the list to the boldest ideas, the marketing team identified 8 factors and selected two different versions of each factor to

TABLE 5.13
E-mail Advertising Experiment: Factors and Levels

Factor	Control ($-$)	New Idea ($-$)
A Link to online catalog	No	Yes
B Design of e-mail	Simple	Stronger brand image
C Partner promotions	None	Offers from two-partner companies
D Navigation bar on side	Current	Additional buttons
E Background color	White	Blue
F Discount offer	15% off	No discount
G Subject line	Exclusive e-mail offer	Special offer for our customers
H Free gift	None	Free pen-and-pencil set

test. For each factor, the company refers to the current setting as the control. Table 5.13 summarizes the control and the new idea to be tested for each factor.

A: *Link to online catalog.* The e-mail included a "Shop our catalog online" button near the bottom of the message. The team felt that an obvious link to the Web site would encourage customers to browse through the selection of products.

B: *Design of e-mail.* In the past, e-mails used a basic font with a small company logo at the top. The team wanted to test a stronger brand image, with a larger logo, more stylized font, and greater use of the company's brand colors.

C: *Partner promotions.* The marketing team believed that promoting several well-known brand-name products would encourage customers to make a purchase. They decided to promote two specific brands in two bright boxes under "Offers from our partners" at the bottom of the e-mail.

D: *Navigation bar on side.* E-mails currently went out with a sidebar similar to the navigation bar on the company Web site, but with a shorter list of links. The company decided to test the current navigation bar versus one with more choices.

E: *Background color.* All e-mails were sent with dark text on a white background. The creative director thought that changing to a blue background might help the e-mail stand out.

F: *Discount offer.* The Internet director had gone back and forth between offering a special e-mail discount or not. He thought the discount helped but never had quantified whether it generated enough sales to justify the lower margin.

G: *Subject line.* The Internet director had been testing different e-mail subject lines. Until this test, "Exclusive e-mail offer from *ABC*" was the winner. Since he knew the subject line was important, he wanted to test another version, "Special offer for our best customers."

H: *Free gift.* They had never before offered a free gift with online orders. They knew other companies were doing so, and decided it was worth trying, selecting an attractive but low-cost pen-and-pencil set as the free gift.

The firm used the 16-run 2_{IV}^{8-4} design in Table 5.14 to study the impact of these 8 factors on the order rate. Each version of the e-mail was sent to 1,000 addresses randomly chosen from

TABLE 5.14
The Design and the Results

			FACTOR					Response:
A	B	C	D Navigation	E	F	G Subject	H	Purchase
Link	Design	Partner	Bar	Color	Discount	Line	Gift	Rate (%)
−	−	−	−	−	−	−	−	2.23
+	−	−	−	+	+	+	−	1.47
−	+	−	−	+	+	−	+	1.81
+	+	−	−	−	−	+	+	2.38
−	−	+	−	+	−	+	+	2.03
+	−	+	−	−	+	−	+	1.62
−	+	+	−	−	+	+	−	1.28
+	+	+	−	+	−	−	−	1.98
−	−	−	+	−	+	+	+	1.78
+	−	−	+	+	−	−	+	2.30
−	+	−	+	+	−	+	−	2.30
+	+	−	+	−	+	−	−	1.53
−	−	+	+	+	+	−	−	1.11
+	−	+	+	−	−	+	−	1.81
−	+	+	+	−	−	−	+	1.93
+	+	+	+	+	+	+	+	1.82

Effect	Estimate
Average	1.8363
A	0.0550
B	0.0850
C	**−0.2775**
D	−0.0275
E	0.0325
F	**−0.5675**
G	0.0450
H	**0.2450**
AB + CE + DF + GH	0.0425
AC + BE + DG + FH	**0.1650**
AD + BF + CG + EH	0.0300
AE + BC + DH + FG	0.0250
AF + BD + CH + EG	0.0600
AG + BH + CD + EF	−0.0325
AH + BG + CF + DE	0.0875

NOTE: Significant effects are shown in boldface.

a list that the firm had purchased. The response variable, the proportion of customers that ordered from the Internet site, is given in the last column. The average response was 1.84%.

The first four columns of the design represent a full 2^4 factorial in the factors A, B, C, and D. The levels of the remaining four factors in the adjacent columns are obtained from the four generators

$$E = ABC \qquad F = ABD \qquad G = ACD \qquad H = BCD$$

These are the generators in Table 5.7, but with F and H interchanged.

The design is resolution IV. Ignoring interactions of order three and higher, this design allows clear (unconfounded) estimates of all eight main effects. The 2-factor interactions are confounded in strings (groups) of four.

In Section 4.6.4 of Chapter 4 we showed how to find the standard error of an estimated effect when the responses are proportions, as is the case in this example. The standard error is given by

$$standard\ error(\text{effect}) = \sqrt{\frac{\bar{p}(1 - \bar{p})}{N/2} + \frac{\bar{p}(1 - \bar{p})}{N/2}} = \sqrt{4\frac{\bar{p}(1 - \bar{p})}{N}}$$

where $\bar{p} = 0.00184$ is the overall success proportion and $N = 16{,}000$ is the total sample size. This results in

$$standard\ error(\text{effect}) = \sqrt{\frac{4(0.00184)(1 - 0.00184)}{16{,}000}} = 0.000678\ \text{or}\ 0.068\%.$$

At the 5% significance level, an estimated effect is significant if its absolute value is greater than $0.068(1.96) = 0.133\%$.

The factors C (partner promotion), F (discount offer), and H (free gift) are significant. Making available offers from two-partner companies reduces the response rate by 0.28 percentage points. The team theorized that additional offers may have confused the message and given customers too many disjointed offers to choose from. Not offering a 15% discount decreases the response by 0.57 percentage points. The team calculated that the loss of margin by offering a 15% discount is more than covered by the increase in the number of orders. A free pen-and-pencil gift set increases the purchasing rate by 0.25 percentage points. Analyzing profitability, the cost of the gift was easily covered by the increase in orders.

The fourth largest effect, $l_{AC} = 0.165$, is an estimate of $AC + BE + DG + FH$. These four 2-factor interactions are confounded, and we only know that their sum is 0.165. However, we do know that factors F and H are significant. Experience has shown that significant 2-factor interactions tend to involve factors that have significant main effects, an experimental design principle that is called *effect heredity*. Since F and H were identified as significant main effects, effect heredity suggests that the estimated effect is most likely due to an interaction between F (discount) and H (free offer).

The interaction diagram for factors F and H is shown in Figure 5.3. The interaction supports the main effects: the 15% discount ($F-$, both points on the left) is always better, and the free gift ($H+$, the top line) increases the response over no free gift. The interaction can be understood by comparing both points on the left with both points on the right. On the right (with no discount), offering the free pen-and-pencil set gives a large jump in response versus offering no free gift; the response changes by 0.41 percentage points from 1.35% to 1.76%. In contrast, the points on the left show that, with the 15% discount ($F-$), the free gift increases response only slightly (from 2.08% to 2.16%). Overall, this interaction shows that the 15% discount is great, the free gift is good, but both together may be overkill—the free gift adds little to the benefit of the discount offer. These data helped the marketing team gain deeper insight into customer behavior, showing that one strong incentive is valuable, but additional incentives are probably unnecessary.

The company decided to offer the 15% discount and avoid the partner promotions. In addition, they planned to run further experiments to study the interaction between smaller discounts and the free gift offer.

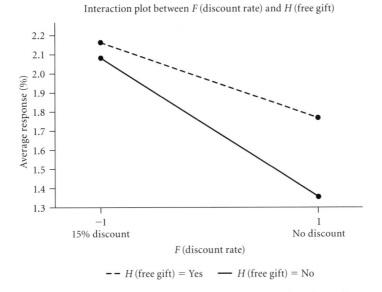

Figure 5.3 Interaction Diagram Between Factors F (discount) and H (free gift)

5.11 NOBODY ASKED US, BUT . . .

A strategy of carrying out one single, all-encompassing experiment is usually ill-conceived because it leaves no room for subsequent experimentation, and shortcuts the accumulation of knowledge. Learning is sequential, and as R.A. Fisher has said so well, the best time to plan an experiment is after you have done it. An approach that spends a portion of the available resources (perhaps 50%) on the initial experiment and saves the remainder for follow-up runs is more efficient. Two-level fractional designs and their foldovers (of a single factor or all factors) are ideally suited for this purpose.

At the outset of a study, one often encounters a large number of conflicting theories and numerous factors that are thought to have an effect on the response. At that stage, 2-level fractional designs are especially useful for screening purposes, to separate what Joseph Juran has called the "vital few" factors from the "trivial many."

In a resolution III fractional factorial design it is possible that a significant main effect and its confounded 2-factor interaction might have about the same magnitude but opposite signs and cancel each other out. In that case the experimenter would miss two significant effects. It is unlikely, but to paraphrase a once-popular bumper sticker, "stuff happens." The approaches in this chapter are powerful but not foolproof. The key is to experiment. Missing something occasionally will be of small consequence compared to the accumulation of insights over time.

In Chapter 4, we discussed estimating the variance (and standard error) of an effect by assuming that higher-order interactions are negligible and represent experimental error. But that approach does not apply in general to fractional factorial designs. The problem is that higher-order interactions are usually confounded with main effects or 2-factor interactions. In the 2_{IV}^{7-3} design, we pointed out that one of the 15 estimated effects is a string of higher-order interactions and could be used to estimate the variance of an effect, but this estimate would only have 1 degree of freedom. Similarly, in the 2_{IV}^{6-2} design shown in Table 5.7, two of the 15 estimates are strings of higher-order interactions and could be used to estimate the variance of an effect with 2 degrees of freedom. With 32-run experiments, there are more opportunities to use this approach. For example, a 2^{6-1} 32-run experiment with generator $6 = 12345$ would be resolution VI with main effects confounded with 5-factor interactions and 2-factor interactions confounded with 4-factor interactions. These leave each 3-factor interaction confounded with one other 3-factor interaction. There are 10 such pairs, which could be used to estimate the variance of an effect with 10 degrees of freedom.

In the 2_{III}^{5-2} fractional design we gave the generators as $D = AB$ and $E = AC$. A natural question is, Why not set one of the generators equal to the column of signs of the 3-factor interaction ABC? For example, we might use $D = AB$ and $E = ABC$. If you work out the confounding pattern for this choice of generators you will see that it is equivalent to the pattern for the generators that we have given. For the 2_{IV}^{6-2} design we gave the generators as $E = ABC$ and $F = BCD$. In this case, you might ask, Why not $E = ABCD$? The defining relation with $E = ABCD$ rather than $E = ABC$ is $I = ABCDE = BCDF = AEF$. The shortest word has three letters, so this design would have resolution III, while the one we listed is resolution IV.

In general, we want to choose generators to achieve the highest possible resolution, but there may be a number of choices that lead to designs with the same resolution. How do we choose among them? For example, consider a 2^{7-2} design for 7 factors in 32 runs. Two of the choices are design 1 with generators $6 = 123$ and $7 = 1245$ and design 2 with generators $6 = 123$ and $7 = 145$. For design 1 the defining relation is $I = 1236 = 12457 = 34567$, while for design 2, it is $I = 1236 = 1457 = 234567$. The length of the shortest word is 4 in each case, so these designs are both resolution IV, with 2-factor interactions confounded with other 2-factor interactions. But design 2 has six pairs of confounded 2-factor interactions: (12, 36), (13, 26), (16, 23), (14, 57), (15, 47), (17, 45), whereas design 1 has only three pairs that are aliases: (12, 36), (13, 26), (16, 23). As a result, design 1 is preferable to design 2. Notice that design 2 has two 4-letter words in its defining relation, whereas design 1 has only one. For design 1, the 4-letter word is 1236, and only pairs of 2-factor interactions involving these 4 factors will be confounded. But design 2 has a second 4-letter word, 1457, and pairs of 2-factor interactions involving these four factors will be confounded as well. So the tie-breaker between these two (or more) resolution IV designs is the number of 4-letter words in each defining relation. Both designs are resolution IV designs, so they each must have at least one 4-letter word, but design 1 has the *fewest* 4-letter words and is called the *minimum aberration design*. Note that all of the designs in Tables 5.5 and 5.7 are minimum aberration designs.

In Chapter 3, we introduced the important idea of *blocking*. The same concept applies to factorial and fractional factorial designs. For example, in a 2^3 experiment (with factors labeled A, B, C), it might be necessary to perform 4 runs on one day and 4 runs on another. Randomizing the assignment of runs to days will result in a valid experiment, but if there is a day effect, it will lead to an increase in experimental error. An alternative is to perform the experiment in two blocks of 4 runs each. The runs in the first block (day 1) would be those for which the signs of the ABC column are plus, with the runs in the second block (day 2) being the runs for which ABC is minus. This arrangement will confound the (possible) day effect with the 3-factor interaction that is very likely to be zero and can be ignored. In the online experiment that we discussed in Section 5.7.1, the two 8-run experiments were run at different times, which also introduces a block effect, as the students might perform better during one time period compared to the other. But in this case, the same 8-run experiment is repeated in the second block. It may well be that the scores in one block are higher than the scores in the other, but any block effect would be added to each of the 8 runs in the block and would have no influence on the estimated effects. The difference between this and the first example is that in the first, a block effect would influence only 4 of the runs. The first situation would be analogous to the online experiment if the complete 2^3 experiment were repeated on the second day.

EXERCISES

Exercise 1   An experiment is to be performed with 4 factors, labeled A, B, C, and D, each at 2 levels. The design consists of 8 runs with the generator chosen to have the highest possible resolution. Write down the design matrix for this design, and show the complete confounding pattern ignoring 3-factor and higher-order interactions.

Raw material, which is not one of the 4 factors being tested, is available from two different suppliers. There is only enough raw material from supplier 1 to do four runs. The other four runs have to use raw material from supplier 2. Management is concerned that the difference in raw material between suppliers might affect the results. Suppose management is certain that the AD and BC interactions are both zero. Which of the 4 runs in your design matrix should use material from supplier 1, and which should use material from supplier 2? Explain why you made this choice.

Exercise 2   Two Master Black Belts, Bill and Karen, meet at an online dating site and go out on a date. Karen brings along a piece of paper showing a 2^{6-2} design with generators $E = ABC$ and $F = BCD$. In a passionate discussion, Bill argues that replacing the generator $E = ABC$ with the generator $E = ABCD$ will result in a better confounding pattern. "We would confound E with a 4-factor interaction so this pattern is obviously going to be better." Karen replies, "No Bill, you are wrong, the pattern is going to be equivalent." Comment on the opinions expressed by these two lovebirds.

Exercise 3   Karen Donegan, a staffer at the business school, is thinking about buying some new golf equipment to improve her game in preparation for the annual golf tournament. One day she comes across some notes on experimental design that she finds on top of a file cabinet near her office. "Hmm," she thinks to herself as she begins to read them, "this looks interesting. Maybe this will help me decide what equipment to purchase."

Karen has been thinking about replacing her current steel-shaft, small-head driver. The company that makes her clubs has three other drivers that she is considering. The first has the same steel shaft but a newly designed very large head. The second has a graphite shaft with the same small head that is on her current driver, while the third has the graphite shaft with the newly designed very large head. She is also wondering whether a new pair of golf shoes might help, as well as switching to an expensive golf ball (at $3.00 per ball) rather than sticking with her discount store special (at $0.75 per ball). She also has her eye on a rather expensive new golf sweater, which should put her in the right frame of mind to hit some really long drives. Finally, she does not wear a golf glove when she plays, and her husband pointed out to her that she is no Fred Couples, one of the few golf professionals who does not wear a glove. After spending an evening reading the notes, Karen concludes that "this is pretty clear" and decides to perform the experiment whose design matrix is shown below. She includes 6 factors and decides to do 16 runs, all to be performed at the local driving range. Each run is performed in random order.

Karen buys the necessary golf balls and the glove, but is able to borrow the shoes and golf clubs from a local store. She has her own steel-shaft, small-head driver, and she has 10 days to return the sweater for a full refund.

The design matrix and the results of the experiment are shown below. Each run is a single drive (shot), and the response variable is the distance the shot carries in yards. Karen's goal is to find the equipment that will maximize her distance, but she doesn't want to spend money on anything that will not help in that regard.

	LEVEL	
Factor	−	+
A	Small club head	Large club head
B	Steel shaft	Graphite shaft
C	Cheap ball	Expensive ball
D	No glove	Glove
E	Old shoes	New shoes
F	Old sweater	Expensive new sweater

DESIGN WITH GENERATORS: $E = ABC$, $F = BCD$

Run	A	B	C	D	E	F	Response
1	−	−	−	−	−	−	182
2	+	−	−	−	+	−	157
3	−	+	−	−	+	+	155
4	+	+	−	−	−	+	226
5	−	−	+	−	+	+	184
6	+	−	+	−	−	+	166
7	−	+	+	−	−	−	177
8	+	+	+	−	+	−	218
9	−	−	−	+	−	+	178
10	+	−	−	+	+	+	152
11	−	+	−	+	+	−	135
12	+	+	−	+	−	−	232
13	−	−	+	+	+	−	173
14	+	−	+	+	−	−	156
15	−	+	+	+	−	+	154
16	+	+	+	+	+	+	223

Estimated effects: Average = 179.25

$l_A = 24.00 \rightarrow A$ $l_{AB} = 45.50 \rightarrow AB + CE$ $l_{ABC} = l_E = -9.25 \rightarrow E$

$l_B = 21.50 \rightarrow B$ $l_{AC} = -5.25 \rightarrow AC + BE$ $l_{ABD} = 6.75 \rightarrow$ 3-factor interactions

$l_C = 4.25 \rightarrow C$ $l_{AD} = 6.75 \rightarrow AD + EF$ $l_{ACD} = 0.50 \rightarrow$ 3-factor interactions

$l_D = -7.75 \rightarrow D$ $l_{BC} = 1.75 \rightarrow BC + AE + DF$ $l_{BCD} = l_F = 1.00 \rightarrow F$
$l_{BD} = -0.25 \rightarrow BD + CF$ $l_{ABCD} = l_{AF} = 0.00 \rightarrow AF + DE$ $l_{CD} = -2.00 \rightarrow CD + BF$

(a) Verify the estimated effects and their confounding patterns (assume interactions of order 3 or higher are negligible).

(b) Find all the effects including higher-order interactions that are confounded with C, with CD, with AB.

(c) Suppose that based on prior extensive experience at the driving range measuring the length of her drives, Karen has determined that the standard deviation of a single drive is 13 yards. What is the 95% confidence interval for an effect? Which effects are significant?

(d) Based on your analysis of the results, what are the most reasonable and likely conclusions you can draw without doing any additional runs? What should Karen do? Show square diagrams, if appropriate. What is the regression prediction equation?

(e) How much additional yardage could Karen expect to get if she followed your advice?

(f) Suppose instead of doing all 16 runs on one day, Karen decides to do 8 of the runs on one day and the other 8 runs the next day. Karen's husband Harold suggests she do the even-numbered runs on one day and the odd-numbered runs on the other. Is this a good idea? Explain. A professor at the business school suggests she do runs 1, 3, 6, 8, 10, 12, 13, and 15 on one day and the others on the next day. Karen asks, why? The professor says, "Look at the column of signs for the ACD interaction. The runs with $-$ signs should be run on one day and the runs with $+$ signs should be run on the other day. That's how I picked the runs for each day." Explain why the professor's suggestion is better than Harold's. It may help to think of the day as the "seventh factor" in the experiment.

Exercise 4   The Natural Delight Food Company makes a range of frozen foods. One product is the Natural Delight Soy Burger, a nonmeat product. The company is interested in testing a number of factors related to this product.

Factor A: Location. The choice is between the natural foods freezer case and the freezer case where beef hamburgers are sold. The supermarket has offered the same amount of shelf space at either location, and the company has to decide which location to choose. In the past, the product has been sold in the natural foods case. The company wonders whether the higher customer flow past the hamburger freezer case might lead to higher sales.

Factor B: Package color. The existing package is green for the environment. A marketing manager suggests red lettering on a white background.

Factor C. An in-store special display located halfway between the two alternative freezer locations. The display would show a happy person eating a Natural Delight Soy

Burger. The company logo would be displayed, and the following words boldly shown: "Oh boy, you will love our Soy Burger!"

Factor D: Free samples. The brand manager feels strongly that setting up a table next to the location of the display, with an extremely attractive store employee cooking and offering shoppers samples of the product, would be very helpful. "If we can just get people to try our product, they will buy it," he says.

Factor E: Sticker. The company has been adding a fancy sticker to the package with the words "stay healthy." They would like to test whether or not the sticker has an effect on sales. The sticker would match the package color.

Factor F: Package lettering. Currently, the lettering on the package uses a modern font. A summer intern, the marketing manager's nephew, suggests testing a more traditional-style font.

The company has identified a group of 16 stores all of very similar size, with about the same weekly sales of the Soy Burger. The experiment will be run over a 4-week period from mid-July to mid-August. The response variable is dollar sales of the product over the 4-week period. As shown below, the design matrix consists of 16 runs, each a specific setting of each of the 6 factors. Each of the 16 runs is randomly assigned to one of the 16 stores.

Based on an extensive analysis of a very large amount of historical data for the test stores, the company estimates that the variance of the response of a single run (the variance of dollar sales at a store over the 4-week period) is $800.

The design matrix and the experimental results are shown below.

Factor	LEVEL	
	−	+
A	Natural foods case	Beef hamburger frozen foods case
B	Green package	Red lettering on a white background
C	No display	Display
D	No free samples table	Free samples table
E	"Stay healthy" sticker	No sticker
F	Current modern lettering	Traditional lettering

DESIGN WITH GENERATORS: $E = ABC$, $F = BCD$

Run	A	B	C	D	E	F	Response
1	−	−	−	−	−	−	1,110
2	+	−	−	−	+	−	1,120
3	−	+	−	−	+	+	970
4	+	+	−	−	−	+	1,025
5	−	−	+	−	+	+	1,000
6	+	−	+	−	−	+	980
7	−	+	+	−	−	−	1,125
8	+	+	+	−	+	−	1,095
9	−	−	−	+	−	+	960
10	+	−	−	+	+	+	975
11	−	+	−	+	+	−	930
12	+	+	−	+	−	−	950
13	−	−	+	+	+	−	930
14	+	−	+	+	−	−	950
15	−	+	+	+	−	+	975
16	+	+	+	+	+	+	960

(a) Obtain the estimated effects and determine the confounding patterns (see Exercise 3).

(b) Based on a 95% confidence interval, which effects are significant?

(c) What settings of the factors would you recommend? Explain why you picked these settings.

(d) What is the regression prediction equation and what are the predicted 4-week sales given your answer to part c?

(e) What additional comments, suggestions, or observations, if any, do you have?

Exercise 5 Consider Exercise 1 in Chapter 4. Eagle Brands studies the effects of 6 factors. Consider the 2^{6-1}, 2^{6-2}, and 2^{6-3} fractional factorial designs. Discuss the confounding patterns and the resolution of these designs.

Exercise 6 Consider Case 6 [Mother Jones (B)] from the case study appendix.

(a) Analyze the results of the experiment. Which effects are statistically significant at the 5% level? At the 10% level?

(b) What settings for the factors would you recommend?

(c) What is the regression prediction equation and the predicted response if significant factors are set at their best levels?

Exercise 7 Consider Case 5 (Office Supplies E-mail Test) from the case study appendix.

(a) In the section on Planning the Test, it is stated that with 35,000 names and an average response rate of 1%, an effect would have to change the response by about 20% (from 1.0% to 1.2%) to have a 50:50 chance of being found significant. Confirm this statement by applying the appropriate sample size calculations. Use computer software such as Minitab or JMP. What magnitude of change could you detect if you wanted to be 80% confident?

(b) Ignoring the three customer segments, discuss the advantages/disadvantages of the 16-run 2^{13-9} and 32-run 2^{13-8} fractional factorial designs. Discuss achievable resolutions and the implied confounding patterns. Use design software such as Minitab or JMP if available.

(c) Explore the differences among the three customer segments. Can you conclude whether or not the effects of the 13 studied factors depend on the customer segment?

(d) Which design (and which run size) would give you unconfounded estimates of all 2-factor interactions among the 13 factors? Is there a smaller design you could use if you were only interested in one specific interaction (say, the *KL* interaction)?

(e) Obtain the standard errors of the estimated effects, using the approach described in Section 4.6.4.

(e1) Note that the sample sizes are not the same. Nevertheless, suppose that they were and assume that the $N = 34,060$ names were divided equally among the 32 cells.

(e2) Use the variance $var(p_i) \cong p_i(1 - p_i)/n_i$ in deriving the variance of the estimated effects. Use the fact that an estimated effect is the difference of average proportions (of size 16) at the low and high settings.

(e3) Compare your results in (e1) and (e2) with those given in the case.

(f) The variance of the observed proportion, $var(p_i) = \pi_i(1 - \pi_i)/n_i$, depends on the true proportion and the sample size, which both vary across the design runs. This violates the assumption that responses are equally reliable, a fact that is needed for the equal weighting of the proportions when calculating estimated effects.

Logistic regression provides the appropriate approach of analyzing categorical response data such as the number of successes among a given number of trials. Abraham and Ledolter (2006, Chap. 11) discusses this approach in detail. Most statistical software packages such as Minitab include routines for logistic regression. Use logistic regression to replicate the results given in this case.

Exercise 8   Consider Case 7 [Peak Electronics: The Broken Tent Problem (B)] from the case study appendix.

(a) Analyze the results. Estimate the effects, and obtain their significance by comparing the estimated effects with their standard error.

(b) Which are the significant effects? What are the best settings? How well did Lou Pagentine predict which variables would be significant? What is the regression prediction equation for the number of broken tents on the panel? Estimate by how much the number of broken tents would be reduced by using the best settings as compared to the current settings.

Exercise 9   Discuss applications of factorial and fractional factorial designs to questions that arise in your field of study (marketing, operations management, management information systems, economics, engineering, etc). Discuss the factors, the response, and the process that you would follow to conduct such experiments.

Search the literature in your field of study and find applications where these design methods have been used.

Exercise 10   Paper helicopter experiment. This experiment was first recommended by George Box. A discussion of this experiment and construction guidelines for paper helicopters are given in Ledolter and Swersey (1997). Also, there are numerous references to this experiment on the Web.

Construct a paper helicopter by varying the length and the width of the blades. Vary the weight of the helicopter by using different paper stock and/or adding paper clips to the helicopter. Drop the helicopter from a high location (say, 12 feet), and determine the flying time. The objective is to maximize the flying time. Carry out the experiment, preferably with replications. Analyze the data and discuss your findings.

Exercise 11 Consider the 8-run 2^3 factorial design. Suppose that you conduct the 8 runs in standard order. However, it turns out that only 4 runs can be conducted on a single day. You are concerned that the experimental conditions change from day to day, and that the mean levels of runs carried out on different days are not the same.

(a) Would this fact affect the estimates that you obtain from your experiment? For example, would it affect the main effect of factor 1, factor 2, factor 3?

(b) Think about another run order that would not affect the main effects.
    Hint: What about running the four experiments with $123 = -$ on day 1, and the four experiments with $123 = +$ on day 2? Why would this be a better strategy?

Exercise 12 Investigate the effects of the following 5 factors on the expansion of pinto beans:

Soaking fluid: water $(-)$ or beer $(+)$

Salinity: no salt $(-)$ or salt $(+)$

Acidity: no vinegar $(-)$ or vinegar $(+)$

Soaking temperature: refrigerator temperature $(-)$ or room temperature $(+)$

Soaking time: 2 hours $(-)$ or 6 hours $(+)$

Carry out the following experiment. Select a pinto bean, measure its "size," put the bean into a soaking fluid, and—after a certain amount of elapsed time—measure its size again. Use five tablespoons of soaking fluid to soak each bean. Make sure that the liquid covers the bean. Use regular beer because light beer might act like water. For salt, add 1/4 teaspoon to the soaking fluid. For vinegar, add 1 teaspoon to the soaking fluid.

(a) Discuss how you measure the "size of a pinto bean" and its "expansion." Give a detailed description of your measurement procedure, so that it can be carried out by other people; that is, give an operational definition.

(b) Design, set up, and execute a 2^{5-1} fractional factorial experiment. Conduct two replications, which may be run concurrently. Analyze the effects of the 5 factors. Write a short report that summarizes your findings. Support your findings with appropriate graphs and calculations. What have you learned? What was the most difficult part of your experiment? If you had to do it over again, what would you change?
    Note: In carrying out this experiment, you need $(16)(2) = 32$ small paper containers for soaking the beans.

Exercise 13 Eibl, Kess, and Pukelsheim (1992) discussed how the response variable, paint coat thickness, depends on a set of 6 input factors. Their objective was to find factor settings that achieve a desired target value for paint coat thickness of 0.8 mm.

They consider the following 6 input factors A through F (listed here in decreasing order of their assumed importance): belt speed, tube width, pump pressure, paint viscosity, tube height, and heating temperature. All factors could be varied continuously. Level 0 stands for the standard operating condition. All factors were scaled so that levels between -3 and $+3$ were technically feasible, without increasing cost.

(a) The first experiment varied the factor levels between -1 and $+1$; it was expected that this experiment could detect the linear effect of these changes. The table given below lists the observed paint thickness (in mm) for a 2-level fractional factorial experiment with four replications at each factor-level combination. The order of the 32 experiments was fully randomized.

(a1) Show that this design is a 2^{6-3} fractional factorial design. Discuss the confounding patterns.

Hint: Notice that factors A, B, and D form a full 2^3 factorial design. Write out the calculation columns, and discuss how the levels of the remaining factors C, E, and F were selected. You will find that $C = BD$, $E = AD$, and $F = AB$.

(a2) Calculate the averages from the replications, and analyze the averages. Find the important effects, calculate their standard errors, and interpret your findings.

A	B	C	D	E	F	Thickness of Paint Coat (mm)			
-1	-1	1	-1	1	1	1.62	1.49	1.48	1.59
1	-1	1	-1	-1	-1	1.09	1.12	0.83	0.88
-1	1	-1	-1	1	-1	1.83	1.65	1.71	1.76
1	1	-1	-1	-1	1	0.88	1.29	1.04	1.31
-1	-1	-1	1	-1	1	1.46	1.51	1.59	1.40
1	-1	-1	1	1	-1	0.74	0.98	0.79	0.83
-1	1	1	1	-1	-1	2.05	2.17	2.36	2.12
1	1	1	1	1	1	1.51	1.46	1.42	1.40

(b) A follow-up experiment focused on the first 4 factors (factors A through D). The results are given below. The levels of the factors were changed because of the findings in the initial experiment. When analyzing the data, you can transform the factor levels into -1 and $+1$; the -1 in your new coding of factor A corresponds to -1.5 on the original scale; the $+1$ in the new coding corresponds to 0.5 on the original scale. You can do the same for the other factors.

(b1) Show that this design is a 2^{4-1} fractional factorial. Determine the generator of the design and the confounding patterns. Can you think of another, and better, 2^{4-1} fractional factorial design?

(b2) Calculate the averages from the replications, and analyze the averages. Find the important effects, calculate their standard errors, and interpret your findings.

A	B	C	D	Thickness	
−1.5	−2	−2	−2	1.51	1.18
0.5	−2	−2	−2	0.64	0.78
−1.5	0	0	−2	1.74	1.98
0.5	0	0	−2	1.33	1.06
−1.5	−2	0	0	1.71	1.60
0.5	−2	0	0	1.15	1.29
−1.5	0	−2	0	1.71	1.61
0.5	0	−2	0	0.91	1.30

(c) Another follow-up experiment was conducted with just the first 3 factors. The results from this 2^3 factorial experiment are given below.

A	B	C	Thickness	
1.0	−2	−2	0.57	0.58
1.5	−2	−2	0.51	0.66
1.0	−1	−2	0.62	0.74
1.5	−1	−2	0.69	0.49
1.0	−2	−1	0.75	0.58
1.5	−2	−1	0.53	0.64
1.0	−1	−1	0.79	1.04
1.5	−1	−1	0.78	0.79

Calculate the averages from the replications, and analyze the averages. Find the important effects, calculate their standard errors, and interpret your findings.

(d) Summarize your findings from all three experiments. Can you find factor settings that achieve the desired target value for paint coat thickness (0.8 mm)?

6 | PLACKETT-BURMAN DESIGNS

6.1 INTRODUCTION

In Chapter 5 we focused on 2-level fractional factorial designs. As we have seen, in those designs the number of runs N is a power of 2 ($N = 4$, 8, 16, 32, etc.). In this chapter, we discuss another important class of fractional designs called Plackett-Burman designs.

In a classic 1946 paper in the journal *Biometrika*, Plackett and Burman showed how to construct 2-level orthogonal designs when the number of runs N is a multiple of 4 ($N = 4$, 8, 12, 16, 20, 24, and so on). If the run size is a power of 2 (for example, $N = 8$, 16, 32, ...), these designs are identical to the fractional factorial designs that we studied in Chapter 5.

The 2-level fractional factorial designs leave large gaps in the run sizes of the available designs. For example, 7 factors can be studied in 8 runs with a 2_{III}^{7-4} design, but if the number of factors is between 8 and 15, 16 runs are needed; and 32 runs are needed for 16 to 31 factors. The Plackett-Burman designs for $N = 12$, 20, and 24 fill in these gaps.

Suppose that we wish to estimate the main effects of 8 factors and want to achieve this through a design with as few runs as possible. We could use the 2_{IV}^{8-4} fractional factorial in Table 5.7. This design does not confound main effects with 2-factor interactions, but it requires 16 runs. A Plackett-Burman design with the smaller run size $N = 12$ is an option if economy of run size is important.

Plackett-Burman designs have resolution III—confounding main effects with 2-factor interactions. Traditionally, they have been used to estimate main effects under the assumption that 2-factor interactions are largely negligible or small in magnitude. More recently, researchers have begun to explore the so-called projective properties of Plackett-Burman designs and have shown that in some circumstances they can be effectively used to identify likely 2-factor interactions. In Section 6.3.3, in our discussion of the results of a case study, we make use of this important property of Plackett-Burman designs.

TABLE 6.1
Plackett-Burman Design for up to 11 Factors in N = 12 Runs

Runs	FACTOR										
	1	2	3	4	5	6	7	8	9	10	11
1	+	+	−	+	+	+	−	−	−	+	−
2	−	+	+	−	+	+	+	−	−	−	+
3	+	−	+	+	−	+	+	+	−	−	−
4	−	+	−	+	+	−	+	+	+	−	−
5	−	−	+	−	+	+	−	+	+	+	−
6	−	−	−	+	−	+	+	−	+	+	+
7	+	−	−	−	+	−	+	+	−	+	+
8	+	+	−	−	−	+	−	+	+	−	+
9	+	+	+	−	−	−	+	−	+	+	−
10	−	+	+	+	−	−	−	+	−	+	+
11	+	−	+	+	+	−	−	−	+	−	+
12	−	−	−	−	−	−	−	−	−	−	−

6.2 THE 12-RUN PLACKETT-BURMAN DESIGN

Table 6.1 shows the design matrix for the 12-run Plackett-Burman design. The rows in this matrix represent the runs ($N = 12$) and the columns represent up to 11 factors. As in all Plackett-Burman designs, the entire design matrix is constructed from an initial row of plus and minus signs that is given in Appendix 6.1. The last entry in row 1 ($−$) is placed in the first position of row 2. The other entries in row 1 fill in the remainder of row 2, by each moving one position to the right. The third row is generated from the second row using the same method, and the process continues until the next to the last row is filled in. A row of all $−$ signs is then added to complete the design.

The design is orthogonal, since for any two factors (columns) the number of runs at each of the four factor-level combinations $(−\,−), (−\,+), (+\,−), (+\,+)$ is the same (3 runs). Because the design is orthogonal, each of the 11 linear contrasts is independently estimated. We obtain each estimate in the usual way, by taking the average of the responses when the column entry is at the plus sign $(\bar{y}_+)$ minus the average of the responses when the column entry is at the minus sign $(\bar{y}_-)$.

Confounding in Plackett-Burman Designs

Plackett-Burman designs are resolution III. Main effects are confounded with 2-factor interactions, but the nature of the confounding is different from fractional factorial designs. (Appendix 6.2 gives a general discussion of confounding in Plackett-Burman designs.) As we have seen, in fractional factorial designs two effects are confounded if for each run the signs in each effect column are either the same or opposite. For confounded effects, the correlation between the signs of the two columns is always perfect (± 1).

In Plackett-Burman designs, a main effect and its confounded 2-factor interaction column contains signs that are correlated, but not perfectly. For example, consider the main effect of factor 1 and the 26 interaction for the $N = 12$-run design. We use $+1$ and $−1$ to represent the column signs, and we multiply the entries in columns 2 and 6 to obtain the entries in column 26. Writing each column as a row to save space, and listing the run numbers above the entries, we have

Run	1	2	3	4	5	6	7	8	9	10	11	12
Column 1:	+1	−1	+1	−1	−1	−1	+1	+1	+1	−1	+1	−1
Column 26:	+1	+1	−1	−1	−1	−1	+1	+1	−1	−1	+1	+1

Both columns have the same number of plus and minus signs, and they add up to zero. Furthermore, the sum of the squares of the entries in each column is 12, the number of runs N. The columns are correlated. In 8 of the 12 runs, the column signs match; whereas for 4 runs, they are opposite. The correlation between these two mean zero columns (let us call them x and z) is given by

$$\rho = \frac{\sum x_i z_i}{\sqrt{\sum x_i^2} \sqrt{\sum z_i^2}} = \frac{8 - 4}{12} = \frac{1}{3}$$

This correlation of 1/3 indicates that there is some linkage between the signs of these two columns, but the correlation is fairly weak.

In fractional factorial designs, if two effects are confounded, the correlation between column entries is either $\rho = +1$ or $\rho = -1$. In Plackett-Burman designs, if two effects are confounded, the absolute value of the correlation between column entries is strictly less than 1. One says that the effects are *partially confounded* or *partially aliased*.

Calculating the correlation between each main effect column and each 2-factor interaction column, we find that for each factor its main effect is partially confounded with all 2-factor interactions not involving that factor. In addition, we find that for all 11 factors, the correlation between column signs for a factor's main effect and each of its confounded interactions is either $+1/3$ or $-1/3$.

The correlation between each main effect column and each 2-factor interaction column that includes the main effect factor is zero. You can see this, for example, by looking at the column for factor 1 and the interaction column 12 that are listed below.

Run	1	2	3	4	5	6	7	8	9	10	11	12
Column 1:	+1	−1	+1	−1	−1	−1	+1	+1	+1	−1	+1	−1
Column 12:	+1	−1	−1	−1	+1	+1	−1	+1	+1	−1	−1	+1

The column signs match in half of the runs, and the correlation between these two columns is given by

$$\rho = \frac{\sum x_i z_i}{\sqrt{\sum x_i^2} \sqrt{\sum z_i^2}} = \frac{6 - 6}{12} = 0$$

Table 6.2 lists all confounding (correlation) coefficients among main effects and 2-factor interactions in the $N = 12$ Plackett-Burman design with 8 factors. The first eight columns in Table 6.1 are used for the levels of the 8 factors, but any other set of eight columns could have been taken. In showing the confounding coefficients, we limit ourselves to 8 factors, because this makes it easier to display the confounding coefficients in a compact table.

TABLE 6.2

Confounding Coefficients Between Main Effects and 2-Factor Interactions in the Plackett-Burman Design with N = 12 Runs and 8 Factors

TWO-FACTOR INTERACTIONS

	1 2	1 3	1 4	1 5	1 6	1 7	1 8	2 3	2 4	2 5	2 6	2 7	2 8	3 4	3 5	3 6	3 7	3 8	4 5	4 6	4 7	4 8	5 6	5 7	5 8	6 7	6 8	7 8
Main effects																												
1			△	△	■	△	△	△	△	△	■	△	△	■	△	△	■	△	■	■	△	△	△	△	△	△	■	■
2		△	△	△	△	■	△					■	△	△	△	△	■	△	■	△	△	■	■	■	△	△	△	△
3	△		■	△	■	△	△			△	△	△	△			△	△	△	△	△	△	△	■	△	△	■	■	■
4	△	■		■	△	△	△	■		△	△	■	■		△	■	△	■		△	■	■	■	△	△	■	△	△
5	△	△	■		△	△	△	△	△		△	△	△	△	△	■	△	△		■	■	△			■	△	△	△
6	■	△	■	△		■	■	△	■	■		△	△	■	△	■			△		△	△	△	△		■		△
7	△	■	△	△	△		■	△	△	■	△		△	■	△	■		△	△	△	■	△	△	■	△	△	△	■
8	△	△	△	△	■	■		△	■	△	△	△		■	△	■	△		△	△	■	■	△	△	■	△	△	△

NOTE: An empty cell indicates zero correlation and no confounding. The symbol ■ represents confounding coefficient +1/3, and the symbol △ represents confounding coefficient −1/3.

Consider the linear contrast for column 1, $l_1 = \bar{y}_+ - \bar{y}_-$, which compares the average responses at the plus level of column 1 (runs 1, 3, 7, 8, 9, 11) with the average responses at the minus level of column 1 (runs 2, 4, 5, 6, 10, 12). The first row of Table 6.2 shows that it is an estimate given by

$$l_1 \rightarrow 1 + \frac{1}{3}(26 + 34 + 37 + 45 + 46 + 68 + 78)$$

$$-\frac{1}{3}(23 + 24 + 25 + 27 + 28 + 35 + 36 + 38 + 47 + 48 + 56 + 57 + 58 + 67)$$

In general, the contrast $(\bar{y}_+ - \bar{y}_-)$ associated with each design column is an estimate of the main effect plus the *weighted* sum of the 2-factor interactions that are confounded with that main effect (we are ignoring 3-factor and higher-order interactions). The weight applied to each 2-factor interaction is the correlation (shown in Table 6.2) between the entries in the main effect column and the entries in that interaction column. (In fractional factorial designs, the weights applied to each confounded 2-factor interaction are also the correlations, which are either $+1$ or -1.)

Implications of Plackett-Burman Confounding

Now let us return to considering the 12-run design for 11 factors. Although each main effect is confounded with many 2-factor interactions, in most applications, we would expect very few 2-factor interactions to be important (effect sparsity) and those that are to be smaller in magnitude compared to the main effects (hierarchical ordering). In many situations, especially in the early stages of an experimental investigation, it is appropriate to ignore 2-factor interactions and assume that each of the 11 estimated effects represents a main effect. This is reasonable as long as the magnitudes of 2-factor interactions are relatively small compared to main effects. But what if this is not the case?

Suppose we analyze the results of the experiment under the assumption that each of the 11 estimates represents a main effect and a single 2-factor interaction (unknown to us) is fairly large. What effect will this have on our main effect estimates? Consider l_1 again. There are 45 two-factor interactions partially confounded with the main effect of factor 1. Suppose that just one of these (say, 37) is large, while the rest are negligible (zero). Then l_1 estimates $1 + 1/3(37)$. If we always take l_1 to be an estimate of 1 (the main effect of 1), then our estimate of 1 will be *biased* by an amount equal to one third of the magnitude of the 37 interaction. If the 37 interaction is positive, our estimate of 1 will be too large, while if the 37 interaction is negative our estimate of 1 will be too small. As a result, we may mistakenly believe that the main effect of 1 is significant when it is not, or that it is not significant when it actually is. And because the 37 interaction is confounded with 8 other main effects (all but the main effects of factors 3 and 7), its presence will bias all of those main effect estimates as well. The presence of a second 2-factor interaction (e.g., 24) would increase or decrease the main effect bias depending on its sign.

In contrast, with a resolution III fractional factorial design, each 2-factor interaction is confounded with a single main effect. [The fact that a 2-factor interaction cannot be confounded with two different main effects is easy to show. Assume that a 2-factor interaction (say, 12) were confounded with two main effects (say, 3 and 4). Then the defining relation would contain the words 123, 124, and (123)(124) = 34, making the fractional factorial a resolution II design.] But in this case, the bias in the main effect estimate will equal the full magnitude of the interaction, not just a fraction as in the Plackett-Burman design.

Because of the extensive partial confounding in Plackett-Burman designs, the option of a complete foldover can be especially useful. Adding 12 more runs to the existing design by switching the signs in all factor columns will result in a 24-run resolution IV design with main effects no longer confounded with 2-factor interactions.

Given the complex confounding patterns of resolution III Plackett-Burman designs, it may seem at first glance that they would not provide any useful information about 2-factor interactions. But as we show in the following example, under some circumstances it may be possible to estimate one or more 2-factor interactions from the results of a Plackett-Burman design.

6.3 PLACKETT-BURMAN DESIGN FOR $N = 20$ RUNS: A DIRECT MAIL CREDIT CARD CAMPAIGN

A leading Fortune 500 financial products and services firm uses direct mail to reach new customers. [For a detailed discussion, see Case 9 (Experimental Design on the Front Lines of Marketing: Testing New Ideas to Increase Direct Mail Sales) in the case study appendix.] But as competition increased over the years, response to the firm's offers had declined steadily. In an effort to reverse this trend, the company hired a consultant to help with the planning and execution of a large mailing of a credit card offer.

The marketing team identified 19 factors to test (Table 6.3), and the consultant specified the 20-run Plackett-Burman design shown in Table 6.4. Here, as in the 12-run design, the entire design matrix is generated from the first row listed in Appendix 6.1. The procedure is exactly the same as before. The last entry in row 1 ($-$) is placed in the first position of row 2. The other entries in row 1 fill in the remainder of row 2, by each moving one position to the right. This process continues until the next to the last row is filled in. A row of all minus signs is then added to complete the design.

Factors $A-E$ were approaches aimed at getting more people to look inside the envelope, while the remaining factors related to the offer inside. Factor G (sticker) refers to the peel-off sticker at the top of the letter to be applied by the customer to the order form. The firm's marketing staff believed that a sticker increases involvement and is likely to increase the number of orders. Factor N (product selection) refers to the number of different credit card images that a customer could chose from, and the term "buckslip" (factors Q and R) describes a small separate sheet of paper that highlights product information.

A total of 100,000 people, randomly chosen from a list of potential customers, participated in the experiment. Each of the 20 runs in Table 6.4 describes a test package that was sent to 5,000 people. The response variable is the fraction of people who respond to the credit card offer.

TABLE 6.3
The 19 Factors and Their Levels

	Factor	(−) Control	(+) New Idea
A	Envelope teaser	General offer	Product-specific offer
B	Return address	Blind	Add company name
C	"Official" ink-stamp on envelope	Yes	No
D	Postage	Preprinted	Stamp
E	Additional graphic on envelope	Yes	No
F	Price graphic on letter	Small	Large
G	Sticker	Yes	No
H	Personalize letter copy	No	Yes
I	Copy message	Targeted	Generic
J	Letter headline	Headline 1	Headline 2
K	List of benefits	Standard layout	Creative layout
L	Postscript on letter	Control version	New postscript
M	Signature	Manager	Senior executive
N	Product selection	Many	Few
O	Value of free gift	High	Low
P	Reply envelope	Control	New style
Q	Information on buckslip	Product info	Free gift info
R	Second buckslip	No	Yes
S	Interest rate	Low	High

TABLE 6.4
The 20-Run Plackett-Burman Design and the Results of the Experiment

							FACTOR												Responses	
A	B	C	D	E	F	G	H	I	J	K	L	M	N	O	P	Q	R	S	(of 5,000)	Rate
+	+	−	−	+	+	+	+	−	+	−	+	−	−	−	−	+	+	−	52	1.04
−	+	−	+	−	+	+	+	+	−	+	−	+	−	−	−	−	+	+	38	0.76
+	−	+	+	−	−	+	+	+	+	−	+	−	+	−	−	−	−	+	42	0.84
+	+	−	+	+	−	−	+	+	+	+	−	+	−	+	−	−	−	−	134	2.68
−	+	+	−	+	+	−	−	+	+	+	+	−	+	−	+	−	−	−	104	2.08
−	−	+	+	−	+	+	−	−	+	+	+	+	−	+	−	+	−	−	60	1.20
−	−	−	+	+	−	+	+	−	−	+	+	+	+	−	+	−	+	−	61	1.22
−	−	−	−	+	+	−	+	+	−	−	+	+	+	+	−	+	−	+	68	1.36
+	−	−	−	−	+	+	−	+	+	−	−	+	+	+	+	−	+	−	57	1.14
−	+	−	−	−	−	+	+	−	+	+	−	−	+	+	+	+	−	+	30	0.60
+	−	+	−	−	−	−	+	+	−	+	+	−	−	+	+	+	+	−	108	2.16
−	+	−	+	−	−	−	−	+	+	−	+	+	−	−	+	+	+	+	39	0.78
+	−	+	−	+	−	−	−	−	+	+	−	+	+	−	−	+	+	+	40	0.80
+	+	−	+	−	+	−	−	−	−	+	+	−	+	+	−	−	+	+	49	0.98
+	+	+	−	+	−	+	−	−	−	−	+	+	−	+	+	−	−	+	37	0.74
+	+	+	+	−	+	−	+	−	−	−	−	+	+	−	+	+	−	−	99	1.98
−	+	+	+	+	−	+	−	+	−	−	−	−	+	+	−	+	+	−	86	1.72
−	−	+	+	+	+	−	+	−	+	−	−	−	−	+	+	−	+	+	43	0.86
+	−	−	+	+	+	+	−	+	−	+	−	−	−	−	+	+	−	+	47	0.94
−	−	−	−	−	−	−	−	−	−	−	−	−	−	−	−	−	−	−	104	2.08

The resulting response rates are shown in Table 6.4. The estimated effects, which are differences between average responses at the plus and minus levels of the factor columns, are listed in Table 6.5. A Pareto chart, where estimated effects are ordered according to their absolute magnitudes, is shown in Figure 6.1. Significance of each effect is determined by comparing the estimate with (twice) its standard error. In Chapter 4, Section 4.6.4, we showed how to calculate the standard error of an estimated effect when the response is a proportion.

TABLE 6.5
The Estimated Main Effects

Factor	Estimate
Average	1.298
A	0.064
B	0.076
C	0.032
D	0.044
E	0.092
F	−0.128
G	**−0.556**
H	0.104
I	**0.296**
J	**−0.192**
K	0.088
L	−0.116
M	−0.064
N	−0.052
O	0.092
P	−0.096
Q	−0.080
R	**−0.304**
S	**−0.864**

NOTE: Effects that exceed twice the standard error $\pm(2)(0.0717) = \pm 0.143$ are indicated in boldface.

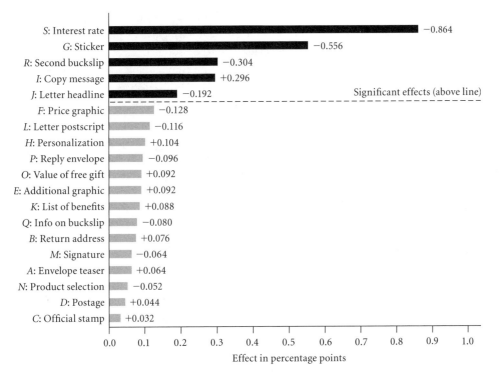

Figure 6.1   Graphical Display of the Estimated Main Effects

In this case, 1,298 people (or 1.298%) placed an order. Using this estimated proportion, the standard error of each estimated effect is

$$standard\ error(\text{effect}) = 100\sqrt{\frac{4(0.01298)(0.98702)}{100,000}} = 0.0717\ (\text{percent})$$

6.3.1 Main Effects Interpretation of the Results

All Plackett-Burman designs have complex confounding patterns, with each main effect partially confounded with many 2-factor interactions. We will discuss the confounding pattern for the 20-run design in the next section. In interpreting the results of this experiment, we initially ignore these possible 2-factor interactions and assume that each estimate is attributable to the main effect.

The following 5 factors had a significant effect on the response rate:

S−: *Low interest rate.* Increasing the credit card interest rate reduces the response by 0.864 percentage points. In addition, it was very clear based on the firm's financial models that the gain from the higher rate would be much less than the loss due to the decrease in the number of customers.

G−: *Sticker.* The sticker (G−) increases the response by 0.556 percentage points, resulting in a gain much greater than the cost of the sticker.

R−: *No second buckslip.* A main effect interpretation shows that adding another buckslip reduces the number of buyers by 0.304 percentage points. One explanation offered for this surprising result was that the buckslip added unnecessary information and obscured the simple "buy now" offer. A more compelling explanation that we discuss in the next section is that the significant effect is not the result of the main effect of factor R, but is due to an interaction between two other factors.

I+: *Generic copy message.* The targeted message (I−) emphasized that a person could chose a credit card design that reflected his or her interests, while the generic message (I+) focused on the value of the offer. The creative team was certain that appealing to a person's interests would increase the response, but they were wrong. The generic message increased the response by 0.296 percentage point.

J−: *Letter headline #1.* The result showed that all "good" headlines were not equal. The best wording increased the response by 0.192 percentage point.

6.3.2 Evidence of a 2-Factor Interaction Between
S (Interest Rate) and G (Sticker)

The main effect of each factor in the 19-factor, 20-run Plackett-Burman design is confounded with all 2-factor interactions not involving that factor, a total of 153 two-factor interactions (18!/2!16!). By enumerating all correlations among design and interaction columns, we find that of the 153 interactions confounded with each main effect, 144 have correlations −0.2 or +0.2, and 9 have correlations −0.6.

Each 2-factor interaction appears in the confounding pattern of 17 main effects. For 16 of these main effects, the correlation with this interaction is −0.2 or +0.2, whereas for a single main effect, the correlation with this interaction is -0.6. This is important because it implies that a large 2-factor interaction will create a large bias (-0.6 times the value of the interaction) in the estimate of one particular main effect.

Factors S (interest rate) and G (presence of a sticker) are by far the largest effects in Table 6.5. The correlation between the main effect of R (second buckslip), and the SG interaction is -0.6. You can check this by calculating the correlation between the column entries, as we did in the previous section. Hence, a significant SG interaction would bias the estimate of the main effect of R by -0.6 times the value of the interaction. This suggests that it may not be the main effect of factor R that is important, but the 2-factor interaction between S and G. This interpretation is supported by the principle of effect heredity as the main effects of S and G are the most important factors. As one might expect, at the high interest rate the effect of having a sticker is small (a change from 0.776% to 0.956% is implied by the results in Table 6.4), but at the low interest rate, the effect of having the sticker is much larger (a change from 1.264% to 2.024%). The sticker is most effective when the customer receives a more attractive offer.

6.3.3 Further Analysis of the Credit Card Experiment

The confounding of main effects and interactions introduces some uncertainty into our interpretation of the results. Of course, a straightforward approach for obtaining unconfounded main effects is a foldover of the original Plackett-Burman design. The combination of a Plackett-Burman design and its complete foldover creates a resolution IV design where main effects are not confounded with 2-factor interactions. Of course, 2-factor interactions are still confounded in a rather complicated fashion. In the credit card experiment, a foldover was not carried out (with 40 runs, it would have greatly increased the operational complexity of the mailing), and hence we can not be certain which combinations of main effects and interactions are responsible for the significant estimates in Table 6.5.

Plackett-Burman designs have traditionally been used as main-effects designs, and generally they should be avoided if the experimenter is concerned about the presence of fairly large interactions. Nevertheless, these designs allow the investigator to identify, in certain circumstances, selected 2-factor interactions. Box and Tyssedal (1996) showed that a Plackett-Burman design produces for any 3 factors a complete factorial arrangement, with some combinations replicated. We discuss this concept, called projectivity, in more detail in Appendix 6.2.

We use this projectivity idea to provide more evidence that the apparent main effect of R (second buckslip) is actually a consequence of the bias created by the SG interaction. Consider the 3 factors S, G, and R. Of the 20 runs in Table 6.4, there is at least one run at each of the eight factor-level combinations of these 3 factors. In specifying each combination, we let the first sign indicate the level of S, the second sign represent the level of G, and the last sign represent the level of R. There are 4 runs at each of the four combinations $(- - -)$, $(- + +)$, $(+ + -)$, $(+ - +)$, and one run at each of the remaining four combinations. Because we have at least one response at each combination, we have a full factorial arrangement in factors S, G, and R (ignoring the other factors). Because the number of runs at each combination is not the same, we must use regression to estimate the effects. Doing so, we

TABLE 6.6

Three Regression Models Relating the Response Rate to Factors S (Interest Rate),
G (Sticker), R (Second Buckslip), I (Copy Message), and
J (Letter Headline) (Minitab Output)

(A) REGRESSION OF RESPONSE RATE ON S, G, R, AND THEIR INTERACTIONS

```
The regression equation is
Rate = 1.33 - 0.386 S - 0.320 G - 0.0613 R + 0.151 SG
        - 0.0700 SR + 0.0762 GR + 0.0450 SGR
```

Predictor	Coef	SE Coef	T	P
Constant	1.32500	0.06602	20.07	0.000
S	-0.38625	0.06602	-5.85	0.000
G	-0.32000	0.06602	-4.85	0.000
R	-0.06125	0.06602	-0.93	0.372
SG	0.15125	0.06602	2.29	0.041
SR	-0.07000	0.06602	-1.06	0.310
GR	0.07625	0.06602	1.16	0.271
SGR	0.04500	0.06602	0.68	0.508

```
S = 0.236185 R-Sq = 90.2% R-Sq(adj) = 84.6%
```

(B) REGRESSION OF RESPONSE RATE ON S, G, AND SG

```
The regression equation is
Rate = 1.30 - 0.432 S - 0.278 G + 0.188 SG
```

Predictor	Coef	SE Coef	T	P
Constant	1.29800	0.05245	24.75	0.000
S	-0.43200	0.05245	-8.24	0.000
G	-0.27800	0.05245	-5.30	0.000
SG	0.18800	0.05245	3.58	0.002

```
S = 0.234585 R-Sq = 87.2% R-Sq(adj) = 84.8%
```

(C) REGRESSION OF RESPONSE RATE ON S, G, I, J, AND SG

```
The regression equation is
Rate = 1.30 - 0.432 S - 0.278 G + 0.151 SG + 0.118 I - 0.0657 J
```

Predictor	Coef	SE Coef	T	P
Constant	1.29800	0.04407	29.46	0.000
S	-0.43200	0.04407	-9.80	0.000
G	-0.27800	0.04407	-6.31	0.000
SG	0.15130	0.04594	3.29	0.005
I	0.11774	0.04501	2.62	0.020
J	-0.06574	0.04501	-1.46	0.166

```
S = 0.197073   R-Sq = 92.1%   R-Sq(adj) = 89.3%
```

find that the three significant effects are S, G, and SG, confirming that it is the SG interaction, not the main effect of R, that is significant.

Table 6.6(a) shows the Minitab output when regressing the response rate on the main and interactions effects of the three factors S, G, and R. The standard errors of the estimated regression coefficients use the pooled variance from the eight factor-level combinations, assuming that the other factors have no effect on the response. The t-ratios and the probability values of the regression coefficients listed in this table indicate that S, G, and SG are significant, while all other effects (including the main effect of factor R) are insignificant. Table 6.6(b) lists the results of the regression on the significant effects S, G, and SG. The regression explains 87.2% of the variability in the response rate.

Cheng (1995) showed that in the 20-run Plackett-Burman design, for any 4 factors, estimates of the four main effects and the six 2-factor interactions involving these four factors can be obtained when their higher-order (3- and 4-factor interactions) are assumed to be negligible. Having eliminated factor R, we apply Cheng's finding and consider a model that includes the four factors that were significant in our initial main effects analysis: S, G, I, and J, together with their six 2-factor interactions. The result of this regression shows that all 2-factor interactions except SG are insignificant, leading to a model with the four main effects and the SG interaction. The fitting results for the model with S, G, SG, and the two main effects of I and J are shown in Table 6.6(c). The five effects explain 92.1% of the variation, a rather modest improvement over the 87.2% that is explained by S, G, and SG. It is clear that factors S (interest rate) and G (sticker) and their interaction SG are the main drivers of the response rate.

6.4 NOBODY ASKED US, BUT . . .

In a Plackett-Burman design, each of the calculated effects estimates a main effect plus a weighted sum of the 2-factor interactions that are confounded with that main effect, with the weight (coefficient) of each 2-factor interaction being the correlation between its column signs and the column signs of the main effect. It is not too difficult to see why this is so. In Section 6.2, we showed the columns of signs of the main effect of 1 and the 26 interaction and found the correlation between them to be 1/3. For simplicity, assume that the main effect of factor 1 is confounded only with the 26 interaction. Looking once again at those two columns of signs, we can write l_1 in the following way,

$$l_1 = \frac{2}{3}\left[\left(\frac{y_1 + y_7 + y_8 + y_{11}}{4}\right) - \left(\frac{y_4 + y_5 + y_6 + y_{10}}{4}\right)\right] + \frac{1}{3}\left[\left(\frac{y_3 + y_9}{2}\right) - \left(\frac{y_2 + y_{12}}{2}\right)\right]$$

The first term in brackets is $(\bar{y}_+ - \bar{y}_-)$ for the 8 runs (2/3 of the total number of runs) when the signs in the two columns are the same. It estimates $1 + 26$ (the main effect of 1 plus the 26 interaction). The second term in brackets is $(\bar{y}_+ - \bar{y}_-)$ for the remaining 4 runs (1/3 of the total number of runs) when the signs in the two columns are opposite. It estimates $1 - 26$ (the main effect of 1 minus the 26 interaction). The linear contrast l_1 combines these two estimates;

$$l_1 \rightarrow \frac{2}{3}[1 + 26] + \frac{1}{3}[1 - 26] = 1 + \frac{1}{3}26$$

giving us an estimate of the main effect plus 1/3 of the 26 interaction.

Plackett and Burman created their designs in response to the problem of reducing the failure rate of an explosive weapon used by the British in World War II. The key to the development of these designs was the fact that the weapon consisted of 22 components. To improve it, the engineering approach was to make a single change to each component and experiment to see which changes reduced the weapon's failure rate. This involved testing 22 factors, each at two levels (old version of the component or new version). A 32-run fractional factorial design would have meant testing 32 different versions of the weapon in each experiment. Using Plackett and Burman's 24-run design, the experiments and improvement of the weapon were carried out more quickly, a great benefit during times of war.

CONSTRUCTION OF PLACKETT-BURMAN DESIGNS IN

$N = 12$, 20, AND 24 RUNS

Table 6.1 lists the Plackett-Burman design for $N = 12$ runs. The design matrix was constructed as follows. Starting from the first row (which is listed below in the row under $N = 12$), you cyclically rearrange the symbols. That is, the sequence of plus and minus signs in row 1 gets pushed to the right by one space to form row 2, and the minus sign in the far-right position of row 1 gets moved to the far-left position in row 2. The plus sign in the far-right position in the second row gets moved to the far-left position in row 3, and so on. The cyclical rearrangement of rows continues until row 11. The 12th row is a row of all minus signs. Alternatively, you can cycle through the 11 rows by pushing the sequence of plus and minus signs to the left and moving the entry in the far-left position of a row to the far-right position of the subsequent row. This only changes the order of the runs.

Similarly, for $N = 20$. You create the first 19 runs by cyclically rearranging the symbols in the row shown below; the 20th row is a row of all minus signs. The resulting design matrix is shown in Table 6.4. The same procedure is used for generating the design matrix for $N = 24$ runs. The initial rows needed to construct Plackett-Burman designs with $N \geq 28$ runs can be found in the original Plackett and Burman (1946) reference and in advanced books on design of experiments.

This procedure results in the standard order of a Plackett-Burman design. As with all experiments, one should randomize the assignments of the experimental units to each run, or if experiments are carried out in time-order, one should randomize the order of the runs.

N	1	2	3	4	5	6	7	8	9	10	11	12	13	14	15	16	17	18	19	20	21	22	23
12	+	+	−	+	+	+	−	−	−	+	−												
20	+	+	−	−	+	+	+	+	−	+	−	+	−	−	−	−	+	+	−				
24	+	+	+	+	+	−	+	−	+	+	−	−	+	+	−	−	+	−	+	−	−	−	−

PROPERTIES OF PLACKETT-BURMAN DESIGNS:

CONFOUNDING PATTERNS AND PROJECTIVITY

Sections 6.2 and 6.3 describe the rather complicated confounding patterns of Plackett-Burman designs. In this appendix we discuss these patterns in more detail, and we show how they are derived. We ignore interactions of order 3 and higher, and focus on the confounding of main effects and 2-factor interactions. Furthermore, we discuss the projectivity properties of Plackett-Burman designs, which make these designs useful for factor screening.

Result

Consider an orthogonal design with k factors at 2 levels each, such as the fractional factorial or the Plackett-Burman design. The confounding coefficient between the main effect of factor i and the 2-factor interaction among factors j and r is given by the correlation coefficient between the design vector $\mathbf{x}_i$ and the interaction (calculation) column $\mathbf{x}_{jr}$. Let us denote this correlation coefficient as $\rho_{i(jr)}$.

Proof

General regression results about the bias of regression estimates when fitting an incorrect model are used to show this result. This approach was employed by Margolin (1968) in his analysis of the confounding patterns in Plackett-Burman designs.

We are fitting the main-effects model:

$$\mathbf{y} = \mathbf{x}_1\beta_1 + \mathbf{x}_2\beta_2 + \cdots + \mathbf{x}_k\beta_k + \boldsymbol{\varepsilon} = X\boldsymbol{\beta} + \boldsymbol{\varepsilon} \qquad (A6.1)$$

where $X = [\mathbf{x}_1, \mathbf{x}_2, \ldots, \mathbf{x}_k]$ is the orthogonal design matrix and $\boldsymbol{\beta} = (\beta_1, \beta_2, \ldots, \beta_k)'$ is the vector of main effects. Table 6.1 lists the design matrix of the Plackett-Burman design in $N = 12$ runs. The design matrix for $N = 20$ is shown in Table 6.4.

Assume that 2-factor interactions are present and that the true model is given by

$$\mathbf{y} = X\boldsymbol{\beta} + X_*\boldsymbol{\beta}_* + \boldsymbol{\varepsilon} \qquad (A6.2)$$

$X_* = [\mathbf{x}_{12}, \mathbf{x}_{13}, \ldots, \mathbf{x}_{23}, \ldots, \mathbf{x}_{k-1,k}]$ is the design matrix consisting of 2-factor interaction (calculation) columns, and $\boldsymbol{\beta}_* = (\beta_{12}, \beta_{13}, \ldots, \beta_{23}, \ldots, \beta_{k-1,k})'$ is the vector of 2-factor interactions. General regression results (see Draper and Smith, 1981, p. 117; Abraham and Ledolter, 2006, p. 208) imply that the calculated main effects, obtained by taking (one-half of) the difference of the response averages at the plus and minus levels of the factors, are estimates of

$$\boldsymbol{\beta} + (X'X)^{-1}X'X_*\boldsymbol{\beta}_* \qquad (A6.3)$$

For orthogonal designs (such as the fractional factorial and Plackett-Burman designs) the matrix $X'X$ is diagonal with diagonal elements given by N, the number of runs. Furthermore, the columns of X and X_* sum to zero, and their squares add up to N. Hence, the matrix $(X'X)^{-1}X'X_*$ is a matrix of correlations between design columns $\mathbf{x}_i$ and calculation columns $\mathbf{x}_{jr}$. This implies that the usual main effect estimate of factor i is an estimate of

$$\beta_i + \sum_j \sum_{j<r} \rho_{i(jr)} \beta_{jr} \qquad (A6.4)$$

The confounding coefficient between the main effect of factor i and the 2-factor interaction among factors j and r is given by $\rho_{i(jr)}$, the correlation coefficient between the design vector $\mathbf{x}_i$ and the interaction (calculation) column $\mathbf{x}_{jr}$.

Discussion

1. The result implies that a main effect is unconfounded with all interactions that contain the main effect factor. The column of products of the elements in a design column $\mathbf{x}_i$ and an interaction (calculation) column $\mathbf{x}_{ir}$ containing the main effect is identical to the column $\mathbf{x}_r$ (as the product of a column with itself leads to a column of ones). The sum of such a column is zero, implying $\rho_{i(ir)} = 0$.

2. A main effect in most Plackett-Burman designs is confounded with all other interactions that do not contain the main effect as a factor. Depending on the run size of the Plackett-Burman design and the particular main and interaction effect being considered, this correlation can take on various values.

Consider the $N = 12$ Plackett-Burman design in Table 6.1, and the design column $\mathbf{x}_1$ (for factor 1) and the interaction column $\mathbf{x}_{23}$ (which one gets by row-wise multiplication of the entries in $\mathbf{x}_2$ and $\mathbf{x}_3$). The correlation between these columns is $-1/3$. The correlation between $\mathbf{x}_1$ and $\mathbf{x}_{26}$ is $+1/3$. Only main effects and interactions that contain the main effect are uncorrelated. All other confounding coefficients are either $-1/3$ or $+1/3$.

For the $N = 20$-run Plackett-Burman design, the confounding coefficients are either -0.2, $+0.2$, or -0.6. For example, the confounding coefficient between the main effect of A and the BC interaction in Table 6.4 is -0.2; it is $+0.2$ for the main effect of A and the BD interaction, and -0.6 for the main effect of R and the SG interaction. Only main effects and interactions that contain the main effect are uncorrelated.

For the $N = 24$ run Plackett-Burman design, the nonzero correlations are either $-1/3$ or $+1/3$. In contrast to the $N = 12$- and $N = 20$-run designs, not every main effect is confounded with interactions that do not contain that main effect; some correlations are in fact zero.

Projectivity Properties

Plackett-Burman designs are useful in screening situations where the objective is to identify important factors for more detailed study. The principle of "effect sparsity" suggests that, most likely, only a few factors among a large pool of potential factors are important. When choosing a design for factor screening, it is important to consider projections of the design into small subsets of factors. Box and Tyssedal (1996) define a factorial design to be of *projectivity p* if it produces, for any subset of p factors, a complete factorial (possibly with some combinations replicated).

Box and Tyssedal (1996) show that (most) Plackett-Burman designs attain projectivity 3. This is true for the $N = 12$ and $N = 20$ designs considered in this chapter. Exceptions to this rule, and projectivity less than 3, are the Plackett-Burman designs in $N = 40$ and $N = 56$ runs.

Convince yourself of this fact by considering the first 3 factors in the 12-run Plackett-Burman design in Table 6.1. Consider the eight factor-level combinations of these 3 factors, and show that each one has at least one run; four of the eight factor-level combinations (the ones at $(- - -), (+ + -), (+ - +), (- + +))$ have two runs. The general result by Box and Tyssedal (1996) implies that this projectivity holds for *any subset* of 3 factors, not just the first 3.

Plackett-Burman designs are main-effects designs, and they should be avoided if we are concerned about possibly large interactions. Nevertheless, the projectivity of these designs allows the investigator to identify, in certain circumstances, selected 2-factor interactions. Assuming that there are no more than 3 active factors, one can estimate the main effects and the interactions among these 3 factors. Furthermore, while the Plackett-Burman design in $N = 20$ runs will not project into full 2^4 factorials, Cheng (1995) shows that the projection of this design onto any 4 factors has the property that all main effects, and 2-factor interactions of these 4 factors can be estimated when their higher-order (order 3 and 4) interactions are assumed negligible. This is a remarkable result, as it shows that one can estimate all main effects and 2-factor interactions of 4 factors, and one can do so without specifying a priori which 4 factors are important.

In terms of their projectivity, Plackett-Burman designs have an advantage over resolution III fractional factorial designs that have projectivity of only 2. Consider, for example, the 2_{III}^{15-11} design generated by associating factors 5–15 with the interaction columns of a full factorial in factors 1–4. It is easy to check that the 16 runs of factors 1, 2, and 5 = 12 in Table 5.7 will not generate runs at all eight factor-level combinations in these three factors.

Because of their complicated alias structures, experimenters have sometimes been reluctant to use Plackett-Burman designs for experimentation (see Draper, 1985). However, the interesting projective properties of Plackett-Burman designs provide a compelling rationale for their use.

EXERCISES

Exercise 1   Consider Case 8 (Experiments in Retail Operations: Design Issues and Application) from the case study appendix.

(a) In this case, we study 10 factors through a 24-run design that consists of a 12-run Plackett-Burman design and its foldover. The 2^{10-5} fractional factorial design in 32 runs would be another potential design. Is it possible to achieve a resolution IV design in 32 runs? If so, discuss the generators and the confounding patterns (assuming that interactions of order 3 or higher are negligible).

(b) Confirm the estimated effects in the test result section.

(c) Obtain standard errors of the estimated effects, using the following approaches:

(c1) Use the percent changes of week 1 and week 2 as independent replications, calculate a variance estimate at each factor-level combination, average the

24 variances, and substitute the pooled variance estimate into the equation for the standard error in Section 4.4. Check whether the observations for weeks 1 and 2 are uncorrelated.

Hint: Calculate the number of runs for which week 2 has larger sales than week 1. Under the null hypothesis, you expect $24/2 = 12$ runs. You can use the binomial distribution with $N = 24$ and $\pi = 0.5$ (expressing the fact that there is a 50:50 chance of increasing sales) to assess the probability value of the sample result.

(c2) Determine the significance of the estimated effects through normal probability plots and Lenth's PSE.

(c3) Explain how regression can be used to calculate the standard errors of the estimated effects. *Hint:* The regression relates the 24×1 vector of responses to a constant and the three design vectors A, D, and F. This leaves $24 - 4 = 20$ degrees of freedom for the standard deviation of the error, $s = 0.1092$. This estimate is used in the covariance matrix $V(\hat{\boldsymbol{\beta}}) = s^2(X'X)^{-1}$. The square roots of the diagonal elements are one-half the standard errors of the estimated effects. Note that none of these calculations are needed when using computer software to execute the regression.

(c4) Discuss whether our earlier conclusion about the significance of the effects is affected by the three different approaches of calculating the standard errors of the effects.

(d) Discuss problems that may arise in the planning and execution of this experiment.

Exercise 2 Consider Case 9 (Experimental Design on the Front Lines of Marketing: Testing New Ideas to Increase Direct Mail Sales) from the case study appendix.

(a) Consider the Plackett-Burman design in Table A9.2. Check that the design is orthogonal. Consider any two factors, say factors A and F, and verify that the design includes 5 runs at each of the four factor-level combinations. Repeat this check for other pairs of factors.

(b) Reanalyze the data from the Plackett-Burman and the factorial designs in Tables A9.2 and A9.6.

(c) A binomial approximation (see Section 4.6.4) was used to determine the standard errors of the estimated effects. Use alternative approaches:

(c1) Construct a normal probability plot of the estimated effects and determine the significance of the factors that way.

(c2) Use Lenth's PSE approach to assess the significance. Discuss the similarities and differences among these three approaches.

Note: The underlying true sign-up proportion π depends on the advertisement scenario for that particular run. Decisions (yes/no) by individuals in that group are the results of Bernouilli trials with success probability π. This implies that the variance of the sample proportion calculated from the n individuals in that group is given by $var(p) = \pi(1 - \pi)/n$.

This derivation assumes that the same proportion π applies to all subjects in the group, an assumption that may not be correct. Sign-up rates may vary across subjects, $\pi_i = \pi + \xi_i$, where ξ_i expresses the subject variability. One can show that the heterogeneity across subjects increases the variance of a proportion and that $var(p) > \pi(1 - \pi)/n$. Discuss the effect of heterogeneity on the standard error of an estimated effect.

(d) The variance of the observed proportion, $var(p) = \pi(1 - \pi)/n$, depends on the true proportion that varies across the design runs. This violates the assumption that responses are equally reliable, a fact that is needed for the equal weighting of the proportions when calculating estimated effects.

It has been shown that the *square root* and the *arcsine* transformations stabilize the variability, and their use has been suggested when working with proportions. Hence, instead of calculating the effects from the proportions p, one transforms the proportions by taking their square roots and obtains the estimated effects from the transformed proportions. Similarly, one can apply the arcsine (or inverse sin, or $\sin^{-1}$) transformation, using calculator keys or commands in statistical software programs. The arcsine is usually given in radians and needs to be transformed into degrees by multiplying the result in radians with the factor $360/(2 \times 3.14159)$.

An even better approach of analyzing binary (yes/no) data is to use logistic regression. Chapter 11 of Abraham and Ledolter, *Introduction to Regression Modeling* (2006) discusses this approach in detail.

(e) Determine the partial confounding in the 20-run Plackett-Burman design in Table A9.2. Follow the approach in Appendix 6.2. Assume a model with 2-factor interactions, and determine the bias of the main effect estimates. For example, consider the estimate that corresponds to factor A (column 1). Consider the interaction between factors G and H, represented by the column of their products (column 2). The correlation between these two columns expresses the confounding factor between GH and the main effect of A. Of course, many interactions will confound the main effect of A. You can use the approach in Appendix 6.2 to determine the complete confounding pattern. Alternatively, using an Excel spreadsheet, you can enumerate the correlations between each main effect and its confounded 2-factor interactions.

(f) In the section on key metrics and sample size, we claim that an overall sample size of 100,000 (and a sample size of 5,000 in each of the 20 cells) implies a certain power (80%) of detecting a 17% shift (from 1% to 1.17%) in the average response rate. How was this determined? Recreate the steps that are involved in this analysis. Use computer software such as Minitab or JMP.

(g) Consider a 32 run 2^{19-14} fractional factorial design to study the 19 factors.

(g1) Is it possible to achieve a resolution IV design?

(g2) If only resolution III is possible, discuss the advantages and disadvantages of the 20-run Plackett-Burman and the 32-run 2^{19-14} fractional factorial designs.

(g3) Is it possible to construct a resolution IV design in $N = 40$ runs? Discuss.

Exercise 3   Obtain a half-fraction of a 2^4 factorial design by associating the levels of factor 4 with (i) the 123 interaction, (ii) the 12 interaction.

(a) Which of the two fractions has the more preferable confounding pattern? Discuss.

(b) Consider the projectivity properties of the two fractions. Assume that you have reasons to believe that only 3 factors are important, but you do not know which 3 factors. Does either fraction result in a full 3-factor factorial of any 3 factors? Discuss.

Exercise 4   Consider the study in Section 6.3. Suppose instead of a 20-run Plackett-Burman design for 19 factors, the experimenters had eliminated 4 factors and had chosen the 2^{15-11}_{III} fractional factorial design shown in Table 5.7 of Chapter 5. Show that in contrast to the 20-run Plackett-Burman design, which has projectivity 3, this fractional factorial design has projectivity less than 3. To do so, you need to show that not all choices of 3 factors have at least one run at each of the eight factor-level combinations.

7 | EXPERIMENTS WITH FACTORS
AT THREE OR MORE LEVELS

7.1 INTRODUCTION

Until now we have discussed experiments with factors at just two different levels. We coded the two levels as "low" and "high," or "−1" and "+1," or simply "−" and "+." A factor may describe two catalysts in a chemical reaction, two ways of displaying information in an ad copy, two cover prices of a magazine, or two different budgets for an advertising campaign. With just two levels, the assignment of the "−" and "+" levels is arbitrary.

In some applications a test factor may have three (or more) levels. There may be three catalysts, three methods, and three prices. It is common to code the three levels as −1, 0, and +1. The factor may be categorical, with no particular order among the categories. In this case, the assignment of the categories to the coded levels is arbitrary as any one of the three categories can be associated with a certain level. This will not be the case if the factor is continuous. The cover price of a magazine may have been set at one, two, or three dollars. Or, the temperature of a chemical reaction may have been studied at 1,500, 2,000, and 3,000 degrees. For continuous factors, the assignment of the actual levels to the coded ones, −1, 0, and +1, carries additional meaning. In addition to studying whether or not the mean responses at the three levels are the same, one can explore the functional relationship between the mean response and the continuous factor. We probably would not expect that the cover price has a linear effect on sales. Linearity of the sales response to changes in price may actually be the hypothesis that needs to be confirmed or refuted from the data. Whereas a linear function of price can be fitted (perfectly) to sales responses at two different price levels, we need at least three price levels to fit a quadratic function. With just two levels for price, it is impossible to check whether a linear relationship is appropriate.

Section 7.2 discusses the general factorial experiment with two factors; the first factor A is studied at a different levels, while the second factor B is studied at b levels. A complete factorial experiment requires runs at all ab factor-level combinations. We show how to estimate and test the main effects of the two factors, and we discuss how to assess the interaction effect. An example is given in Section 7.3. Section 7.4 discusses additional useful analyses when factors are continuous and explains how to test the linearity of the response relationship. Section 7.5 considers an experiment with three continuous factors where two factors are studied at two levels each, and one factor is studied at three levels.

Experiments with $k > 2$ factors at three or more levels require many runs. Three-level factorial experiments, for example, require 3^k runs; 9 runs are required for $k = 2$ factors; 27 runs, for $k = 3$ factors; 81 runs, for $k = 4$ factors; and so on. Such designs become nonparsimonious very quickly, because they require more runs than the experimenter is usually able or willing to carry out. Hence, it is important that the experimenter has screened the factors beforehand and has reduced the number of factors so that a few important ones can be studied at more than two levels. If there are still too many factors, one can reduce the number of runs in 3-level designs by considering orthogonal fractions. We introduce fractional 3-level factorial designs in Section 7.6.

7.2 THE ANALYSIS OF THE GENERAL 2-FACTOR FACTORIAL EXPERIMENT

Consider 2 factors A and B. Factor A is studied at a levels, while factor B is studied at b levels. We assume that the same number of runs, n, is carried out at all ab factor-level combinations. Factor-level combinations are sometimes referred to as *cells*. The experiment results in a total of abn responses, y_{ijr}, for $i = 1, 2, \ldots, a$ (factor A); $j = 1, 2, \ldots, b$ (factor B); and $r = 1, 2, \ldots, n$ (replication).

The analysis that follows assumes that the abn observations are independent and that the standard deviation of the experimental error is the same for each observation. It is important that the experiment is carried out in a way such that these assumptions are satisfied. Assume that you study the sales of a supermarket item as a function of its price and the level of its advertising. If you want to use the analysis discussed in this section, it is important that the observations come from abn different supermarkets, and that the assignment of the supermarkets to the treatment combinations is done at random. The analysis of the data changes if the same stores are used across all levels of one factor (say, advertising) as a "store effect" may make the observations dependent.

It is important to recognize when these assumptions are not met and when it is not appropriate to use the methods that are discussed in this section. For example, in agricultural experiments on the yield effect of type of seed and fertilizer, one typically randomizes seed and fertilizer within subplots of several larger (whole) plots of land. Such experiments are referred to as *split-plot experiments*. Although one obtains a total of abn observations from a different seeds and b different fertilizers, the analysis explained in this section, which assumes equal precision and independence of the observations will, most likely, not be appropriate. One can expect the responses to be more alike within each whole plot, and less alike from one whole plot to another. While the correct analysis of such data is straightforward, it is more complicated than the analysis we describe in this section.

We write the response as

$$y_{ijr} = \bar{y}_{\ldots} + (\bar{y}_{i\ldots} - \bar{y}_{\ldots}) + (\bar{y}_{\cdot j \cdot} - \bar{y}_{\ldots}) + (\bar{y}_{ij\cdot} - \bar{y}_{i\ldots} - \bar{y}_{\cdot j \cdot} + \bar{y}_{\ldots}) + (y_{ijr} - \bar{y}_{ij\cdot})$$

where $\bar{y}...$ is the overall mean of the abn observations, $\bar{y}_{i}..$ is the mean of the bn responses when factor A is at its ith level, $\bar{y}._{j}.$ is the average of the an observation when factor B is at its jth level, and $\bar{y}_{ij}.$ is the average of the n observations at the (i, j) factor-level combination. Recall the dot notation introduced in Section 3.3 of Chapter 3. Dots in place of subscripts indicate that we average the observations over these subscripts.

The difference $(\bar{y}_{i}.. - \bar{y}...)$ measures the effect of factor A, while averaging over all levels of factor B. There is no main effect to factor A if the differences $(\bar{y}_1.. - \bar{y}...)$, $(\bar{y}_2.. - \bar{y}...), \ldots, (\bar{y}_a.. - \bar{y}...)$ are zero, or nearly so. Large differences imply a main effect of factor A.

The difference $(\bar{y}._{j}. - \bar{y}...)$ measures the effect of factor B, while averaging over all levels of factor A.

The third component in the expression on the right-hand side, $\bar{y}_{ij}. - [\bar{y}... + (\bar{y}_i.. - \bar{y}...) + (\bar{y}._{j}. - \bar{y}...)] = (\bar{y}_{ij}. - \bar{y}_i.. - \bar{y}._{j}. + \bar{y}...)$, measures the interaction between factors A and B. It is the difference between the observed mean response at the (i, j) factor-level combination and the predicted mean response that is implied by the main effects of factors A and B alone. Interaction is negligible if these differences are small. Large differences imply an interaction between factors A and B.

The last component $(y_{ijr} - \bar{y}_{ij}.)$ represents the error. It is the deviation of the observation from its respective cell mean.

Similar to our discussion of the analysis of variance (ANOVA) in Chapter 3, we can partition the total sum of squares into several (four) components: the sum of squares of factor A (main effect of A), the sum of squares of factor B (main effect of factor B), the sum of squares of the interaction AB, and the sum of squares of error.

$$SST = \sum_{i=1}^{a} \sum_{j=1}^{b} \sum_{r=1}^{n} (y_{ijr} - \bar{y}...)^2$$

$$= bn \sum_{i=1}^{a} (\bar{y}_i.. - \bar{y}...)^2 + an \sum_{j=1}^{b} (\bar{y}._{j}. - \bar{y}...)^2 + n \sum_{i=1}^{a} \sum_{j=1}^{b} (\bar{y}_{ij}. - \bar{y}_i.. - \bar{y}._{j}. + \bar{y}...)^2$$

$$+ \sum_{i=1}^{a} \sum_{j=1}^{b} \left\{ \sum_{r=1}^{n} (y_{ijr} - \bar{y}_{ij}.)^2 \right\}$$

$$= SS(A) + SS(B) + SS(AB) + SS(error)$$

We obtain this decomposition by squaring $y_{ijr} - \bar{y}...$ and summing over i, j, and r. For an equal number of observations in each cell, all sums of cross products of the components are zero. Similarly, we can partition the degrees of freedom of the total sum of squares into the degrees of freedom of its components,

$$abn - 1 = (a - 1) + (b - 1) + (a - 1)(b - 1) + ab(n - 1)$$

These entries are displayed in the ANOVA in Table 7.1. The mean squares are obtained by dividing the sums of squares by their respective degrees of freedom. The F-ratios in the ANOVA table, $MS(AB)/MS(error)$, $MS(A)/MS(error)$, and $MS(B)/MS(error)$, are used to test the presence of interaction, main effect of A, and main effect of B. The probability

TABLE 7.1
ANOVA Table for the Two-Factor Factorial Experiment

Source of Variation	Sum of Squares SS	Degrees of Freedom df	Mean Squares MS	F	Probability Value
Factor A	SS(factor A)	$a - 1$	SS(factor A)/$(a - 1)$	MS(factor A)/ MS(error)	Prob value (factor A)
Factor B	SS(factor B)	$b - 1$	SS(factor B)/$(b - 1)$	MS(factor B)/ MS(error)	Prob value (factor B)
Interaction AB	SS(interaction)	$(a - 1)(b - 1)$	SS(interaction)/ $[(a - 1)(b - 1)]$	MS(interaction)/ MS(error)	Prob Value (interaction)
Error	SS(error)	$ab(n - 1)$	SS(error)/$[ab(n - 1)]$		
Total	SS(total)	$abn - 1$			

values of the F-ratios, which computer programs usually list in the last column, express the statistical significance of these components. The probability values are obtained from the F-distributions as follows:

probability value(interaction)
$$= P[F((a - 1)(b - 1), ab(n - 1)) > MS(AB)/MS(error)]$$

$$\textit{probability value}(\text{main effect of } A) = P[F(a - 1, ab(n - 1)) > MS(A)/MS(error)]$$

$$\textit{probability value}(\text{main effect of } B) = P[F(b - 1, ab(n - 1)) > MS(B)/MS(error)]$$

A main effect or an interaction is considered significant if its associated probability value is smaller than the significance level that is usually taken as 0.05. The significance of the interaction needs to be established first, as main effects have little meaning if interactions are present. Interactions are visualized through interaction plots, similar to the ones we considered for the 2-level designs in the previous chapters. Main effects are displayed by graphing response averages against the levels of factors A and B. Neither the ANOVA calculations nor the main effects and interaction plots need to be carried out by hand as commonly available statistical software such as Minitab include convenient functions that perform these analyses without much effort.

Establishing the statistical significance of interactions and main effects is a useful first step, but by no means the only important one. It is essential to know that the magnitude of an observed effect is larger than its chance variation. Otherwise, one would start attaching practical significance to differences that are not well estimated. However, once the significance of the effects has been established, it becomes important to learn how the groups differ. If interactions are small, it is appropriate to look at the main effects separately and compare the response averages across the considered factor levels. Simple plots of the averages, together with their standard errors and confidence intervals, tell us how response averages differ across the levels. For example, the average response at each level of factor A is an average of bn observations, and the standard error of such an average is given by $s/\sqrt{bn}$, where $s = \sqrt{SS(error)/ab(n - 1)}$. For continuous factors, it also makes sense to explore whether the relationship is linear. This is discussed in Section 7.4. Of course, an interpretation of main effects alone is not meaningful if interactions are present. In such a situation, one needs to interpret the interaction plot. We illustrate this in the following example.

7.3 EXAMPLE: BAKING A CAKE

Results of a 3^2 factorial experiment in a study on how to best bake a cake are given in Table 7.2. A commercially available cake mixture is baked by varying the baking temperature (factor A) and the baking time (factor B). Three different *temperatures* are chosen: level 0 represents the temperature recommended by the instructions on the package, whereas levels -1 and $+1$ represent temperatures 10% below and 10% above the recommended level. Three different *times* are chosen, with level 0 representing the recommended time setting, and levels -1 and $+1$ representing the time settings 10% below and 10% above the recommended level. Three independent replications are made at each temperature and time combination, and the finished cakes are tasted by experts and rated on a $0-6$ quality scale.

The ANOVA table is shown in Table 7.3. The entries in the table can be calculated from the equations in Section 7.2. Although this is not particularly difficult, the calculations are tedious given that they involve calculating factor A averages, factor B averages, and average ratings for each of the ab factor-level combinations, as well as the calculation of the various sums of squares. Fortunately, computing software is available for the calculations. Statistical software such as Minitab and JMP includes routines that compute the ANOVA table and conduct tests of significance for main effects and interaction. All that is needed is for the user to enter the information into a spreadsheet; responses into one column (say, column 1), the levels of factor A into a second (column 2), and the levels of factor B into a third (column 3). In this example, there are 27 rows. The first row contains 0 (the response), -1 (level of A), and -1 (level of B). The second row contains $0, -1, -1; \ldots$; the last (27th) row contains 2, 1, and 1. The particular order of the rows does not matter.

There is quite a strong interaction effect between time and temperature. The F-statistic for interaction, $F = 7.34$, is highly significant when compared to percentiles of the $F(4, 18)$

TABLE 7.2
Cake Ratings

Temperature	Time	Response
-1	-1	0, 0, 3
0	-1	0, 2, 4
1	-1	4, 5, 6
-1	0	2, 3, 4
0	0	3, 6, 6
1	0	1, 2, 3
-1	1	4, 5, 6
0	1	1, 3, 5
1	1	0, 1, 2

TABLE 7.3
ANOVA Table: Cake Ratings

Source	DF	SS	MS	F	P
Temperature	2	2	1.0000	0.47	0.630
Time	2	2	1.0000	0.47	0.630
Interaction	4	62	15.5000	7.34	0.001
Error	18	38	2.1111		
Total	26	104			

Interaction diagram

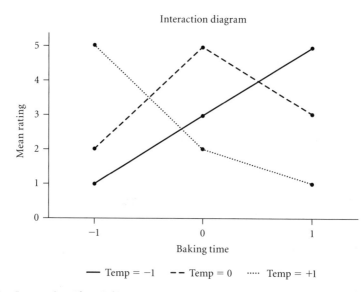

Figure 7.1 Interaction Plot: Cake Ratings

distribution; the probability value (0.001) is small, much smaller than the usually adopted significance level of 0.05. The interaction diagram shown in Figure 7.1 demonstrates the nature of the interaction. For a temperature lower than the recommendation (level -1), the quality of the cake is increased by increasing the baking time. For a higher-than-recommended temperature (level 1), the baking time, not surprisingly, should be reduced. At the recommended temperature, quality suffers if the baking time is either lower or higher than the recommended value. The cake turns out best if the recommendations for time and temperature are followed. But, the cake mix is "robust" in the sense that two other settings (lower temperature, but longer time; and higher temperature, but shorter time) are equally acceptable. A cake baked at a lower temperature has to be baked longer, whereas a cake baked at a higher temperature requires a shorter baking time.

7.4 A USEFUL INTERPRETATION OF EFFECTS IN FACTORIAL EXPERIMENTS WITH CONTINUOUS FACTORS

7.4.1 Orthogonal Polynomials

The sum of squares of a main effect in the 3^2 factorial experiment has two degrees of freedom; see Table 7.3. The F-test for the significance of a main effect assesses whether the three means at the low, middle and high levels of a factor are the same (i.e., $\mu_{-1} = \mu_0 = \mu_1$). This amounts to testing whether the three differences between the group means and the overall mean $\mu = (\mu_{-1} + \mu_0 + \mu_1)/3$ are zero; that is, $\mu_{-1} - \mu = \mu_0 - \mu = \mu_1 - \mu = 0$. Since deviations from a mean always sum to zero, two zero differences are sufficient to establish that the three means are the same. This fact explains the two degrees of freedom in the main effects test in Table 7.3. The restrictions among the means that are tested by the F-test (here there are two, $\mu_{-1} - \mu = 0$ and $\mu_1 - \mu = 0$) are commonly referred to as the contrasts.

Three group means can be represented in many different, equivalent ways. One can express them by their differences from the overall average (this was done above). Or, one can express them by their differences from the mean of a reference group (e.g., the group where the factor is at its low level). Or, one can express the means in a way that helps us test whether the functional relationship between the means and the continuous factor is linear. Statisticians refer to the different ways of representing the means as "parameterizations."

Consider the following two equivalent parameterizations of the three group means:

$$
\begin{bmatrix} \mu_{-1} \\ \mu_0 \\ \mu_1 \end{bmatrix} = \begin{bmatrix} 1 & 0 & 0 \\ 1 & 1 & 0 \\ 1 & 0 & 1 \end{bmatrix} \begin{bmatrix} \beta_0 \\ \beta_1 \\ \beta_2 \end{bmatrix} = \begin{bmatrix} 1 & -1 & 1 \\ 1 & 0 & -2 \\ 1 & 1 & 1 \end{bmatrix} \begin{bmatrix} \alpha_0 \\ \alpha_1 \\ \alpha_2 \end{bmatrix}
$$

We have written them in matrix form. In the first representation, $\mu_{-1} = \beta_0$, $\mu_0 = \beta_0 + \beta_1$, and $\mu_1 = \beta_0 + \beta_2$. This implies $\beta_0 = \mu_{-1}$, $\beta_1 = \mu_0 - \mu_{-1}$ and $\beta_2 = \mu_1 - \mu_{-1}$. The parameter $\beta_0 = \mu_{-1}$, the mean of the low group, becomes the standard against which the other means are compared to. The parameters β_1 and β_2 are the differences between μ_0 and μ_1, and this standard. A test of the equality of the three means amounts to testing whether $\beta_1 = \beta_2 = 0$. Here the low group is taken as the reference group, but any other group could have been used.

In the second representation, the α coefficients are such that $\alpha_0 = (\mu_1 + \mu_0 + \mu_{-1})/3$, $\alpha_1 = (\mu_1 - \mu_{-1})/2$, and $\alpha_2 = (\mu_1 - 2\mu_0 + \mu_{-1})/6$. We obtain these expressions by solving the equations that relate the means to the alpha coefficients. The alpha coefficients have a nice interpretation if the factor is continuous, and when the distances between the levels are meaningful. The coefficient $\alpha_0 = (\mu_1 + \mu_0 + \mu_{-1})/3$ is the overall mean, and α_1 and α_2 represent the linear and quadratic components of the relationship among the mean response and the coded factor levels. [The coded levels $(-1, 0, 1)$ may represent unequally spaced levels in the original metric, say, temperatures at 1,500, 2,000, and 3,000 degrees.] One can see this as follows. Assume a quadratic model between the mean and the factor levels, $\mu_i = a + bi + ci^2$, with mean responses at the three levels ($i = -1, 0, 1$) given by $\mu_{-1} = a - b + c$, $\mu_0 = a$ and $\mu_1 = a + b + c$. The coefficient $\alpha_1 = (\mu_1 - \mu_{-1})/2 = [(a + b + c) - (a - b + c)]/2 = b$ represents the linear term in the relationship, while the coefficient $\alpha_2 = (\mu_1 - 2\mu_0 + \mu_{-1})/6 = [(a + b + c) - 2(a + 0 + 0) + (a - b + c)]/6 = c/3$ is proportional to the quadratic component. Factor A has no effect if $\alpha_1 = \alpha_2 = 0$. The association is linear if $\alpha_2 = 0$.

The second representation of the group means in terms of linear and quadratic components is useful when describing the relationship between the mean response and the levels of a continuous factor. It has the additional advantage that the columns in the matrix that relates the means μ_{-1}, μ_0, μ_1 to the coefficients $\alpha_0, \alpha_1, \alpha_2$,

$$
\begin{bmatrix} \mu_{-1} \\ \mu_0 \\ \mu_1 \end{bmatrix} = \begin{bmatrix} 1 & -1 & 1 \\ 1 & 0 & -2 \\ 1 & 1 & 1 \end{bmatrix} \begin{bmatrix} \alpha_0 \\ \alpha_1 \\ \alpha_2 \end{bmatrix}
$$

are *orthogonal*; you can check that all pairwise vector products formed with the three columns of the matrix are zero. [The vector product of the first two columns is $(1)(-1) +$

$(1)(0) + (1)(1) = 0$; of the first and third column, $(1)(1) + (1)(-2) + (1)(1) = 0$; and of the second and third column, $(-1)(1) + (0)(-2) + (1)(1) = 0$.] The columns $(-1, 0, 1)$ and $(1, -2, 1)$ are known as the *orthogonal linear* and *quadratic polynomials* for a factor with three levels.

Here, we have discussed the situation when each factor has three levels, and we partition the effect into a linear and a quadratic component. Orthogonal polynomials for continuous factors with 4 and 5 levels are shown in Appendix 7.1. For 4 levels, one parameterizes the 4 means in terms of an overall mean and linear, quadratic and cubic components.

7.4.2 Partitioning Sums of Squares into Interpretable Components

Assume that both factors in the 3^2 factorial design are continuous. We use the orthogonal linear and quadratic polynomials to partition the sum of squares of each factor into a linear and a quadratic component, each with one degree of freedom. This becomes useful for testing whether the relationship between the response and the levels of a continuous factor is linear.

For that, we construct columns for the linear and quadratic components. The column of the linear component, $A(\text{lin})$, is assigned the value -1 when A is at the low level, value 0 when A is at the middle (0) level, and value $+1$ when A is at the high level; these are the coefficients in the linear polynomial. The column of the quadratic component, $A(\text{qua})$, is assigned the coefficients in the quadratic polynomial; the value $+1$ when A is either at the low or high level, and value -2 when A is at the middle level. The same procedure is used to construct $B(\text{lin})$ and $B(\text{qua})$, the linear and quadratic components for factor B. The four columns are listed in Table 7.4. The length of these columns is determined by the number of observations (runs).

The AB interaction sum of squares has four degrees of freedom. Also this sum of squares can be partitioned into four orthogonal components—the linear by linear, the linear by quadratic, the quadratic by linear, and the quadratic by quadratic interaction components.

TABLE 7.4

Regression Formulation of the 3^2 Factorial Design, with Linear and Quadratic Main and Interaction Effects

						REGRESSOR COLUMNS			
DESIGN FACTORS		MAIN EFFECT OF FACTOR A		MAIN EFFECT OF FACTOR B		INTERACTION AB			
A	B	A (lin)	A (qua)	B (lin)	B (qua)	AB (lin × lin)	AB (lin × qua)	AB (qua × lin)	AB (qua × qua)
-1	-1	-1	1	-1	1	1	-1	-1	1
0	-1	0	-2	-1	1	0	0	2	-2
1	-1	1	1	-1	1	-1	1	-1	1
-1	0	-1	1	0	-2	0	2	0	-2
0	0	0	-2	0	-2	0	0	0	4
1	0	1	1	0	-2	0	-2	0	-2
-1	1	-1	1	1	1	-1	-1	1	1
0	1	0	-2	1	1	0	0	-2	-2
1	1	1	1	1	1	1	1	1	1

We create the columns $AB(\text{lin} \times \text{lin})$, $AB(\text{lin} \times \text{qua})$, $AB(\text{qua} \times \text{lin})$ and $AB(\text{qua} \times \text{qua})$. The elements in column $AB(\text{lin} \times \text{lin})$ are the products of the elements in $A(\text{lin})$ and $B(\text{lin})$, the elements in the column $AB(\text{lin} \times \text{qua})$ are the products of the elements in $A(\text{lin})$ and $B(\text{qua})$, and so on. These columns are also shown in Table 7.4.

We consider a regression model that relates the response vector to these orthogonal columns. That is,

$$y = \beta_0 + \beta_1 A(\text{lin}) + \beta_2 A(\text{qua}) + \beta_3 B(\text{lin}) + \beta_4 B(\text{qua}) + \beta_5 AB(\text{lin} \times \text{lin})$$
$$+ \beta_6 AB(\text{lin} \times \text{qua}) + \beta_7 AB(\text{qua} \times \text{lin}) + \beta_8 AB(\text{qua} \times \text{qua}) + \varepsilon$$

Standard regression software can be used to obtain the estimates and the regression sums of squares that are explained by each of these regressor columns. The orthogonality of the regressor columns has important consequences. We pointed out in Appendix 4.5 that each regression estimate and the regression sum of squares of each column are not affected by the presence of other components in the model, and that the individual regression sums of squares are additive.

Exercise 7 in Chapter 4 shows how to calculate the regression sum of squares that is explained by a single column, say, x with entries $x_1, x_2, \ldots, x_n$. The regression sum of squares is given by $SSR(\mathbf{x}) = [\sum x_i y_i]^2 / [\sum x_i^2]$. Given the responses $y_1, y_2, \ldots, y_n$, it is easy to calculate the regression sum of squares for each of the regressor columns in Table 7.4. For example, the column $A(\text{lin})$ is used to calculate $SSR(A(\text{lin}))$, and the column $A(\text{qua})$ is used to calculate $SSR(A(\text{qua}))$. The regression sum of squares of the main effect of factor A, $SS(A)$, which is calculated in Sections 7.2 and 7.3, turns out to be the sum of these two components: $SS(A) = SSR(A(\text{lin})) + SSR(A(\text{qua}))$. This shows how much of the sum of squares of the main effect of A can be attributed to the linear association, and how much to the quadratic one.

The same procedure can be applied to the main effect of B, as well as the interaction between A and B. We find that $SS(B) = SSR(B(\text{lin})) + SSR(B(\text{qua}))$, and $SS(\text{Interaction}) = SSR(AB(\text{lin} \times \text{lin})) + SSR(AB(\text{lin} \times \text{qua})) + SSR(AB(\text{qua} \times \text{lin})) + SSR(AB(\text{qua} \times \text{qua}))$.

The following example makes use of this decomposition.

7.4.3 Example: Sales of Apple Juice

A medium-sized supermarket was selected to study the impact of price and display on the sales of several store-brand products. Products with stable sales (i.e., no trends) and limited seasonality were selected for this study. Here we focus on the sales of the store-brand apple juice.

A complete 3^2 factorial experiment is carried out to assess the effects of price and display. Three price levels are considered: The cost price (low level -1), which is the cost to the supermarket; the regular price (high level $+1$), which is the recommended retail price to customers as listed in the regional warehouse price manual; and the reduced price (level 0), which is the price halfway between the recommended retail price and the cost to the supermarket. The three display choices are normal display space (level 0), as determined at

TABLE 7.5
Sales of Apple Juice for Changing Price and Display

Display	Price	Sales	
−1	−1	40.8	34.2
0	−1	44.2	53.5
1	−1	91.5	70.5
−1	0	32.0	31.4
0	0	50.2	34.9
1	0	85.7	59.3
−1	1	9.0	18.0
0	1	24.9	24.9
1	1	55.9	31.9

the beginning of the experiment on the stock manager's recommendation; a reduced display (level −1), which amounts to one-half of the normal display; and an expanded display (level 1), which amounts to twice the normal display area.

With three display options and three price levels, the design calls for nine treatment combinations. The design is replicated once. Eighteen weeks are needed for this study, and the time arrangement of the experimental conditions is randomized. Furthermore, each experimental week is preceded and followed by a base week (which is a week where the product is priced at its regular price and displayed at the normal shelf position). For this reason, and because of holiday weeks that are not used, the experiment spans roughly 40 consecutive weeks. The response is the number of units (divided by 10) that sold between Wednesday noon and Sunday 9 p.m. of each experimental week. The design and the observations are shown in Table 7.5.

The interaction plot and the two main effects plots are given in Figure 7.2. The interaction plot reveals very little interaction because the lines connecting average sales for different prices but from the same display are almost parallel. The absence of an interaction makes it appropriate to study the main effects. The sales effects of both price and display are (roughly) linear.

The graphical analysis is supported by the ANOVA in Table 7.6. The sums of squares for display, price, interaction, and error can be obtained from the expressions in Table 7.1 and Section 7.2. The calculations are tedious. It is much simpler to obtain the ANOVA through the Minitab "ANOVA > Two-Way" command. For this, one enters the data into three columns of a spreadsheet. Row 1, for example, contains the entries 40.8, −1, −1; the last (18th) row contains 31.9, 1, 1. The result is shown in Table 7.6.

A decomposition into linear and quadratic components is appropriate in this example as the factors are continuous and the distances between the levels are meaningful. The Minitab command "ANOVA > Two-Way" does not provide the decomposition into orthogonal components automatically. For that one needs to construct the spreadsheet shown in Table 7.7. It contains the regressor columns of Table 7.4 and the responses of Table 7.5. The data for week 1 are in rows 1–9, while the data for week 2 are in rows 10–18. Rows 10–18 are identical to rows 1–9, except that the responses are different.

Below, we illustrate the calculation of $SSR(D(\text{qua}))$, the regression sum of squares that is explained by the quadratic component of display. $D(\text{qua})$ is the fourth column in Table 7.7,

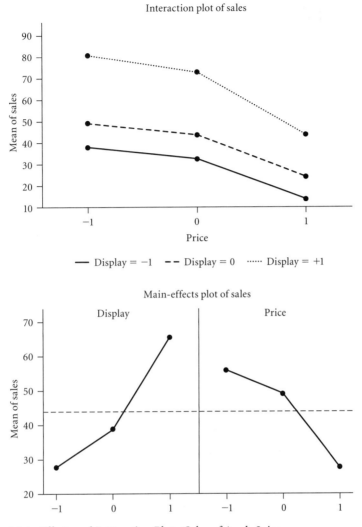

Figure 7.2   Main Effects and Interaction Plots: Sales of Apple Juice

TABLE 7.6
ANOVA Table: Sales of Apple Juice

Source	DF	SS	MS	F	P
Display	**2**	**4636.1**	**2318.0**	**19.32**	**0.001**
D(lin)	**1**	**4385.4**	**4385.4**	**36.55**	**0.000**
D(qua)	1	250.7	250.7	2.09	0.182
Price	**2**	**2624.8**	**1312.4**	**10.94**	**0.004**
P(lin)	**1**	**2411.2**	**2411.2**	**20.10**	**0.002**
P(qua)	1	213.6	213.6	1.78	0.215
Interaction	4	130.1	32.5	0.27	0.889
DP(linxlin)	1	85.8	85.8	0.72	0.420
DP(linxqua)	1	9.9	9.9	0.08	0.781
DP(quaxlin)	1	29.0	29.0	0.24	0.634
DP(quaxqua)	1	5.3	5.3	0.04	0.838
Error	9	1079.7	119.97		
Total	17	8470.7			

NOTE: Significant effects are shown in boldface.

T A B L E 7.7
Regression Formulation: Sales of Apple Juice

DESIGN FACTORS		DISPLAY		PRICE		DISPLAY BY PRICE INTERACTION				
D	P	D (lin)	D (qua)	P (lin)	P (qua)	DP (lin × lin)	DP (lin × qua)	DP (qua × lin)	DP (qua × qua)	Response
−1	−1	−1	1	−1	1	1	−1	−1	1	40.8
0	−1	0	−2	−1	1	0	0	2	−2	44.2
1	−1	1	1	−1	1	−1	1	−1	1	91.5
−1	0	−1	1	0	−2	0	2	0	−2	32.0
0	0	0	−2	0	−2	0	0	0	4	50.2
1	0	1	1	0	−2	0	−2	0	−2	85.7
−1	1	−1	1	1	1	−1	−1	1	1	9.0
0	1	0	−2	1	1	0	0	−2	−2	24.9
1	1	1	1	1	1	1	1	1	1	55.9
−1	−1	−1	1	−1	1	1	−1	−1	1	34.2
0	−1	0	−2	−1	1	0	0	2	−2	53.5
1	−1	1	1	−1	1	−1	1	−1	1	70.5
−1	0	−1	1	0	−2	0	2	0	−2	31.4
0	0	0	−2	0	−2	0	0	0	4	34.9
1	0	1	1	0	−2	0	−2	0	−2	59.3
−1	1	−1	1	1	1	−1	−1	1	1	18.0
0	1	0	−2	1	1	0	0	−2	−2	24.9
1	1	1	1	1	1	1	1	1	1	31.9
Sum of squares		4,385.4	250.7	2,411.2	213.6	85.8	9.9	29.0	5.3	

and its elements are used in the following calculation. We find:

$$\sum x_i^2 = (1)^2 + (-2)^2 + (1)^2 + (1)^2 + (-2)^2 + (1)^2 + (1)^2 + (-2)^2 + (1)^2$$
$$+ (1)^2 + (-2)^2 + (1)^2 + (1)^2 + (-2)^2 + (1)^2 + (1)^2 + (-2)^2 + (1)^2$$
$$= 36$$

$$\sum x_i y_i = 1(40.8) - 2(44.2) + 1(91.5) + 1(32.0) - 2(50.2) + 1(85.7) + 1(9.0)$$
$$- 2(24.9) + 1(55.9) + 1(34.2) - 2(53.5) + 1(70.5) + 1(31.4) - 2(34.9)$$
$$+ 1(59.3) + 1(18.0) - 2(24.9) + 1(31.9)$$
$$= 95$$

$$SSR(D(\text{qua})) = [\,\sum x_i y_i]^2 / [\,\sum x_i^2] = (95)^2/36 = 250.7.$$

This is the number that is reported at the bottom of Table 7.7 and in the ANOVA in Table 7.6. The sums of squares of the other components can be obtained in a similar fashion.

The F-statistic for interaction in Table 7.6, $F = (130.1/4)/(1,079.7/9) = 0.27$, is small and insignificant (probability value 0.889), confirming what we had seen in the interaction plot. Price ($F = 10.94$ with probability value 0.004) and display ($F = 19.32$ with probability value 0.001) are both highly significant. Furthermore, the ANOVA shows insignificant quadratic components for both price ($F = (213.6/1)/(1,079.7/9) = 1.78$ and probability value 0.215) and display ($F = 2.09$ and probability value 0.182). We conclude that the relationships between sales and price and between sales and display are linear.

7.5 A FACTORIAL EXPERIMENT WITH TWO FACTORS
AT TWO LEVELS, AND ONE FACTOR AT THREE LEVELS

Two-level and 3-level designs can be combined. Consider, for example, the experiment where factors A and B have two levels each, while factor C is studied at three levels. We call this a $2^2 3^1$ factorial experiment.

Such a $2^2 3^1$ design was used to improve the consistency of a bottle-filling operation (Montgomery, 2005, p. 184). Process engineers controlled three factors: the operating pressure in the filler (factor A), the number of bottles produced per minute (line speed, factor B), and the percent carbonation (factor C). All three factors are continuous. For the purposes of this experiment, the engineer selected 2 levels for pressure [25 and 30 pounds/ (inch)^2], 2 levels for line speed (200 and 250 bottles/minute), and 3 levels for carbonation (10, 12, and 14%). The response is the (average) deviation of the actual fill height from the targeted fill height. Positive deviations represent fill heights above the target, and negative numbers are fill heights below the target. The average deviation is calculated from all bottles within the same production run. The experiment was replicated once.

The design and the resulting data are given in Table 7.8. The levels of the 12 runs are shown in columns 2–4. The design is orthogonal; you can check that each level combination of any two factors is studied with the same number of runs.

With the data from such an experiment, we can obtain the sum of squares of the main effects of factors A and B (with 1 degree of freedom each), the main effect of factor C (with 2 degrees of freedom), the AB interaction (with 1 degree of freedom), the AC and BC interactions (each with 2 degrees of freedom), the ABC interaction (with 2 degrees of freedom), and—since there are replications—the sum of squares of error. We have not provided the detailed calculation equations for the sum of squares in the 3-factor experiment, as we expect you to use computer software for the computations. The ANOVA table in Table 7.9, without breaking down the 3-level factor C (carbonation) into its linear and quadratic components, can be obtained from the Minitab command "ANOVA > General Linear Model." (Here we have 3 factors, more than the 2 factors allowed in the Minitab command

TABLE 7.8
Bottle-Filling Operation

Run	Pressure (factor A)	Speed (factor B)	Carbonation (factor C)	Response (deviation)	
1	−1	−1	−1	−3	−1
2	1	−1	−1	−1	0
3	−1	1	−1	−1	0
4	1	1	−1	1	1
5	−1	−1	0	0	1
6	1	−1	0	2	3
7	−1	1	0	2	1
8	1	1	0	6	5
9	−1	−1	1	5	4
10	1	−1	1	7	9
11	−1	1	1	7	6
12	1	1	1	10	11

TABLE 7.9
ANOVA Table: Bottle-Filling Operation

Analysis of Variance for Response

Source	DF	SS	MS	F	P
A: Pressure	1	45.375	45.375	64.06	0.000
B: Speed	1	22.042	22.042	31.12	0.000
C: Carbonation	2	252.750	126.375	178.41	0.000
C(lin)	1	248.062	248.062	350.21	0.000
C(qua)	1	4.687	4.687	6.62	0.024
AB	1	1.042	1.042	1.47	0.249
AC	2	5.250	2.625	3.71	0.056
AC(lin)	1	5.063	5.063	7.15	0.020
AC(qua)	1	0.187	0.187	0.26	0.620
BC	2	0.583	0.292	0.41	0.671
BC(lin)	1	0.563			
BC(qua)	1	0.021			
ABC	2	1.083	0.542	0.76	0.487
ABC(lin)	1	0.063			
ABC(qua)	1	1.021			
Error	12	8.500	0.708		
Total	23	336.625			

"ANOVA > Two-Way." The command "ANOVA > General Linear Model" provides the ANOVA for the factorial experiment with more than 2 factors. We must specify three columns that identify the levels of the 3 factors, and request the sum of squares contributions for the main effects of A, B, C, and the interactions AB, AC, BC, and ABC.)

For the additional decomposition into linear and quadratic components, we construct the spreadsheet in Table 7.10. It contains orthogonal polynomials for the main effect and the interactions involving the 3-level factor C. Its effect on the response is expressed through the columns C(lin) and C(qua) that use the coefficients of the linear $(-1, 0, 1)$ and the quadratic $(1, -2, 1)$ polynomials. The interactions that involve the 3-level factor C can be parameterized similarly. The columns that represent the linear and quadratic components of the 2-factor interactions AC and BC are obtained by multiplying the elements in columns A and B with the elements in columns C(lin) and C(qua). The columns representing the linear and quadratic components of the 3-factor interaction are obtained by multiplying column A with BC(lin) and BC(qua). The columns in Table 7.10 are of length 24 as there are 24 observations. The procedure employed in the previous section is used to obtain the regression sums of squares that are associated with each of these columns. The regression sums of squares are listed at the bottom of the table, and they have been added to the ANOVA in Table 7.9.

The 3-factor interaction (ABC) and the three 2-factor interactions (AB, AC, BC) are insignificant. The largest 2-factor interaction is between pressure and carbonation (AC, with $F = 3.71$ and borderline significant probability value $= 0.056$). The lines in the interaction plot in Figure 7.3 that connect averages for different carbonation (C) arising under identical pressure (A) are almost parallel. This is another indication that the AC interaction is negligible. The main effects of all 3 factors are highly significant. Increased pressure, speed, and carbonation increase the average deviation from the target fill height. Since carbonation is studied at 3 levels, we can assess whether the effect is linear or whether a quadratic component is needed. The main-effects plots in Figure 7.3 show that the effect of carbonation is

TABLE 7.10
Regression Formulation of the Mixed $2^2 3^1$ Factorial Experiment, with Linear and Quadratic Main and Interaction Effects of the 3-Level Factor

REGRESSOR COLUMNS

DESIGN FACTORS					MAIN OF C			INTERACTION AC		INTERACTION BC		INTERACTION ABC		
A	B	C	A	B	C(lin)	C(qua)	AB	AC(lin)	AC(qua)	BC(lin)	BC(qua)	ABC(lin)	ABC(qua)	Response
−1	−1	−1	−1	−1	−1	1	1	1	−1	1	−1	−1	1	−3
1	−1	−1	1	−1	−1	1	−1	−1	1	1	−1	1	−1	−1
−1	1	−1	−1	1	−1	1	−1	1	−1	−1	1	1	−1	−1
1	1	−1	1	1	−1	1	1	−1	1	−1	1	−1	1	1
−1	−1	0	−1	−1	0	−2	1	0	2	0	2	0	−2	0
1	−1	0	1	−1	0	−2	−1	0	−2	0	2	0	2	2
−1	1	0	−1	1	0	−2	−1	0	2	0	−2	0	2	2
1	1	0	1	1	0	−2	1	0	−2	0	−2	0	−2	6
−1	−1	1	−1	−1	1	1	1	−1	−1	−1	−1	1	1	5
1	−1	1	1	−1	1	1	−1	1	1	−1	−1	−1	−1	7
−1	1	1	−1	1	1	1	−1	−1	−1	1	1	−1	−1	7
1	1	1	1	1	1	1	1	1	1	1	1	1	1	10
−1	−1	−1	−1	−1	−1	1	1	1	−1	1	−1	−1	1	−1
1	−1	−1	1	−1	−1	1	−1	−1	1	1	−1	1	−1	0
−1	1	−1	−1	1	−1	1	−1	1	−1	−1	1	1	−1	0
1	1	−1	1	1	−1	1	1	−1	1	−1	1	−1	1	1
−1	−1	0	−1	−1	0	−2	1	0	2	0	2	0	−2	1
1	−1	0	1	−1	0	−2	−1	0	−2	0	2	0	2	3
−1	1	0	−1	1	0	−2	−1	0	2	0	−2	0	2	1
1	1	0	1	1	0	−2	1	0	−2	0	−2	0	−2	5
−1	−1	1	−1	−1	1	1	1	−1	−1	−1	−1	1	1	4
1	−1	1	1	−1	1	1	−1	1	1	−1	−1	−1	−1	9
−1	1	1	−1	1	1	1	−1	−1	−1	1	1	−1	−1	6
1	1	1	1	1	1	1	1	1	1	1	1	1	1	11
Sum of squares			45.375	22.042	248.062	4.687	1.042	5.063	0.187	0.563	0.021	0.063	1.021	

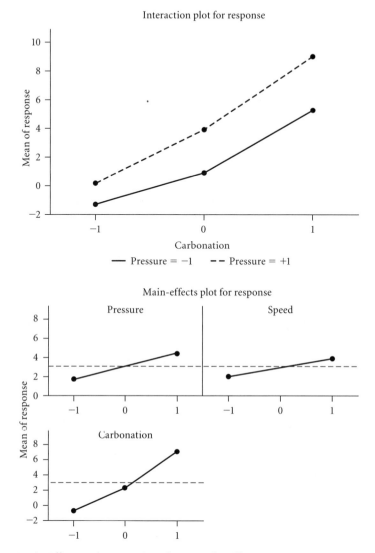

Figure 7.3 Main Effects and Interaction Plots: Bottle-Filling Operation

roughly linear; this assessment is confirmed by the very large and highly significant linear component of factor C ($F = 350$, with probability value = 0.000). The quadratic component of C is much smaller and not nearly as significant ($F = 6.62$ and probability value = 0.024).

7.6 THREE-LEVEL FRACTIONAL FACTORIAL DESIGNS

The number of runs in 3-level factorial experiments grows quite rapidly with the number of factors k. Even with only 3 factors, 27 runs are required. Orthogonal fractions of 3^k factorial experiments can be constructed. This reduces the number of runs but, depending on the particular fraction, confounds certain main and interaction effects. We describe a few simple 3^{k-p} fractional factorial designs in this section. In particular, we discuss the 3^{3-1} and 3^{4-2} fractional factorials for studying 3 and 4 factors in 9 runs.

TABLE 7.11
The Graeco-Latin Square Design with Four Factors
at Three Levels Each

		FACTOR B		
		Level -1	Level 0	Level 1
Factor A	Level -1	$a\,\alpha$	$b\,\beta$	$c\,\gamma$
	Level 0	$b\,\gamma$	$c\,\alpha$	$a\,\beta$
	Level 1	$c\,\beta$	$a\,\gamma$	$b\,\alpha$

TABLE 7.12
The 3^{4-2} and 3^{3-1} Fractional Factorial Designs

THE 3^{4-2} DESIGN				THE 3^{3-1} DESIGN		
A	B	C	D	A	B	C
-1	-1	-1	-1	-1	-1	-1
0	-1	0	1	0	-1	0
1	-1	1	0	1	-1	1
-1	0	0	0	-1	0	0
0	0	1	-1	0	0	1
1	0	-1	1	1	0	-1
-1	1	1	1	-1	1	1
0	1	-1	0	0	1	-1
1	1	0	-1	1	1	0

A 3^{4-2} fractional factorial in 9 runs can be constructed from the Graeco-Latin square design in Table 7.11. The design involves 4 factors (A, B, C, and D), at 3 levels each. The rows and columns in this table represent the coded factor levels ($-1, 0, +1$) for factors A and B. Factor levels for factor C are given by the Latin letters a, b, and c, where a represents level -1, b represents level 0, and c represents level 1. The factor levels for factor D are given by the Greek letters α, β, γ, where α corresponds to level -1, β corresponds to level 0, and γ corresponds to level 1. Graeco-Latin square designs have the property that each letter (Latin, as well as Greek) appears exactly once in each row and each column. Also, each Latin letter appears with each Greek letter exactly once.

The 9 runs of the 3^{4-2} fractional factorial design are listed in Table 7.12. The first two columns list the runs in a full 3^2 factorial experiment. The levels for factors C and D result from the Graeco-Latin letters in Table 7.11. For example, the first run with letter combination $a\,\alpha$ is described by ($A = -1, B = -1, C = -1, D = -1$); the next run with letter combination $b\,\gamma$ is ($A = 0, B = -1, C = 0, D = +1$); . . . ; the last run with letter combination $b\,\alpha$ is ($A = +1, B = +1, C = 0, D = -1$). The 3^{3-1} fractional factorial design in Table 7.12 is obtained by omitting the factor D. Both designs are orthogonal. You can check that each level-combination of any two factors is studied with the same number of runs (namely, one).

The 9 runs of the 3^{4-2} design in Table 7.12 allow us to estimate the sums of squares of all four main effects. Each sum of squares has 2 degrees of freedom, because comparisons of 3 levels are involved. Since the design is orthogonal, one can obtain the sum of squares of a factor by averaging over the other 3 factors. (Minitab's "Stat > ANOVA > General Linear Model" provides the ANOVA table. Replications are needed to obtain an error sum of squares and test the significance of the four main effects.) Of course, the main effects are

TABLE 7.13

A 3^{4-1} Fractional Factorial Designs in 27 Runs

	FACTOR		
A	B	C	D
−1	−1	−1	−1
0	−1	−1	0
1	−1	−1	1
−1	0	−1	0
0	0	−1	1
1	0	−1	−1
−1	1	−1	1
0	1	−1	−1
1	1	−1	0
−1	−1	0	0
0	−1	0	1
1	−1	0	−1
−1	0	0	1
0	0	0	−1
1	0	0	0
−1	1	0	−1
0	1	0	0
1	1	0	1
−1	−1	1	1
0	−1	1	−1
1	−1	1	0
−1	0	1	−1
0	0	1	0
1	0	1	1
−1	1	1	0
0	1	1	1
1	1	1	−1

confounded with 2-factor interactions, which limits the usefulness of such designs to initial screening experiments, where the aim is to identify important factors for further study. A 3^{4-1} fractional factorial design in 27 runs is needed if the experimenter wants a 3-level design that does not confound the four main effects with 2-factor interactions. This resolution IV design is shown in Table 7.13.

Similar to 2^{k-p} fractional factorial designs, we can construct 3^{k-p} designs by first writing down a full 3-level factorial in $(k - p)$ factors and generating the remaining p columns of the design matrix from specified generators. The generators imply a defining relationship, and from the defining relationship, we can obtain the confounding structure. However, the steps are more complicated in 3-level designs, because they involve modulus-3 arithmetic. Modular arithmetic is a system of arithmetic for integers, where numbers "wrap around" after they reach a certain value—the modulus. Two integers a, b are said to be the same modulo 3 if their difference is divisible by 3. In this case, we write $a = b$ (mod 3). Consider the sum S of the (coded) levels of the 3 factors A, B, and C in the full 3^3 factorial design in Table 7.13. The sum can take any one of eight possible values, $S = -3$ (if all three factors are at −1), −2, −1, 0, 1, 2, 3 (if all three factors are at +1). We have selected the level of the fourth factor D as $D = -1$ if $S = 0$ (mod 3); that is, if S is −3, 0, or 3. We have selected $D = 0$ if $S = 1$ (mod 3); that is, if S is −2 or 1. And we have selected $D = +1$ if $S = 2$ (mod 3); that is, S is −1 or 2. These are the levels shown in Table 7.13.

7.7 NOBODY ASKED US, BUT . . .

Designs that study continuous factors at more than two levels are useful if one wishes to explore the functional relationship between the response and the factors. Such designs allow us to fit quadratic models, which will tell us the levels of the factors that maximize (or minimize) the response. The literature refers to this area as *response surface analysis*.

The 3-level factorial and fractional factorial designs are useful, but other designs, such as central composite and simplex designs, have been employed. The central composite design in 2 factors, for example, adds to the four runs of the 2^2 factorial design [i.e., $(-1, -1)$, $(+1, -1)$, $(-1, +1)$, $(+1, +1)$], a center point $(0, 0)$ and four "star" points [i.e., $(-w, 0)$, $(w, 0)$, $(0, -w)$, $(0, w)$]. This design studies each factor at 5 different levels, namely $-w$, -1, 0, 1, and w, with w specified by the experimenter. These designs are included in most experimental design software programs such as Minitab and JMP. For more on these designs, see Box, Hunter, and Hunter (2005).

TABLE OF ORTHOGONAL POLYNOMIALS

For factors with two levels:

	Linear
Level 1	-1
Level 2	1

For factors with three levels:

	Linear	Quadratic
Level 1	-1	1
Level 2	0	-2
Level 3	1	1

For factors with four levels:

	Linear	Quadratic	Cubic
Level 1	-3	1	-1
Level 2	-1	-1	3
Level 3	1	-1	-3
Level 4	3	1	1

For factors with five levels:

	Linear	Quadratic	Cubic	Quartic
Level 1	-2	2	-1	1
Level 2	-1	-1	2	-4
Level 3	0	-2	0	6
Level 4	1	-1	-2	-4
Level 5	2	2	1	1

EXERCISES

Exercise 1 Consider the data in Section 7.3.

(a) Use an available computer program to obtain the ANOVA table in Table 7.3. Obtain the interaction plot in Figure 7.1. Obtain the main effects plots of factor A and factor B, and comment on whether or not these plots are useful.

(b) For this rather small data set, calculate the nine cell averages, the three averages for factor A and the three averages for factor B. Use the expressions in Table 7.1 to calculate the sums of squares and convince yourself that the results coincide with the ones given in Table 7.3.

(c) Discuss in detail your experimental procedure. How would you carry out the baking experiment if you had to use your home oven, and the rating procedure if you had access to a set of friends? Does your procedure satisfy the assumptions in Section 7.2 that justify the analysis in Section 7.3? Can you expect that the observations are independent and of equal precision?

Exercise 2   Consider Case 10 (Piggly Wiggly) from the case study appendix.

(a) For each of the two products, complete the ANOVA table. In particular:

   (a1) Specify the degrees of freedom and calculate the mean squares and the appropriate F-statistics.

   (a2) Calculate an estimate of the standard deviation of the error.

   (a3) Assess whether one needs to consider the 3-factor interaction.

   (a4) Assess whether one needs to consider any of the 2-factor interactions.

(b) Give graphical interpretations of the results. If you find significant 2-factor interactions, show the relevant interaction diagrams. Display the main effects of factors that do not interact. Summarize your conclusions. How do price, display, and advertising affect sales?

   Hints: For White House apple juice, only main effects are relevant. For Mahatma rice, you will also notice a Price $\times$ Display interaction. For White House apple juice, it makes sense to calculate average responses for the levels of price (and for the levels of display, and advertising). You can obtain these averages from the given cell averages (each average comes from an equal number of observations; here two). Use the estimate of the standard deviation in (a2) to obtain an estimate of the standard error of these averages. Display the averages on a dot diagram and superimpose on these graphs the distribution of these averages. The distribution of the averages is sometimes referred to as a *sliding reference distribution*. One can imagine sliding the distribution along the x-axis trying to "cover" the averages. The main effect of a factor is insignificant if all averages can be covered by the reference distribution.

   A similar graph can be made for Advertising in the Mahatma rice data as Advertising does not appear to interact with either Price or Display. Also, one can collapse the table of averages and obtain cell averages for Price and Display. These averages can be represented in an interaction diagram. Error bands can be calculated after obtaining the standard error of the averages (which, similar to the earlier discussion, can be calculated from the estimate in (a2); one just needs to keep track of the number of observations that go into each average).

(c) Discuss the following issues:

   (c1) Any weaknesses of the study. Do you think that the experimenters have done a good job?

   (c2) How seasonality (if present) would affect the results.

   (c3) How one could guard against seasonality at the design stage.
         Hint: Randomization of assignment; but even better, blocking for season.

   (c4) How one could check for seasonality at the analysis stage.

Hint: Construct indicator variables to denote season (quarter) and incorporate the indicator variables into the ANOVA/regression analysis.

(c5) Keeping track of the data: Should experimenters have kept track of the number of weekly customers, and should they have analyzed unit sales per customer? Why or why not?

(d) The experiment used just one store. How would you proceed if you had more stores available? What about if the stores were of different sizes?

(e) Discuss the practical difficulties of carrying out such an experiment. Did this study do a reasonable job?

Exercise 3   Consider Case 11 (United Dairy Industries) from the case study appendix.

(a) Consider the Latin square design for the test markets in Part 1. Show that the Latin square is orthogonal (i.e., same number of runs at each level-combination of any two factors). Orthogonality implies that one can ignore time and location when obtaining the effects of advertising. Ignoring these two factors, the observations for the four advertising groups are as follows.

```
0 cents (A): 7360 13153 11852 7557 Ave =  9981
3 cents (B): 7364 11258 12089 7900 Ave =  9653
6 cents (C): 8049 13880 11800 8501 Ave = 10558
9 cents (D): 9010 13147 11450 7776 Ave = 10346
```

Use the ANOVA calculations for the completely randomized one-factor experiment in Section 3.2, and calculate the sum of squares due to Advertising. Check that it is identical to the one listed in the ANOVA table in Part 1 of the case study.

Repeat this for the factor City. Ignoring advertising and time, obtain the ANOVA for the completely randomized one-factor experiment (with factor City). Show that the sum of squares due to City is the same as the one listed in the ANOVA table in Part 1. Discuss.

(b) Compare the ANOVA results in Part 1 of the case study and the results in Table 3.10 of Chapter 3. Discuss the significance of advertising. Why are the two tests different? Which test is more relevant?

(c) The analysis in Part 2 is the same as the analysis in the randomized complete block experiment in Section 3.3. Repeat the analysis using your statistical software of choice.

Exercise 4   Orthogonal fractions of 3-level factorial designs are discussed in Section 7.6. We used Graeco-Latin square arrangements to construct these designs. This strategy can be extended to factors with more than 3 levels.

Consider factors with 5 levels. Construct orthogonal fractional factorial designs that allow you to study the main effects of four 5-level factors in just 25 runs.

(a) Write down a table of 5 rows and 5 columns. Add Latin letters to the 25 cells. The first row consists of letters a, b, c, d, and e. Rearrange the letters cyclically, similar to the strategy we used to construct the runs in Plackett-Burman designs (Chapter 6). Shift the row of letters to the left, and move the letter in the far-left position of a row to the far-right position of the subsequent one. Similarly, for the Greek letters $\alpha, \beta, \gamma, \delta, \varepsilon$. But, now shift the letters to the right, and move the letter in the far-right position of a row to the far-left position of the subsequent one. Write down the 25 runs that you obtain from this arrangement. Check that the numbers of runs at each factor level combination of any 2 factors are the same, making this an orthogonal design.

(b) From this orthogonal design, you can obtain the sums of squares of 4 factors. How many degrees of freedom are associated with each sum of squares? How do you obtain the sum of squares of error, and how many degrees of freedom are associated with it? Discuss how you would determine the significance of a main effect.

(c) Assume that 2-factor interactions are important. Would this affect the main effects analysis?

Exercise 5 Peter W.M. John (1990) described an 18-run experiment that involves 5 controllable factors, each studied at three levels.

A	B	C	D	E	Average
1	1	1	1	1	39.08
1	2	2	2	2	41.82
1	3	3	3	3	39.77
2	1	2	3	2	42.15
2	2	3	1	3	46.82
2	3	1	2	1	43.05
3	1	3	2	2	46.28
3	2	1	3	3	46.80
3	3	2	1	1	45.67
1	1	3	3	1	39.30
1	2	1	1	2	42.65
1	3	2	2	3	41.37
2	1	1	2	3	39.91
2	2	2	3	1	45.21
2	3	3	1	2	45.51
3	1	2	1	3	43.47
3	2	3	2	1	46.07
3	3	1	3	2	46.67

(a) Show that this is an orthogonal design.

(b) Consider a main effects model and determine the sums of squares of the five main effects (each with 2 degrees of freedom). Obtain the main effects plots. Assume that you want to maximize the response. Determine the best values for each of the five factors. Statistical software such as Minitab's "Stat > ANOVA > General Linear Model" can be used for the calculation of the ANOVA table and the plots of the main effects.

8 NONORTHOGONAL DESIGNS AND COMPUTER SOFTWARE FOR DESIGN CONSTRUCTION AND DATA ANALYSIS

8.1 INTRODUCTION

The designs that we have considered up to this point are orthogonal. The 2-level factorial, fractional factorial, and Plackett-Burman designs in Chapters 4 through 6 are orthogonal; they share the property that each factor-level combination of any 2 factors is studied with the same number of runs. Also, our analysis of the general 2-factor factorial experiment in Section 7.2 assumes that the numbers of runs at the ab factor-level combinations are the same, making it an orthogonal design. The same is true for the 3^k and 3^{k-p} factorial designs in Chapter 7.

Orthogonality of the design simplifies the analysis of the resulting data considerably. Main effects and interactions can be estimated by averaging over all other factors. The estimates are independent, and the sum of squares that is explained jointly by the studied factors can be partitioned into individual, unconditioned sums of squares that ignore all other factors. For example, the total sum of squares in the 2-factor factorial experiment in Section 7.2 can be partitioned into the individual, unconditioned sums of squares for factor A, factor B, the interaction, and the error component.

Such an additive unconditioned decomposition is no longer possible if the design is not orthogonal. Orthogonality, for example, is no longer present if observations are missing from an orthogonal design. More importantly, an orthogonal design may simply not be available in situations that involve many factors with different numbers of factor levels. Consider, for example, 7 factors with 2 factors at 2 levels, 1 factor at 3 levels, 3 factors at 4 levels, and 1 factor at 5 levels. A full factorial with $(2)(2)(3)(4)(4)(4)(5) = 3,840$ runs is certainly orthogonal. Also, a few special orthogonal fractions in fewer than 3,840 runs are possible, but the number of runs of these orthogonal fractions is still quite large. It is simply not possible to find an orthogonal fraction with a moderate number of runs. Other design criteria need to be adopted if one wants to select a good design that is able to study the main effects of these 7 factors in, say, $N = 30$ runs. In this chapter, we discuss useful guiding principles for constructing such nonorthogonal designs. Design concepts such as D- and A-optimality become useful, and we discuss them in Appendix 8.1.

Section 8.2 discusses an interesting case study that involves a nonorthogonal design with many factors and different numbers of factor levels. Section 8.3 talks about useful computer software for design construction and the analysis of the resulting data.

8.2 THE PHONEHOG CASE

We would like to thank Mark Wachen, CEO of Optimost, for providing the data and for sharing his modeling insights with us. Optimost (www.optimost.com) is a technology and services company based in New York that specializes in comprehensive real-time testing and conversion-rate marketing. We would also like to thank Phil Nadel, CEO of Gulfstream Internet (the parent company of PhoneHog), for allowing us to use this case.

PhoneHog.com®, owned and operated by Gulfstream Internet (http://www.phonehog .com), is a subscription-based service through which consumers get free long-distance phone calls. Participants sign up for the program, then earn phone minutes by visiting Internet sites, entering sweepstakes, or trying new products and services. As of November 2005, the system had more than 3 million members.

Since PhoneHog is a subscription-based service, it is important that its Web site can attract customers to sign up for its program. PhoneHog needs to learn which advertising copies, offers, and images increase the sign-up rate of Internet surfers who come in contact with its Web site. Experiments are run continually to determine modifications to the current standard (baseline) strategy with the hope of improving the sign-up rate. This case focuses on 10 different areas of the PhoneHog Web site displayed in Figure 8.1: A-top and A-bottom image of the headline, B (subheadline), C (main copy), D (form), E (privacy copy), F (submit button), G (how it works section), H (main image on right-hand side), and I (footer). The choices for each area are described in Figure 8.2. For example, there are four different choices for the top image of the headline (A-top): the baseline showing pictures of five people making calls and the word *PhoneHog* written next to them, and three experimental versions (picture of a hog's head peaking through the "O" on white background; same picture on blue background; pictures of five people calling with PhoneHog's logo to the right). There are 10 choices for the bottom area of the headline (A-bottom): the baseline "Let our advertisers pay for your long distance calls," and nine experimental versions. Area H describes the main image on the right-hand side of the Web page; in addition to the baseline picture showing a rotating flash image of a woman on the telephone, three experimental pictures are considered: a hog standing in a telephone booth extending a phone, no image at all, and a picture showing a brief summary of the highlights (yes/no) of the program.

8.2.1 Design and Resulting Data

PhoneHog experiments with its Web page continually, and many studies, called "waves," have been carried out in the past. In the specific experiment that is discussed here, several of the test levels in Figure 8.2 were not studied. For example, only the baseline 1 and the five test versions 4, 6, 8, 9 and 10 were considered for the bottom area of the main headline (A-bottom). The active levels for the other factors are indicated in the last column of Table 8.1.

Figure 8.1 PhoneHog's Web Page: Test Area Diagram

A factorial experiment that considers all level combinations of these 10 factors requires $(6)(4)(3)(4)(6)(4)(6)(5)(4)(2) = 1,658,880$ runs. Of course, it is impossible to study all combinations. Only a small fraction of the factorial experiment can be considered. An experiment with just 45 runs was carried out. The runs (which are referred to as "creatives") are listed in Table 8.2. Creative 45 uses the baseline level for each of the 10 areas; the other 44 runs are test versions where one or more baselines are changed to test versions.

The 44 test runs were offered to Internet users randomly and with equal probabilities, while the probability of the baseline (creative 45) was four times as large. The 45 different creatives were made available over a 2-week period. During this period, PhoneHog recorded the number of distinct visitors to the PhoneHog Web site (VISITS) and the number of times visitors clicked the subsequent page to obtain additional information (CLICKS). The click-through rate, CTR = CLICKS/VISITS, measures the success of a run. The results are shown in Table 8.2.

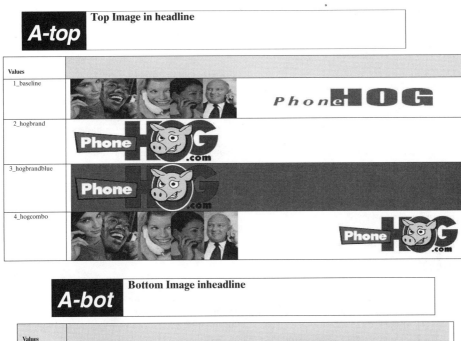

Figure 8.2   Baseline and Test Versions for the 10 Test Areas on PhoneHog's Web Page

In Chapter 5 we learned how to find good orthogonal fractions of 2-level factorial designs. However, the situation is different here as factors with many different levels are involved. This makes it difficult to find orthogonal fractions that have a reasonable number of runs. Other design criteria, different from orthogonality, must be used to determine the fractions.

The experiment involves 10 categorical factors, with varying numbers of factor levels. The numbers of factor levels are given in the last column of Table 8.1. Considering main

B	Subheadline

Values	
1_baseline	**Earn free long distance calls by visiting our advertisers' sites, entering sweepstakes and trying new products and services.**
2_test1	It's easy to earn free long distance calls by visiting our advertisers' sites
3_test2	**Earning free long distance calls with PhoneHog is easy. Just click, register or try a new product. Then start calling for free!**
4_test3	**Get free long distance calls by visiting our advertisers' sites, entering sweepstakes and trying new products and services.**
5_test4	**Earn free long distance calls by visiting our advertisers' sites, entering sweepstakes and trying new products and services**
6_blank	

C	Main Copy

VALUES	
1_baseline	Please complete this brief form so that we can email your free calling card to you and provide you with free membership in PhoneHog.com. As a member of PhoneHog.com, you will receive many exciting opportunities to earn free long distance calls on your new calling card. Thank you.
2_test1	Please take a minute to join now.
3_blank	
4_test3	Complete the brief form below to join PhoneHog today for free. We'll instantly email you a free long distance calling card. As a member of PhoneHog, you will receive many exciting opportunities to earn free long distance calls on your new calling card.
5_test4	Please complete this brief form so that we can email a free long distance calling card to you and provide you with free membership in PhoneHog.com. As a member of PhoneHog.com, you will receive many exciting opportunities to earn free long distance calls on your new calling card.
6_test5	Start earning free long distance today. It takes less than 30 seconds.

Figure 8.2 Continued

effects alone, we need to estimate $1 + (6 - 1) + (4 - 1) + (3 - 1) + (4 - 1) + (6 - 1) + (4 - 1) + (6 - 1) + (5 - 1) + (4 - 1) + (2 - 1) = 35$ effects. Considering creative # 45 as the baseline, the regression formulation of the main-effects model includes a constant, 5 indicator variables for A-bottom ($ab4 = 1$ if test version 4 is used and 0 otherwise, $ab6 = 1$ for version 6, $ab8 = 1$ for version 8, $ab9 = 1$ for version 9, and $ab10 = 1$ for version 10), 3 indicators for A-top ($at2 = 1$ for version 2, $at3 = 1$ for version 3, $at4 = 1$ for version 4), and so on. A minimum of 35 runs is needed to estimate the 35 coefficients. Of course, more runs are required to get better estimates of the main effects, but the cost of too many additional runs may be prohibitive.

The experiment in Table 8.2 contains 45 runs: a control run (creative # 45) and 44 test runs. We want to select the levels of the test runs such that the runs "cover" the region of interest and allow us to estimate the main effects with maximum precision. D-optimality and A-optimality are two design criteria that select the levels of the design matrix with exactly this objective in mind. We discuss these two criteria, the connections between them,

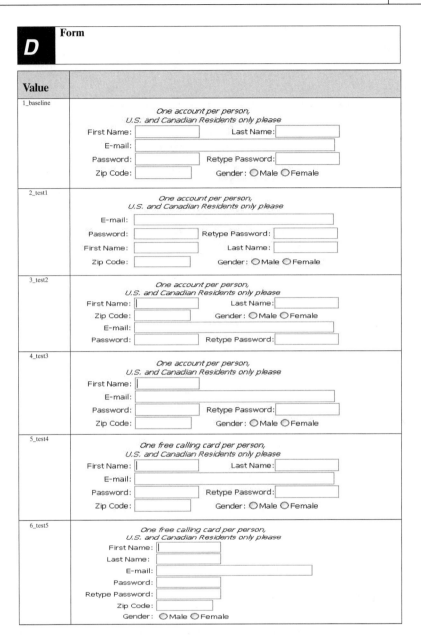

Figure 8.2 Continued

and computer programs to construct such designs in Appendix 8.1. JMP, through its custom design feature, is able to generate a D-optimal design for a given number of runs (see Section 8.3).

Observe that the resulting design in Table 8.2 is no longer orthogonal, and that this fact has consequences for the subsequent data analysis. Also, note that the main-effects model ignores all interactions. The main-effects interpretation could be seriously wrong if interactions are present.

E	**Privacy Copy**

Values	
1_baseline	By clicking below I agree that I have read and accept the Membership & Privacy Agreement. I understand I will receive emails from PhoneHog with opportunities for more free minutes.
2_smallfont	By clicking below I agree that I have read and accept the Membership & Privacy Agreement. I understand I will receive emails from PhoneHog with opportunities for more free minutes.
3_copychange	Send me a free calling card! By clicking below, I agree that I have read and accept the Membership & Privacy Agreement. I understand I will receive emails from PhoneHog with opportunities for free long distance calls.
4_copybold	**Send me a free calling card!** I agree to the Membership & Privacy Agreement. I understand I will receive emails from PhoneHog with opportunities for more free minutes.

F	**Submit Button**

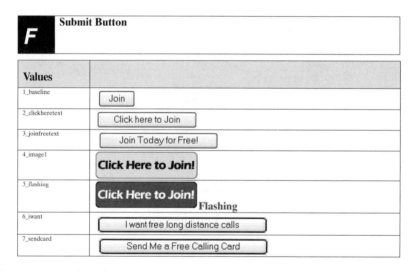

Values	
1_baseline	Join
2_clickheretext	Click here to Join
3_joinfreetext	Join Today for Free!
4_image1	**Click Here to Join!**
5_flashing	**Click Here to Join!** Flashing
6_iwant	I want free long distance calls
7_sendcard	Send Me a Free Calling Card

Figure 8.2 Continued

8.2.2 Analysis of the Data

We use Minitab's "Stat > ANOVA > General Linear Model" command to estimate a *main-effects model* for the click-through rate. Other statistical software programs (such as JMP) work in a similar fashion. In this particular example the data set consists of 11 columns: the column of the response (click-through rate), and 10 columns that contain the categories (levels) for each of the 10 factors (A-bottom, A-top, ..., I). For example, the column for A-bottom contains the six categories (levels) 1, 4, 6, 8, 9 and 10, just like the one in Table 8.2. Minitab produces the analysis of variance table in Table 8.3.

The analysis of variance table looks slightly different from the tables in Chapters 3 and 7, as it shows two different sums of squares. The *sequential sums of squares* (Seq SS) measure the explanatory contribution of each factor as factors are added sequentially to the model. For example, A-bottom explains 8.8764 of the total variation in the click-through rate ($SST = 101.2383$) when it is the only factor in the model. The model with A-bottom and

G	**How it works section**	

VALUES	
1_baseline	**How it works:** Joining PhoneHog is fast, easy and FREE. Once you're a member, we'll instantly email your free PhoneHog Calling Card to you. Make your free calls from any phone, anytime, anywhere.
2_test1_MovetoBC	**How it works:** Joining PhoneHog is fast, easy and FREE. Once you're a member, we'll instantly email your free PhoneHog Calling Card to you. Make your free calls from any phone, anytime, anywhere. **Now between variables B and C**
3_test2_imageversion	**How it works:** Joining PhoneHog is fast, easy and FREE. Once you're a member, we'll instantly email your free PhoneHog Calling Card to you. **Make your free calls from any phone, anytime, anywhere.**
4_test3_longcopy	**How it works:** Joining PhoneHog is fast, easy and FREE. Once you're a member, we'll instantly email your free PhoneHog Calling Card to you. Make your free calls from any phone, anytime, anywhere in the world. Never pay a long distance bill again!
5_movetoBCimage	**How it works:** Joining PhoneHog is fast, easy and FREE. Once you're a member, we'll instantly email your free PhoneHog Calling Card to you. **Make your free calls from any phone, anytime, anywhere.** **Now between variables B and C**

I	**Footer**	

VALUES	
1_baseline	No footer links
2_moreinfo	 **How it works** **Member Login** **FAQ** **Contact**

Figure 8.2 Continued

A-top explains 9.7248 of the variation, implying that the additional contribution of A-top is 0.8484. Or, to say this differently, A-top explains 0.8484 of the variation when it is added to the model with A-bottom. The factor B explains an additional 10.9232 when factor B is added to the model with A-bottom and A-top, and so on. Sequential sums of squares always add up to the regression sum of squares that is explained by the largest model with all factors; the regression sum of squares of the model that includes all factors is $SST - SS(\text{error}) = 101.2383 - 7.4203 = 93.8180$.

For orthogonal designs, the sum of squares of a factor does not change if other factors are present in the model. Hence, we can assess the importance of a factor by comparing its regression contribution (which is unconditional of other factors) to the total variability. In the nonorthogonal situation, this is no longer possible, as the contribution of a factor changes with the factors that are already present in the model.

Main Image on Right Side

1_baseline **Rotating Flash Image**	2_hog	3_center	4_yes/no

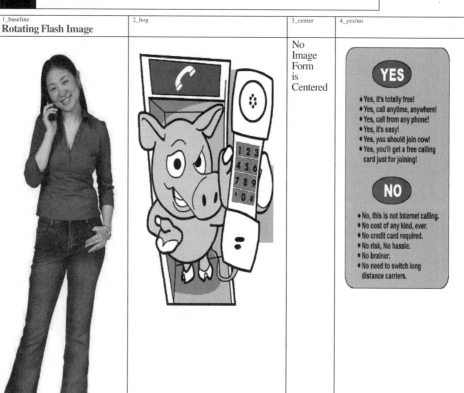

No Image Form is Centered

Figure 8.2 Continued

TABLE 8.1
Test Areas (the Factors) on PhoneHog's Web Page and Their Levels

Area	Number of Levels	Number of Levels and Levels (in parentheses) Used in the Experiment
A-bottom (main headline)	10	6 (1, 4, 6, 8, 9, 10)
A-top (main headline)	4	4 (1–4)
B (sub-headline)	6	3 (1, 3, 6)
C (main copy)	6	4 (1, 2, 5, 6)
D (form)	6	6 (1–6)
E (privacy copy)	4	4 (1–4)
F (submit button)	7	6 (1, 2, 4, 5, 6,7)
G (how it works section)	5	5 (1–5)
H (main image on right side)	4	4 (1–4)
I (footer)	2	2 (1, 2)

NOTE: Level 1 corresponds to the currently used baseline.

TABLE 8.2

Description of the 45 Creatives and the Resulting VISITS, CLICKS, and Click-Through Rate,
CTR=CLICKS/VISITS

			FACTOR										
Creatives (runs)	A-bottom	A-top	B	C	D	E	F	G	H	I	Visitors	Clicks	CTR (%)
1	6	3	3	2	2	2	4	3	2	1	2,177	405	18.60
2	10	3	6	1	1	4	2	4	4	2	2,150	420	19.53
3	10	4	1	2	4	1	5	3	4	1	1,988	376	18.91
4	10	3	6	6	6	2	1	1	2	1	2,163	412	19.04
5	6	2	3	6	1	3	7	1	4	1	2,239	391	17.46
6	6	4	6	1	6	2	2	2	1	1	2,204	404	18.33
7	9	2	3	1	6	4	1	3	2	2	2,119	410	19.34
8	9	1	6	2	1	2	5	5	1	2	2,124	434	20.43
9	6	4	1	5	4	4	7	4	2	1	2,088	360	17.24
10	9	2	1	1	4	2	1	2	4	2	2,208	452	20.47
11	6	1	6	1	4	1	4	4	3	2	2,131	413	19.38
12	4	2	1	5	5	2	4	2	4	2	2,262	479	21.17
13	8	3	1	2	6	4	7	1	3	2	2,151	420	19.52
14	10	4	1	1	1	3	4	3	2	1	2,242	393	17.52
15	10	2	6	5	3	1	2	5	3	1	2,134	391	18.32
16	6	3	1	6	3	3	5	2	3	2	2,101	356	16.94
17	9	2	1	2	2	3	2	4	1	1	2,089	353	16.89
18	10	1	1	2	6	3	6	5	4	1	2,144	385	17.95
19	8	2	6	6	2	1	6	3	3	1	2,090	360	17.22
20	4	4	3	6	6	1	5	4	4	2	2,068	413	19.97
21	6	4	6	2	5	4	1	5	1	2	2,054	379	18.45
22	10	1	6	2	3	3	7	2	2	2	2,202	453	20.57
23	4	3	6	1	2	3	1	5	4	1	2,087	374	17.92
24	10	1	3	5	3	2	1	4	3	1	2,087	393	18.83
25	1	3	1	6	5	2	6	4	1	2	2,123	433	20.39
26	8	2	6	5	5	3	5	4	2	1	2,077	347	16.70
27	9	4	6	5	3	4	6	1	4	1	2,059	339	16.46
28	9	3	3	6	5	1	2	2	2	1	2,211	413	18.67
29	1	2	3	2	3	1	4	1	1	2	2,123	426	20.06
30	1	2	1	6	4	4	2	5	2	2	2,162	406	18.77
31	4	1	1	6	3	4	2	3	1	1	1,649	289	17.52
32	8	1	3	5	5	2	2	3	4	2	2,257	513	22.72
33	4	1	1	5	2	1	6	1	2	2	2,123	384	18.08
34	4	3	3	1	4	1	7	5	1	1	2,188	389	17.77
35	10	1	3	5	2	4	5	2	1	2	2,244	406	18.09
36	8	4	1	5	1	1	1	2	1	2	2,202	434	19.70
37	1	3	6	5	4	3	6	3	1	2	2,241	448	19.99
38	4	4	3	2	4	3	2	1	3	2	2,185	454	20.77
39	4	2	3	2	1	4	6	2	3	1	2,166	479	22.11
40	1	4	6	6	2	2	7	3	3	2	2,194	451	20.55
41	1	1	1	1	5	4	5	1	3	1	2,094	364	17.38
42	8	1	6	6	4	4	4	2	4	1	2,214	397	17.93
43	9	4	3	5	6	3	4	5	3	2	2,061	390	18.92
44	8	4	3	1	3	2	6	5	2	2	2,072	441	21.28
45	1	1	1	1	1	1	1	1	1	1	7,698	1,282	16.65

NOTE: The numbers under areas *A*-bottom, *A*-top, *B* through *I* refer to the available levels in each of the 10 test areas listed in Table 8.1. Level 1 represents the baseline level.

TABLE 8.3
ANOVA (Regression) Results for CTR= CLICKS/ VISITS

Analysis of Variance for CTR=CLICKS/VISITS(%), using Adjusted SS for Tests

Source	DF	Seq SS	Adj SS	Adj MS	F	P
A-bot	5	8.8764	8.6401	1.7280	2.33	0.120
A-top	3	0.8484	0.0146	0.0049	0.01	0.999
B	2	10.9232	5.9607	2.9803	4.02	0.052
C	3	6.3376	9.3502	3.1167	4.20	0.036
D	5	7.0235	7.3990	1.4798	1.99	0.165
E	3	27.1948	14.9875	4.9958	6.73	0.009
F	5	1.3796	3.5688	0.7138	0.96	0.484
G	4	7.3321	5.6330	1.4083	1.90	0.187
H	3	4.8428	3.6671	1.2224	1.65	0.240
I	1	19.0596	19.0596	19.0596	25.69	0.000
Error	10	7.4203	7.4203	0.7420		
Total	44	101.2383				

S = 0.861413   R-Sq = 92.67%   R-Sq(adj) = 67.75%

The *partial (or adjusted) sum of squares* (Adj SS) measures the explanatory contribution of a factor as this factor is added *last* to the model. For example, the adjusted sum of squares of factor B (5.9607) in Table 8.3 is the contribution of factor B when it is added to the model that does not include factor B (i.e., the model with A-bottom, A-top, C through I). Because of the nonorthogonality, the partial (adjusted) sum of squares is different from the sequential sum of squares (which is 10.9232 for factor B).

The degrees of freedom of each factor sum of squares correspond to the number of indicator variables that are needed to represent the levels of that factor. For a factor with a levels, the degrees of freedom are $a - 1$. The adjusted mean squares are obtained by dividing the partial (adjusted) sums of squares by their degrees of freedom.

A sensible strategy for assessing the importance of the various factors is to consult the *adjusted* mean squares and their associated F-statistics and probability values. We notice that the factors A-bottom, B, C, E, and I affect the sign-up rate. The weakest component among these five factors is A-bottom, with probability value 0.12.

We need to find out which levels of these factors are beneficial. The main-effects plots in Figure 8.3 show that the best level for A-bottom is 4 ("Stop paying for long-distance calls"). Level 3 works best for subheadline B ("Earning free long distance calls with PhoneHog is easy. Just click, register or try a new product. Then start calling for free!"). The simple invitation, "Please take a minute to join now," works best for the main copy (level 2 of factor C). The small-font privacy line (level 2 of E) works better than all others, perhaps because it does not highlight the "fine print." Level 2 of factor I (providing link buttons to get to more information) is quite successful in enticing viewers to visit the subsequent Web pages. Hence the *best* factor-level combination is given by

$$(A\text{-bottom} = 4, B = 3, C = 2, E = 2, I = 2)$$

The main-effects plots were obtained as an option in Minitab's "Stat > ANOVA > General Linear Model" command. Minitab displays the fitted means from the main-effects regression model with an intercept and 0/1 indicator variables for the absence/presence of the

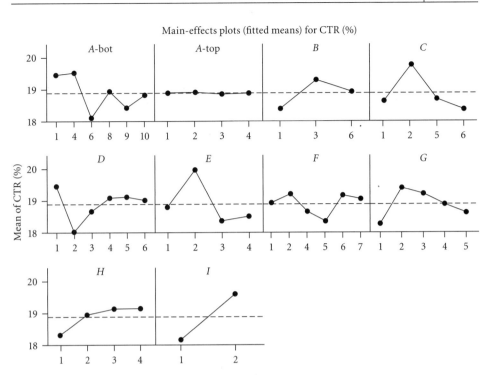

Figure 8.3 Main-Effects Plots for Click-Through Rate

various levels (35 coefficients in total; see Section 8.2.1). The fitted means are different from the means that are obtained by averaging over all other factors. However, the differences are usually minor as long as the design is not too different from an orthogonal design.

The results of the main-effects regression model with the 5 identified factors are shown in Table 8.4. Substituting (A-bottom = 4, B = 3, C = 2, E = 2, I = 2) into the estimated equation leads to the predicted click-through rate

$$17.3671 + 0.5336 + 0.8932 + 0.7673 + 1.4663 + 1.5314 = 22.56.$$

This represents a 35% improvement over the click-through rate of the current base level (creative # 45; CTR = 16.65).

Comment. Experimentation at PhoneHog is a continual activity that tries to improve on the current best results. Our best factor-level combination becomes a baseline for the next set of runs. Also note that the *winner* among the 45 studied runs is creative # 32. Its click-through rate (22.72%), and its factor levels are quite similar to the predicted click-through rate and the factor levels of the best strategy. It certainly makes sense to also include this winning strategy (A-bottom = 8, B = 3, C = 5, E = 2, I = 2) as one of the creatives in the next wave.

There is a lot of uncertainty in our analysis. There are many factors, the factors are categorical, and there are many factor levels. Our assumed main-effects model may be incorrect. It may well be that nothing works except for one single specific combination of factor-levels. If we are lucky to have this particular combination as part of the experiment, its result will stand out as the clear winner. A situation such as the one described here introduces

TABLE 8.4

Model Coefficients: Model with A-bottom, B, C, E, and I (baseline 1 selected as reference level)

The regression equation is
CTR(%) = 17.4 + 0.534 ab4 - 0.905 ab6 + 0.243 ab8 - 0.515 ab9 + 0.056 ab10
          + 0.893 b3 + 0.375 b6 + 0.767 c2 - 0.160 c5 - 0.165 c6 + 1.47 e2
          - 0.195 e3 + 0.035 e4 + 1.53 i2

Predictor	Coef	SE Coef	T	P	
Constant	17.3671	0.5368	32.35	0.000	
Factor A-bottom (baseline 1)					
ab4	0.5336	0.5120	1.04	0.306	(largest)
ab6	-0.9047	0.5244	-1.73	0.095	
ab8	0.2427	0.5220	0.46	0.645	
ab9	-0.5153	0.5287	-0.97	0.338	
ab10	0.0556	0.5146	0.11	0.915	
Factor B (baseline 1)					
b3	0.8932	0.3576	2.50	0.018	(largest)
b6	0.3752	0.3472	1.08	0.288	
Factor C (baseline 1)					
c2	0.7673	0.4111	1.87	0.072	(largest)
c5	-0.1604	0.4037	-0.40	0.694	
c6	-0.1648	0.4056	-0.41	0.687	
Factor E (baseline 1)					
e2	1.4663	0.4118	3.56	0.001	(largest)
e3	-0.1946	0.4132	-0.47	0.641	
e4	0.0347	0.3977	0.09	0.931	
Factor I (baseline 1)					
i2	1.5314	0.2968	5.16	0.000	(largest)

S = 0.938145   R-Sq = 73.9%   R-Sq(adj) = 61.7%

complicated interactions, and a main-effects analysis and its implied best levels could be seriously flawed. So, it is a good idea also to include the winner as one of the creatives in the next wave.

Alternatively, interactions may not matter and the result may be affected only by main effects. Furthermore, because the experiment represents a very small portion of all possible level-combinations, it is very likely that the best combination is not part of the studied runs of the experiment. In this case, our main-effects model and its implied best factor-level combination will outperform the winner in the experiment, supporting a strategy of including the best creative in the next waive of experiments.

To be on the safe side, we recommend that both the *best* and the *winning* combinations are included in the next stage of experimentation. Our approach to designing experiments is powerful but not foolproof. The key is to experiment. Missing something occasionally or including something that eventually turns out to be unnecessary will be of small consequence compared to the accumulation of insights over time.

8.3 COMPUTER SOFTWARE FOR DESIGN CONSTRUCTION AND DATA ANALYSIS

Useful tools for the construction of designs and the analysis of the resulting data are included in most statistical software packages. Our discussion in this section focuses on two general statistics software programs with strong process control and design components: JMP

(The Statistical Discovery Software, http://www.JMPdiscovery.com) is SAS Institute's package for exploratory data analysis. Minitab (http://www.minitab.com) is distributed by Minitab Statistical Software. The emphasis of the following discussion is on design-related aspects of these programs.

Many other useful software packages specifically targeted to the construction of experimental designs and the analysis of the resulting data are also available. Design-Ease and Design-Expert, distributed by Stat-Ease (htpp://www.statease.com), share many of the features found in JMP and Minitab.

8.3.1. Minitab

Minitab makes it easy for the user to construct *2-level factorial* and *fractional factorial designs*. The user enters the number of factors and is then presented with a list of full and fractional designs and their run sizes. After deciding on the number of runs, the user can construct the design either through default generators that optimize the resolution of the design, or by stipulating specific generators. In either situation, Minitab obtains the design columns, displays—if desired—a randomized arrangement of the runs, and indicates the confounding patterns of the particular fraction that is selected. If default generators are used, Minitab displays the adopted generators. The user can add center points, replicate the design, block the experiment by specifying blocking generators, and modify the design by considering foldovers. These can be complete (full) foldovers where the signs of all factors are changed, or foldovers of individual factors. Minitab displays the confounding pattern of the modified design.

Minitab can also construct Taguchi *orthogonal array designs* for designs with factors at three or more levels. These designs include 3-level factorial and fractional factorial designs, Latin square and Graeco-Latin square designs, and mixed-level designs such as an 8-run design with two 2-level factors and one 4-level factor. However, Minitab does not specify the implied confounding patterns for these designs. The user needs to understand that the Latin and Graeco-Latin square designs as well as the mixed-level design mentioned here are orthogonal main-effects designs that leave main effects and 2-factor interactions confounded; refer to the discussion in Section 7.6.

Minitab also constructs *response surface designs* (central composite and Box-Behnken designs) and *mixture designs* (simplex and lattice designs). Response surface designs are useful for fitting quadratic models that are subsequently used to determine the optimum conditions of the response. For further discussion, see Box and Draper (1987).

Once the design has been carried out, Minitab facilitates an efficient analysis of the data. The analysis of *2-level factorial* and *fractional factorial designs* includes estimates of the effects (as well as the regression coefficients, which are one-half of the effects), standard errors of the estimated effects if replications are available, and the ANOVA table. In the unreplicated situation the user can omit model coefficients (e.g., by specifying a model that contains only certain selected main effects and interactions). Minitab combines the omitted effects into an estimate of the variance and calculates standard errors of the estimates. Normal probability plots and Pareto plots for assessing the importance of the effects, as well as main-effects and interaction plots for assessing the nature of the relationships, are read-ily available.

Programs for determining the sample size are also available in Minitab. In addition to the sample size determination for 2-sample comparisons of means and proportions (which we discussed in Appendix 2.1), one can obtain the sample sizes in the one-way ANOVA model (see Section 3.2) and 2-level factorial and Plackett-Burman designs.

8.3.2 JMP

JMP is an equally versatile and useful software package for the construction and the analysis of a variety of experimental designs. It can construct *full factorial designs* for a speci-fied number of factors with different numbers of factor levels. The analysis of the resulting experimental data includes the ANOVA table for testing the significance of main and inter-actions effects. JMP's output is very similar to Minitab's ANOVA output.

The *screening designs* in JMP are particularly useful. JMP allows the user to construct 2-level factorial and fractional factorial as well as Plackett-Burman designs. For a specified number of 2-level factors, the software offers a list of available 2-level arrangements with their implied numbers of runs. The software allows the user to review and change the gen-erators of the fractions and the available blocking arrangements. The software displays the confounding structure that is implied by the selected generators. Center points can be added, the design can be replicated, the order of the runs can be randomized, and the de-sign or parts of it can be augmented by various foldovers.

JMP facilitates the estimation of user-specified models once the data from the experi-ment have become available. Similar to Minitab, the user can omit certain main effects or interactions from the model, calculate a variance estimate by pooling the omitted effects, and compute standard errors of the estimates. JMP has excellent graphical capabilities; the prediction profiler allows the study of main and interaction effects, and normal probability plots of the effects are easily constructed.

Screening designs provide the user the option to construct full and fractional 3-level, and mixed 2- and 3-level designs. Similar to Minitab, JMP does not provide information on the confounding structure of the 3-level (or mixed 2- and 3-level) fractional factorial designs.

JMP includes procedures for constructing and analyzing *response surface designs*, *mixture designs*, and *Taguchi designs*. It includes programs for *sample size determination* in the one-way ANOVA situation, but it does not determine the sample size in 2-level factorial and Plackett-Burman designs (as is done by Minitab).

Another useful feature of JMP is the construction of *custom designs*. After entering the number of factors and their levels (which can be either categorical or continuous), and af-ter specifying the desired model (in terms of its desired main effects and interaction com-ponents), JMP determines the minimum number of runs that are needed to estimate the model coefficients (this is referred to as the minimum solution). It also calculates the num-ber of runs that are needed when combining all possible factor levels into a factorial arrangement (this is called the grid solution). Furthermore, for a given specified number of runs (not less than the minimum number of runs needed) JMP constructs designs that are optimal with respect to certain optimality criteria (such as D-optimality). A D-optimal design minimizes the determinant of the covariance matrix of the estimated model coefficients; it guarantees efficient estimation of the model coefficients; see Appendix 8.1 for further discussion.

8.4 NOBODY ASKED US, BUT . . .

Web site design provides an ideal area for applying experimental design methods. In the PhoneHog case, the consultant and the company decided to test 10 factors with each factor at many levels, leading to a nonorthogonal design that required a more complex analysis compared to 2-level fractional factorial or Plackett-Burman designs. In the online setting, this was a sensible approach. Although it meant creating 45 different Web pages, doing so was not prohibitively difficult or expensive. But 2-level designs would be useful also in this setting. Many factors could be screened in a large resolution III design to identify the likely few important ones. Then supplementary 2-level experiments could be carried out to test additional alternatives for key factors, or the methods of Chapter 7 could be used to test some of these factors at more than two levels while maintaining an orthogonal design.

Computer software has simplified the construction of suitable designs and the efficient analysis of the resulting data. Computer software avoids tedious hand (or calculator) computations and it simplifies the construction of graphical displays. Use the computer to your advantage, but do not trust its outcomes blindly. Check the reasonableness of the results, as results depend on model assumptions that may be violated. There is no substitute for common sense.

Excel has become the standard computer software for business analysis, and you probably use it. Excel is appropriate and useful for simple analyses, but it is deficient in situations that require a more sophisticated approach. Fortunately, good statistical software packages (such as Minitab, SAS, JMP, SPSS, and R) are available.

Much can be learned by studying a textbook, reading case studies, and solving end-of-chapter exercises. However, it has been our experience that to really master the material, you must apply the methods in the real world. Get out and experiment! Discovering the unexpected is more important than confirming what you know.

DESIGN OPTIMALITY

This appendix is written for readers with a substantial statistics background. We study the effects of k design factors, $x_1, x_2, \ldots, x_k$, on the response y and estimate the $k + 1$ coefficients in the main-effects (first-order) model.

$$y = \beta_0 + \beta_1 x_1 + \beta_2 x_2 + \cdots + \beta_k x_k + \varepsilon$$

The factors may be the price in a marketing study, or the temperature and the concentration of an input factor in an engineering problem. A total of $N \geq k + 1$ experimental runs are needed to estimate the $k + 1$ coefficients. The $N \times (k + 1)$ design (regression) matrix X consists of a column of ones and k columns of factor levels that need to be selected at the design stage. At issue is the optimal selection of the elements in X.

Our interest is in the precise estimation of the regression coefficients $\boldsymbol{\beta} = (\beta_0, \beta_1, \ldots, \beta_k)'$. Least squares theory (see Appendix 4.4) implies that the variance of the estimate $\hat{\boldsymbol{\beta}}$ is given by

$$V(\hat{\boldsymbol{\beta}}) = \sigma^2 (X'X)^{-1}$$

Several optimality criteria have been proposed in the design literature.

> *A-optimality.* We look for a design that leads to the smallest average variance of the resulting estimates. The design that allows us to estimate the parameters with the smallest average error must minimize the *trace* of $(X'X)^{-1}$. This criterion is called A-optimality.

> *D-optimality.* Alternatively, we look for a design that minimizes the volume of the joint confidence region of the parameters. It can be shown that the volume is proportional to the square root of the determinant of $(X'X)^{-1}$. Hence, we want to select the levels of the factors in the design matrix X such that the *determinant* of $(X'X)^{-1}$ is minimized. This criterion is called D-optimality.

The two design criteria are similar, involving two closely related functions of the reciprocals of the eigenvalues of $X'X$. A-optimality minimizes their sum, whereas D-optimality minimizes the sum of their logarithms.

Before one can apply these criteria to determine an optimal design, one must specify the permissible experimental region of the design factors. Also, one needs to remember that the coefficients β_i ($i = 1, 2, \ldots, k$) are affected by the choice of the scale for x_i. If x_i denotes the price measured in dollars, we can change β_i by a factor of 100 by measuring the price in cents. It is often desirable to scale the factors uniformly. We have done so in the 2-level designs in Chapters 4–6 by adopting the scaling -1 and $+1$, implying uniformity across the k factors.

One can show that an orthogonal $N \times (k + 1)$ design matrix X guarantees both A- and D-optimality of the main-effects design. For a proof, see John (1971, p. 194). It is in situations where orthogonal designs cannot be found that A- and D-optimality become important. For example, assume that one desires to study the main effects of three factors, each with two levels (-1 and $+1$). It is straightforward to write down an orthogonal design in $N = 8$ runs, and among 8-run designs, this design allows us to estimate the main effects with the least variability. However, assume that one wants to estimate the main effects from the results of just $N = 6$ runs. An orthogonal design in 6 runs does not exist, and one needs another criterion to determine the optimal design. One can select a D-optimal design and use available computer software (such as the custom design feature in JMP) to determine the levels. The note below on computer software explains the algorithms that are used by the programs to find them. Using JMP, we obtain the following levels for the 6 runs:

Run	Factor 1: x_1	Factor 2: x_2	Factor 3: x_3
1	-1	1	-1
2	-1	1	1
3	1	-1	-1
4	-1	-1	1
5	1	1	-1
6	1	1	1

The matrix X is obtained by adding to the three columns of levels a column of ones. Check that the matrix $X'X$ is no longer diagonal, and verify that the determinant of its inverse is $1/1,024$. Convince yourself that any other arrangement will result in a larger determinant. For example, change the level of the first factor in the first run from -1 to 1. Repeat the matrix algebra, and you will find that the determinant of the resulting $(X'X)^{-1}$ is $1/448$ and larger than $1/1,024$.

Our discussion emphasizes the important role of orthogonality in design of experiments. An orthogonal design is also a D- (and A-) optimal main-effects design. However, for certain problems and run sizes orthogonal designs may not exist, and in these situations D- and A-optimality become useful design criteria.

Observe that A- and D-optimality criteria are model-specific, as they look at the precision of the coefficients in a certain specified model. Here we discuss the main-effects model, but extensions to models that include interactions or quadratic components of factors with more than two levels are possible, and software for finding the optimal designs is readily available.

Categorical Design Factors with More Than Two Levels

A main-effects design with categorical factors at more than two levels can be parameterized in terms of a regression model with an intercept term and indicator variables that express the absence/presence of the various levels of the design factors. Consider the special situation of 2 categorical factors with 3 and 4 levels. One pair of levels (e.g., the first level of factor 1 and the first level of factor 2) becomes the standard against which all other factor levels are compared. The regression model includes $k = (3 - 1) + (4 - 1) = 5$ indicator variables that express the presence/absence of factor levels 2 and 3 for factor 1, and factor levels 2 through 4 for factor 2. The vector of regression coefficients $\boldsymbol{\beta}$ is of dimension 6, and we need a minimum of 6 runs ($N = 6$) to estimate the coefficients. Of course, better estimates of the main effects could be obtained if more runs were available. A 12-run full factorial design with a single run at each level-combination would be an excellent choice. This design is orthogonal, and it is optimal in terms of minimizing the variability of the resulting estimates. But, let us assume that our resources allow for only $N = 8$ runs. An orthogonal design in 8 runs is not possible, and hence one needs another criterion to select the runs. D-optimality is a reasonable criterion as it leads to the most precise main-effects estimates.

Consider the PhoneHog example in Section 8.2 as a second illustrative example. There we study 10 factors with 6, 4, 3, 4, 6, 4, 6, 5, 4, 2 levels, respectively. A full factorial is orthogonal and hence D-optimal, but its number of runs is prohibitive. Assume that we want to estimate the main effects of these factors as precisely as possible and are looking for a D-optimal design that uses a certain small number of runs. We can write down a regression formulation for the main-effects model. It includes an intercept and 34 regressor columns. One particular combination of 10 levels (one for each factor) becomes the standard against which the other levels are compared. The 34 indicators express the absence/presence of the other levels in the considered runs. At a minimum, we need 35 runs to estimate the main effects. Additional runs would help estimate the parameters more precisely. PhoneHog was looking for a design that estimated the parameters (i.e., the main effects) as accurately as possible. D-optimality turned out to be a reasonable criterion for generating the design.

Note on Software

Several iterative algorithms for determining A and D-optimal designs are proposed in the literature and they have been implemented in easy-to-use software packages. The custom designer in JMP, for example, starts with a random design of the desired run size with each of the runs satisfying the restrictions of the design. An iterative algorithm called *cyclical coordinate exchange* (Meyer and Nachtsheim, 1995) is used to improve the design. Each iteration of the algorithm involves testing every value of every factor in the design to determine if replacing that value increases the optimality criterion. If so, the new value replaces the old. Iteration continues until no replacement occurs in an entire iterate. To avoid converging to a local optimum, the whole process is repeated several times using a different random start. The custom designer displays the best of these designs.

A recent article by Kuhfeld and Tobias (2005) describes combinatorial and heuristic optimization methods for constructing D-optimal factorial designs. The authors discuss Federov's approach of iteratively exchanging candidate/design pairs (where runs from the list of possible design runs are swapped) and the coordinate exchange algorithm of Meyer and Nachtsheim (1995) (where coordinates of runs are swapped), and they apply simulated annealing optimization techniques to improve the performance of these two methods. Useful SAS Macros are available from the first author's Web site. An earlier paper by Kuhfeld, Tobias, and Garratt (1994) describes useful marketing applications of D-optimal designs.

EXERCISES

Exercise 1   Consider Case 12 (Almquist & Wyner) from the case study appendix.

(a) Consider the design in Table A12.2 of that case. Show that it is balanced (i.e., same number of runs at each level of every factor) and orthogonal (i.e., same number of runs at each level-combination of any two factors). Show that the main effects of Message and Promotion are not confounded with the Message by Promotion interaction.

   Hint: Use the regression approach and show that the interaction column x_{12} is orthogonal to x_1 (message) and x_2 (promotion).

(b) Using statistics software of your choice, confirm the regression output in Table A12.4. Using results in Appendix 4.4 (Brief Primer on Regression), explain how the program obtains $s = 0.0707$ and the standard errors in Table A12.4. Fit the regression on just Message and Promotion, and explain why the regression coefficients for Message and Promotion are unchanged.

(c) Consider the design in Table A12.5. Formulate a regression model with an intercept, three main effects for the 2-level factors (Subject, Action, and Closing), and linear and quadratic components for the two 3-level factors (Salutation, Promotion); see Section 7.4. Specify the 16×8 design matrix. Imagine fitting two regressions. One model considers all eight regressors, while the second contains the intercept and only the three main effects of the 2-level factors. Would the estimates of the three main effects stay the same? Why or why not?

(d) Using a design software of your choice, obtain a 16-run D-optimal design with three 2-level factors and two 3-level factors. Discuss the software's approach of obtaining such designs.

Exercise 2   Consider Case 13 (PhoneHog) from the case study appendix.

(a) Use Minitab or any other statistics software such as JMP or SPSS, fit the main-effects model with all 10 factors, and confirm the results for the click-through rate in Table 8.3. Second, enter the 10 factors in a different order and observe that the sequential regression sums of squares change. This is a consequence of nonorthogonality. For orthogonal designs, the sums of squares would stay the same.

(b) Focus on the model with just three factors, B, E, and I. Define indicators for the levels of the 3 factors; 3 indicators for B, 4 for E, and 2 for I, for a total of 9 indicator columns, b1, b3, b6, e1, e2, e3, e4, i1, i2. Estimate the model that includes a constant and the indicators b3, b6, e2, e3, e4, and i2. The constant represents the mean response when all factors are at their baseline values (level 1 of B, level 1 of E, level 1 of I). Obtain the least squares estimates and their standard errors. Find the best levels of each factor and obtain an estimate of the click-through rate of the best factor-level combination. Compare this to the prediction (22.56) that you got in Section 8.2.2.

(c) Analyze the action rates, using the same approach that we used for click-through rates.

APPENDIX

CASE STUDIES

This appendix contains thirteen case studies. The following table shows for each case the sections of the book that contain relevant material.

Case No.	Title	Relevant Chapters
1	Eagle Brands	4 and 5
2	Magazine Price Test	4
3	Mother Jones (Part A)	4 and 5
4	Peak Electronics (Part A)	4 and 5
5	Office Supplies E-mail Test	5
6	Mother Jones (Part B)	5
7	Peak Electronics (Part B)	5
8	Experiments in Retail Operations	6
9	Experimental Design on the Front Lines of Marketing	6
10	Piggly Wiggly	7
11	United Dairy Industries	6 and 7
12	Almquist & Wyner	8
13	PhoneHog	8

ACKNOWLEDGMENT

Gordon H. Bell (President, LucidView, 80 Rolling Links Blvd., Oak Ridge, TN 37830, USA, (865) 693–1222, (865) 220–8410 (fax), gbell@lucidview.com) contributed to Cases 2, 5, 8, and 9. We are very grateful to him for allowing us to include these case studies in our text.

CASE 1

EAGLE BRANDS

INTRODUCTION

Bill Evans, Director of Marketing at Eagle Brands, was worried. Eagle, a national producer of packaged sandwich meats, was facing increased competition and declining market share. Looking over the latest quarterly supermarket sales numbers, Evans observed that the situation was not improving. He realized that drastic action was needed to turn things around.

Evans had recently read an article in the *Wall Street Journal* about a statistical approach to product testing and was intrigued by the idea of using it to try out some new marketing initiatives. The article called the approach multivariable testing (MVT), and its proponents claimed it could be used to devise an efficient in-store test of multiple variables that might influence sales and profits. Evans had in the past led a project to test market a new package design, and although the experiment provided very useful results, it had been a major undertaking. He was concerned that testing a number of variables in one experiment might be prohibitively expensive to carry out.

The *Journal* article mentioned QualTest, a management consulting firm specializing in applications of MVT. Evans contacted the firm and arranged to have QualTest give a presentation at Eagle Brands. Evans assembled a group of 10 key people, including the heads of sales, finance, and accounting. Steve Gardner, a senior QualTest consultant, made the presentation, explaining the approach and illustrating it with several case examples of successful experiments for QualTest clients.

DESIGNING THE EXPERIMENT

The response to the QualTest presentation was a positive one, and Eagle Brands hired the firm to help them design and evaluate an in-store marketing experiment. QualTest consultants led by Steve Gardner began a series of meetings with Eagle managers.

One of Eagle's major customers was Zip Stores, a national supermarket chain, and the plan was to select a group of the chain's stores to participate in the test. Bill Evans realized that input from the chain was important to the success of the experiment, and a merchandising manager from Zip agreed to join the team.

TABLE A1.1
Factors and Levels

FACTOR LEVEL	
−	+
Regular package	Deluxe package
No in-store samples	In-store samples
No coupons	Coupons
Current fat percentage	Reduced fat percentage
No gold sticker	Gold sticker
Current package lettering	New package lettering

In the first phase of the work, the group developed an initial list of about 20 factors for possible inclusion in the experiment. Gradually the list was reduced to its final form consisting of 6 factors (Table A1.1).

Arriving at the final list was not an easy matter because each manager had personal favorites among the list of potential changes. Al Douglas (finance) felt that the current deluxe package only added to the cost without affecting sales. Bill Evans (marketing) strongly disagreed. "Al, our lunch meat line of bologna, ham, and turkey has the finest products on the market, and the classy package enhances our quality image." A heated argument ensued.

Gloria Johnson of Zip Stores, the supermarket chain, suggested simply adding a gold star to the package. "It will catch the shopper's eye, and that is important in a crowded display case."

The advertising manager felt that coupons were an important way to increase sales. Others argued that sales would increase but not enough to offset the discount provided by the coupon.

Eagle had in the past occasionally set up in-store displays where customers were offered free samples of bologna, ham, and turkey. There was general agreement that this led to more sales ("if they try it, they'll like it!"), but these free sample displays were expensive, and it was unclear if the increase in sales was sufficient to justify the cost.

Eagle had recently developed a new version of its cold cuts with a reduced level of fat. Extensive testing had shown that taste and appearance was unchanged, and the firm felt that the lower fat product would appeal to health-conscious customers, and therefore sales would increase. The downside was that the low-fat version had a higher production cost.

Several team members felt that a change in package lettering to a bolder look would increase sales. The proposed new lettering would result in a small increase in packaging costs.

HOW MANY STORES SHOULD BE INCLUDED IN THE TEST?

There was general agreement among the members of the team that the shorter the test period, the better. It was agreed that the test would be run for one week across a random sample of stores. The Zip Stores supermarket chain had approximately 500 stores all located in the Midwest. There was extensive data available on sales of the three Eagle Brand products by store and week. The team eliminated weeks from the database that were not considered average. The not-average weeks included the week of the Super Bowl and weeks with special promotional activities. For average weeks, average weekly sales per store of Eagle cold cuts was $1,200 with a standard deviation of $150.

The team identified two consecutive weeks in April that were average. The test would be performed in the second week, with the previous week used to provide a sales baseline.

QUESTIONS

1. Assume that there is one factor that increases average store sales by $100. You want to be 80% confident that a 5% significance test can detect such a large increase. Determine the sample size. That is, determine the number of stores to be used in the test, with half of the stores assigned to the minus level and half of the stores assigned to the plus level of the factor. What if you wanted to be 70% confident to detect a change as large as $80? (See Exercise 1 in Chapter 4.)

2. Now that you know the sample size, discuss the advantages of a multifactor experiment (factorial, or fractional factorial) over the experimentation approach of changing one factor at a time.

3. Eagle Brands wants to learn about the effects of 6 factors. A full 2^6 factorial in 64 runs or various fractions of the full factorial such as 2^{6-1}, 2^{6-2}, and 2^{6-3} could be considered. Discuss the advantages and disadvantages of fractional factorials in this particular setting. Which design would you chose? (See Exercise 5 in Chapter 5.)

4. Discuss the protocol that you would use to carry out the experiment. Discuss whether one should analyze absolute or relative changes in sales.

CASE 2

MAGAZINE PRICE TEST

Gordon H. Bell

INTRODUCTION

The publishing industry has seen a continual decline in magazine sales over the last few years. More magazine titles, free content on the Internet, and lower readership have all led to industry difficulties. Publishers push subscriptions through direct mail and online sales, and they try to advance single-copy newsstand sales in supermarkets and other retail outlets.

One leading publisher wanted to find new ways to increase profitability by testing new price points. However, profit depends not only on the magazine cover price but also on the number of new subscribers and the cost of unsold copies. So, the publisher focused on three factors to test:

Cover price. This is the price paid for a single newsstand copy; it is considerably higher than the per-copy subscription price.

Subscription price. This is the price shown on the subscription card inside each magazine. While publishers like the higher per-copy profit from single copy sales, they want to get as many long-term subscribers as possible since a larger subscriber base increases their advertising revenues.

Number of copies on the newsstand. The publisher looks at the balance between having enough copies available for every customer, yet minimizing the number of leftover magazines. The marketing team wanted to test if a larger (or smaller) excess of copies might increase sales. With a few magazine racks in each store, they wondered if more copies in every rack would lead to higher sales, or if fewer copies in each rack—even an empty rack or two—might encourage customers to "buy now while supplies last."

PRICE TEST

With the high cost of printing different covers and subscription cards, the publisher wanted to minimize the number of test cells. But the marketing director also expected to see some interactions among studied factors and "curvature" in the relationship between sales and price.

TABLE A2.1

Factors, Factor Levels, and the Test Design

Factor	(−) Low Level	(0) Centerpoint	(+) High Level
A Cover price	$3.99	$4.99	$5.99
B Subscription price	$1.00	$2.00	$3.00
C Number of copies on newsstand	1/3rd less	Current	1/3rd more

Test Cell	A *Cover Price*	B *Subscription Price*	C *Number of Copies*	AB	AC	BC	ABC	*Sales**	*Subscriptions**
1	−	−	−	+	+	+	−	−2.31%	15%
2	+	−	−	−	−	+	+	−5.54%	112%
3	−	+	−	−	+	−	+	−1.62%	1%
4	+	+	−	+	−	−	−	−3.10%	30%
5	−	−	+	+	−	−	+	18.30%	18%
6	+	−	+	−	+	−	−	1.41%	106%
7	−	+	+	−	−	+	−	22.60%	1%
8	+	+	+	+	+	+	+	−0.73%	37%
9	0	0	0	0	0	0	0	2.08%	20%

* Percent change versus baseline, averaged over 5 weeks

Price often shows a "stair-step" type of relationship. Price changes tend to have a small effect within a certain range, but sales drop precipitously beyond a certain price point. This is the reason why you see many more products sold at $9.99 than at $10.19.

With some statistical guidance, the marketing team decided to run a full-factorial test design with center points, testing for main effects, interactions, and curvature using only nine test cells. The factors and test design are shown in Table A2.1.

The experiment included all combinations of the 3 factors at 2 levels (the 2^3 factorial experiment) and one test cell with all factors set at the center point (0) level. Each of the nine combinations was run in four different regions over a 5-week period. Historical averages for sales and number of subscriptions were available for each region, and by adding the numbers for the four regions, the publisher obtained a baseline estimate for each cell. Using this baseline, percent changes in sales and subscriptions were calculated for each of the 5 weeks. Weekly average percent changes for sales and subscriptions are shown in Table A2.1. Week-to-week variation was used as a measure of experimental error. Variances among weekly percent changes, calculated for each of the 9 runs, were averaged, resulting in an estimated standard deviation of 5% for weekly sales changes and 15% for weekly subscription changes.

TEST RESULTS

Data analysis results are summarized in Figures A2.1 and A2.2.

Analyzing *sales*, we find significant main effects of A (cover price) and C (number of copies), along with a significant AC interaction. The main effect of A is −11.2%. A $1.00 price reduction from $4.99 to $3.99 increases sales by $11.2/2 = 5.6\%$. The main effect of C is 13.5. Placing one-third more copies on the newsstand increases sales by $13.5/2 = 6.8\%$. The

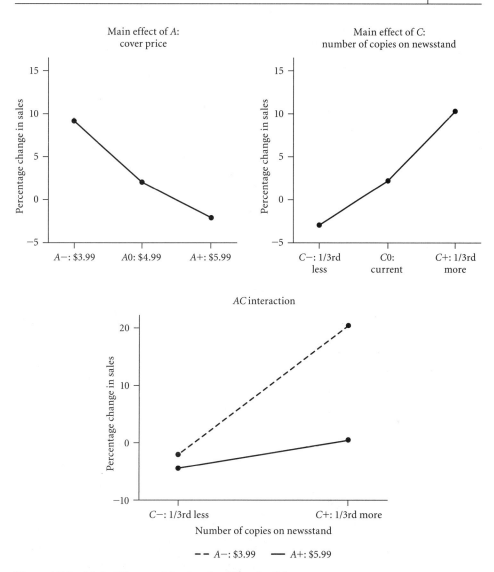

Figure A2.1 Main Effects and Interaction Plots for Sales

sales change at the center point (2.1%), where each of the 3 factors is at the midlevel, falls about in line between the low and high levels; for sales, there is no appreciable curvature.

The *AC* interaction adds further insight to the sales analysis. It shows that the increase in sales at a lower cover price (*A*−) is much smaller if the number of copies is reduced (*C*−). Alternatively, the increase in newsstand copies (*C*+) has a minimal impact if the cover price is high (*A*+). Furthermore, combining the levels that are optimal individually—low cover price and high number of copies—increases sales by more than what can be expected by summing the two individual main effects. Depending on the team's profitability analysis (which is not discussed here), the publisher should increase the number of copies on the newsstand only if it plans to lower the cover price. Otherwise, cost savings from a reduction in newsstand copies may have little impact on sales.

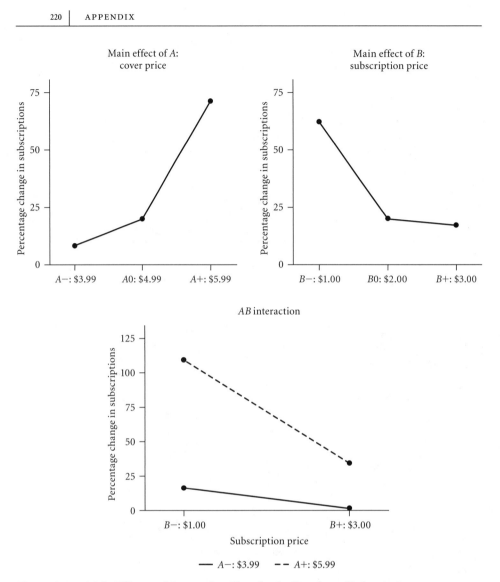

Figure A2.2 Main Effects and Interaction Plots for the Number of Subscriptions

Analyzing the change in the *number of new subscribers* (people who mail in the subscription card taken from the magazine and who pay the bill), we find significant main effects of cover price (A) and subscription price (B), along with significant AB interaction and significant curvature.

Cover price (factor A) has the largest impact on subscriptions, with a high cover price (A+: $5.99) increasing subscriptions 71.3% over historic levels. The low subscription price (B−: $1.00) also has a large impact, increasing subscriptions by 62.8%. But before jumping to the conclusion that a high cover price and a low subscription price are best, these data need to be analyzed along with the data on sales (shown earlier) and profit data (not shown). A high newsstand price along with a low subscription price increases the number of subscribers, but a high cover price also reduces newsstand sales.

The *AB* interaction shows that the two significant main effects are even greater in combination. A high cover price $(A+)$ and a low subscription price $(B-)$ result in more new subscribers than the implied number that is obtained by adding the two individual main effects.

Analysis of curvature leads to important insights. Significant curvature for subscriptions means that, within some range, cover and subscription prices have negligible impact. Yet beyond a certain level, price changes result in a big jump in subscriptions. However, with only one center-point test cell it is not possible to determine which factor (or perhaps both) is causing the curvature. With significant curvature, the next step for the marketing team was to run a new test design with more combinations at different price and subscription levels.

Following this test, the publisher's marketing team ran additional tests to pinpoint the optimal price points. They ended up increasing the cover price and increasing the subscription price, while maintaining the same level of newsstand copies. These price increases ultimately increased profit while maintaining the number of copies sold on the newsstand and through subscriptions.

QUESTIONS

Exercise 2 in Chapter 4

CASE 3

MOTHER JONES (PART A)

INTRODUCTION

Jay Harris, publisher of *Mother Jones* magazine, was constantly searching for ways to increase circulation. *Mother Jones* is named for Mary Harris "Mother Jones" (1830–1930), whom the magazine describes as "orator, union organizer, and hell-raiser."

Mother Jones is published six times a year and is known for its political muckraking. Two articles from the March/April 2000 issue are representative of the magazine's style and substance. The table of contents describes an article called "Slick W" in the following way: "Meet one of the more ordinary failures ever bailed out of the oil business—George W. Bush, a man who lost millions of other people's money and walked away with a tidy profit." The article "Ballots and Bones" has this description: "Chile's newly elected Socialist president wants to avoid Pinochet and focus on prosperity. But survivors of the general's reign insist that the nation cannot move forward until it confronts its bloody past."

In the fall of 1999, Harris got a call from Arthur Swersey, professor of operations research at the Yale School of Management. Harris had gotten to know Professor Swersey well while he was a student at the School of Management. "I can help you increase circulation," Swersey proclaimed. He went on to explain how statistical experimental design methods, long used in manufacturing, could increase the effectiveness of a mailing to potential subscribers. Harris was excited by the concept and quickly put in place a plan to test the experimental design approach in the Spring 2000 mailing.

THE MAILING

A typical mailing was sent to about 400,000 potential subscribers. Most of these people were sent the same package, called the control. The control represented the combination of outer envelope design and other factors that had yielded the best response in the past. Traditionally, a fraction of the 400,000 pieces mailed was used to test a single change in the control, for example, the inclusion of a free *Mother Jones* bumper sticker.

Under the experimental design approaches, numerous changes would be tested simultaneously. After a series of discussions among Harris, Swersey, and members of the market-

TABLE A3.1
Factors and Assigned Levels

| | FACTOR LEVEL | |
Factor	−	+
A	No act now insert	Act now insert
B	No credit card	Credit card
C	Hard offer	Harder offer
D	Strong guarantee	Stronger guarantee
E	No testimonials	Testimonials
F	No bumper sticker	Bumper sticker
G	Gutsy	Ballsy

ing firm that organized the mailings, the seven factors listed in Table A3.1 were identified for the test.

The "act now" insert, a small separate note in color, urged people to act now and to respond/pay today. The front of the insert showed the cover of the next issue and urged the recipients to act immediately to make sure they received this very special issue. The back of the insert in bullet point style and vibrant language described the articles that would appear in the next issue. If the act now insert was included, the words *act now* were also boldly written on the reply card.

The second factor gave people the option of paying by credit card rather than only by personal check.

The factor "hard offer/harder offer" refers to the language in the offer described in the letter from Harris, the publisher. For example, "hard" encourages the person to "get the next issue of *Mother Jones*," and "harder" says "to get the next issue hot off the press, send your reply and subscribe today."

"Guarantee" allows the person to cancel the subscription at any time during the first year, and receive money back for the issues not yet received, while the "stronger guarantee" means that the full subscription price would be refunded, as long as the subscription is canceled before the end of one year.

"Testimonials" refer to an insert in which typical subscribers and notable people make positive comments about the magazine.

"Gutsy/ballsy" refers to a single word that is printed on the outer envelope. Previously, *Mother Jones* had effectively used the word "ballsy," but had received some complaints about this language. Harris was interested in finding out whether softening the language to "gutsy" would produce similar results.

THE MAILING PROCESS

The mailing process consisted of two stages: printing and insertion. Printing produced the addressed outer envelope with either "gutsy" or "ballsy" printed on it. The testimonials were printed on a single sheet of paper, and the reply card included information on the price, whether a credit card could be used, the type of guarantee, and the phrase "act now" included or not.

A firm specializing in mailings was responsible for carrying out the logistics of the experiment. The firm's production line consisted of automated insertion equipment that put the letter from the publisher, reply card, and other materials in the outer envelope. Changing from one combination of factors to another required setting up the equipment for the specified run. A great deal of care was needed to ensure that each run was exactly right—that is, all of the packages in a run containing the correct information. This meant that the fewer the number of runs in the experiment, the easier it would be to carry out.

Harris wanted a design that would not only show main effects but also, if possible, identify 2-factor interactions as well.

With the foregoing information in hand, Professor Swersey sat down to develop the design. He knew that a typical response rate for a mailing was about 2%. Harris had told him that a ¼ percentage increase in response was worth learning about. With this information, he would be able to determine how large this test mailing needed to be to obtain statistically significant results.

QUESTIONS

1. Devise an appropriate design for the mailing. Label each of the factors given in the case as A, B, C, ... What is the confounding pattern for your design (ignoring interactions of order 3 or higher)? Briefly explain why you picked this design.

2. Recall that a ¼ percentage increase from the current subscription rate of 2% was considered important. If we call each run a package, how many people should be sent each package (i.e., how many mailings are there for each run)? What is the total number of mailings (i.e., the total sample size)?

CASE 4

PEAK ELECTRONICS:

THE BROKEN TENT PROBLEM (PART A)

In the late fall of 1991, Peak Management started to get concerned with "broken tents," the number one cause of rework. A *tent* is a piece of photoresist (or film) that covers a hole that is not to be plated with copper. If the tent brakes before copper plating, then the hole is plated and the panel needs to be reworked by scraping the copper from the hole.

THE PROCESS

The production process consists of many steps. Steps 4 and 5 are relevant for this discussion. In step 4 (laminate photoresist), a thin photosensitive film or resist is laminated (bonded) onto a copper panel. The film is applied by rolling a sheet of film onto the panel and putting the panel between two rollers with the heat and pressure causing the film to adhere to the panel. In step 5, a film negative showing the circuitry is placed over the photoresist covered panel, and the panel is exposed to ultraviolet light. The circuitry on the negative is opaque and blocks the UV light. The rest of the resist on the panel is polymerized (hardened) by the UV light. In the developer, the panel moves on a conveyer through an 8-foot-long chamber. A developing solution is sprayed onto the panel and the resist, which is not polymerized (hardened), is washed away, exposing the circuitry. At the end of this stage, each hole that is not to be plated should be tented (covered by film). However, a tent may be broken at this point.

HERCULES, INC.

A quality improvement team at Peak, using a fishbone diagram, identified the photoresist as a likely contributor to the broken tent problem. Peak had been using Dupont 4215 resist. Scientists at Hercules, Inc., a competitor of Dupont, suggested that their film could significantly reduce broken tents. Peak ran a test with the new resist, using 40 lots of 36 panels each, interspersing 10 lots that used the current Dupont 4215 resist. They found no statistically significant differences in the number of broken tents per panel between the Dupont and Hercules resists.

Hercules was obviously unhappy with these results and suggested that Peak run a designed experiment that might improve the process. The Hercules representative and Lou

TABLE A4.1
Five Factors and Their Assigned Levels

	LEVEL	
Factor	Low (−) Level	High (+) Level
Lamination roll thickness	0.062 inch	0.125 inch
Lamination exit temperature	110°F	125°F
Developer spray pressure	15 psi	25 psi
Developer breakpoint	40%	60%
Post lamination hold time	3 hours	24 hours

Pagentine, the engineering manager at Peak, identified 5 factors to include with 2 levels for each factor (see Table A4.1).

The *exit temperature* refers to the temperature of the panel as it leaves the rollers. A higher temperature could improve the bonding, but to achieve a higher temperature, the speed of the rollers would have to be reduced by 50%. This would result in a serious loss of productivity.

The *spray pressure* is the force of the spray hitting the panels. The *developer breakpoint* is the distance that the panel must move on its journey through the 8-foot-long chamber before the resist starts to come off the panel. A 60% setting means that the panel is 60% of its way through the chamber before the resist starts to come off. The breakpoint depends on the speed of the conveyor, as well as the solution pH and temperature. A short breakpoint could make the remaining polymerized resist brittle. Pagentine did not expect that spray pressure or breakpoint would be significant variables.

The lamination rolls consist of a steel center surrounded by rubber. The *thickness* variable is the thickness of the rubber. The expectation is that a thinner roll (0.062 inch) would reduce the number of broken tents. The idea was that the currently used larger rollers would tend to force too much film into the holes, which could cause them to break.

The current settings of the first 4 factors were (0.125 inch, 115 F, 25 psi, and 60%). The last factor, *hold time*, refers to the time between lamination and developing. Peak observed that broken tents tended to occur more often in panels that were laminated at the end of one day, but not developed until the next day.

For the experiment, Peak designed a single test panel with 1,350 holes. It also included holes of larger dimensions than usual, which would likely increase the occurrence of broken tents. The panel had no circuits, and all holes were to be tented. Pagentine felt that the handling of the panels in the plating operation could contribute to broken tents. To eliminate this variable, he designed his test panel with no holes within 3 inches of the panel edge.

The expectation was that under the current conditions about 2% or 3% of the holes on a given test panel would have broken tents. The cost per test panel including labor and inspection was estimated to be about $20.

QUESTIONS

1. Devise an appropriate experimental design for the broken tent problem. Label each of the factors given in the case as *A, B, C, ...* What is its confounding pattern (ig-

noring interactions of order 3 or higher)? Briefly explain why you have selected this design.

2. Would it be beneficial to replicate the design? Why? How would replication help you determine which effects are significant? If you choose not to replicate, how would you decide which effects were significant?

CASE 5

OFFICE SUPPLIES E-MAIL TEST

Gordon H. Bell

INTRODUCTION

Marketing know-how does not always translate well from one channel to another, as one office-supplies retailer came to realize. With consistent growth and solid profit from their retail stores, one industry leader decided to expand into direct marketing channels, mailing out catalogs and sending e-mails to direct small-business customers to their Web site and stores. Two of their biggest challenges were building solid mail and e-mail lists of prospective customers and translating the in-store experience onto a two-dimensional page. A year after starting these new programs, the marketing vice president wanted to speed the learning curve with a more disciplined approach.

Talking with other executives, he decided to bring in an outside consultant to strengthen their marketing testing efforts. Both the catalog and Internet programs had room for improvement, but the flexibility and low cost of e-mail (versus printing multiple catalogs) became the deciding issue on where they would first apply scientific testing techniques.

With fast response, low costs, and flexible production, e-mail was a great place to start testing. In addition, what worked in e-mail could then be tested in the catalog and even in retail stores. However, in this early stage of their business-to-business e-mail program, the Internet marketing director had very few e-mail addresses he could use. Retail sales associates had begun asking for e-mail addresses, and online orders were growing, but at this point the marketing team had only about 35,000 names. Moreover, these names included three distinct customer segments, each with different buying behavior. With so few names, the Internet director had tested one or two new ideas in each monthly drop, but had a difficult time trying to get a statistically significant read on his results.

PLANNING THE TEST

The consultant agreed with the Internet director that sample size could be a problem. He explained that there was no magic shortcut—no way to reduce the natural variation in the marketplace—so it was necessary to overcome variability with bold factors and a sufficiently large sample size. He explained how simple rules of thumb, like "100 orders in each test cell," oversimplified the sample size issue and often led to a weak test with no significant results. In this case, with only 35,000 names and an average response rate of 1%, an effect would have to change the response by about 20% (from 1.0% to 1.2%) to have a 50:50 chance of being found significant.

The consultant offered encouragement. With the right multifactor test design and a focus on bold changes, they could create a strong test with useful results. The consultant summarized the requirements:

- Use one experimental design to test numerous variables, maintaining the same test power no matter how many variables were tested.

- Use all available names, but design the test so differences among segments can be quantified.

- Take advantage of the flexibility of e-mail by using a higher-resolution test design with more test cells yet less confounding among the effects.

TEST FACTORS

After brainstorming ideas and trimming the list down to the boldest ideas, the marketing team identified 13 variables and selected two different versions of each variable to test. These 13 factors could be tested simultaneously in a 16-run design, but for reasons outlined below, the consultant selected a 32-run fractional factorial design instead.

The 32-run design requires greater effort for the marketing team to construct 32 different e-mails, but it has important statistical advantages. Where a 16-run design is only of resolution III (with main effects fully confounded with 2-factor interactions), the 32-run design is of resolution IV (with main effects confounded with 3-factor interactions, but independent of 2-factor interactions). Since higher-order interactions are unlikely, this design reduces the potential confounding error and also helps identify key 2-factor interactions.

The three customer segments also had to be considered. Including a 3-level factor in the test would have led to an unbalanced design, in which each factor level would not have appeared in the same number of runs. Instead, the three segments were defined as a factor with 4 levels, with segment 1 (the largest segment) taking up 2 levels. Just as a 2-level factor requires one column in the test design, a 4-level factor requires three columns. The A, B, and AB interaction columns in Table A5.1 were used to define the three segments.

After creating the test design, one of the 13 factors was eliminated. The team planned to test a search box at the top of the e-mail message, but this was too difficult to execute and column E was left empty. The remaining 12 factors plus the 4-level segment factor are listed in Table A5.2.

TABLE A5.1
Three Customer Segments Treated as a 4-Level Factor

Combinations	A	B	AB	Segment	Available Names
1	−	−	+	Segment 2	11,586
2	+	−	−	Segment 1	13,508
3	−	+	−		
4	+	+	+	Segment 3	8,966

TABLE A5.2

Factors and Their Levels

Factor	(−) Control	(+) New Idea
A Segment	−	+
B Segment	−	+
C Link to online catalog	No	Yes
D Background color	White	Blue
E (empty)		
F Design of e-mail	Simple	Stronger brand image
G Partner promotions	None	Offers from two partner companies
H Navigation bar on side	Current	Additional buttons
J Special-offer starburst	No	"Special e-mail offer" starburst
K Discount offer	15% off	No discount
L Free gift	None	Free pen-and-pencil set
M Products pictured	Few	Many
N "Valued customer" copy	Current	Stronger
O Cross-sell copy	Current	New copy
P Subject line	"Exclusive e-mail offer . . ."	"Special offer for our customers . . ."

A and B: Segment

The marketing team had defined three key customer segments, based on behavioral variables and the timing of recent purchases:

Segment 1 consisted of customers who had made a purchase online or in a store within the last 3 months. Segment 2 had made a purchase within the last 3–6 months. Segment 3 had made a purchase within the last 6–12 months.

C: Link to Online Catalog

The e-mail included a "Shop our catalog online" button towards the bottom of the e-mail. The team felt that a link to the Web site would encourage customers to browse the available products.

D: Background Color

All e-mails were sent with dark text on white background. The creative director thought that a blue background might help the e-mail stand out.

F: Design of E-Mail

E-mails used a basic font with a small company logo at the top. The team wanted to test a stronger brand image, with a larger logo, more stylized font, and greater use of the company's brand colors in the e-mail.

G: Partner Promotions

With brand-name products, the marketing team believed that promoting several brands could help convince customers to make a purchase. They decided to promote two specific brands in two bright boxes under "Offers from our partners" at the bottom of the e-mail.

H: Navigation Bar on Side

E-mails currently went out with a sidebar similar to the navigation bar on the company Web site, but with a shorter list of links. They didn't want to test an e-mail without any sidebar, so instead they decided to test the current navigation bar versus one with more choices.

J: Special-Offer Starburst

Since e-mails were sent to a "select group of customers," they wanted to play up the exclusivity with an eye-catching red star at the upper right stating "Special e-mail offer."

K: Discount Offer

The Internet director had gone back and forth between offering a special e-mail discount or not. He thought the discount helped, but had never quantified whether it pulled in enough sales to justify the lower margin.

L: Free Gift

They had not offered a free gift with online orders before, but wondered whether it was worth a try, as other companies were doing it. They selected an attractive, but low-cost, pen-and-pencil set that they could offer for free. At first, they wanted to offer it only for orders of $50 or more, but choose instead to be bold and offer it with every order.

M: Products Pictured

Every e-mail focused on a selection of products—with pictures and prices—but they never knew how many were best. Some people on the marketing team thought that a simple offer with just a few products would get people to respond faster. Others thought that a larger selection would give more people something of interest. They decided to test a few products versus many products. For the test, every picture was the same size, so e-mails featuring "many products" had additional rows of product pictures.

N: "Valued Customer" Copy

Their standard e-mail copy stated, "As a valued [company] customer, we would like to offer you these Internet-only specials." They tested this against a copy with a stronger message, adding a second sentence about how only their best customers get these special offers.

O: Cross-Sell Copy

The second copy change was designed to sell more products. An additional sentence was added to encourage people to order a variety of office supplies at once to lower shipping costs.

P: Subject Line

The Internet director had been testing different e-mail subject lines. Currently, "Exclusive e-mail offer from [company]" was the winner. Since he knew the subject line was important, he wanted to test another version, "Special offer for our best customers."

TEST DESIGN

The consultant developed a 2^{15-10} fractional factorial test design, based on a 32-run, 5-factor, full-factorial design in factors A–E, with factors F–P assigned to 10 of the 26 interaction columns using the design generators:

$$F = ABC, G = ABD, H = ABE, J = ACD, K = ACE, L = ADE,$$
$$M = BCD, N = BCE, O = BDE, P = CDE$$

Minitab statistical software was used to generate the design columns. The test matrix for the 15 design columns (the 13 factors plus the 2 factors, A and B, that specify the three

TABLE A5.3
Test Design and Test Results

Test Cell	Segment — A	Segment — B	Link to Online Catalog — C	Background Color — D	(empty) — E	Design of E-mail — F	Partner Promotions — G	Navigation Bar on Side — H	Special-Offer Starburst — J	Discount Offer — K	Free Gift — L	Products Pictured — M	"Valued Customer" Copy — N	Cross-Sell Copy — O	Subject Line — P	Names	Orders	Response Rate
1	−	−	−	−	−	−	−	−	−	−	−	−	−	−	−	1,515	21	1.39%
2	+	−	−	−	−	+	+	+	+	+	+	−	−	−	−	883	6	0.68%
3	−	+	−	−	−	+	+	+	−	−	−	+	+	+	−	883	9	1.02%
4	+	+	−	−	−	−	−	−	+	+	+	+	+	+	−	1,169	8	0.68%
5	−	−	+	−	−	+	−	−	+	+	−	+	+	−	+	1,515	9	0.59%
6	+	−	+	−	−	−	+	+	−	−	+	+	+	−	+	781	9	1.15%
7	−	+	+	−	−	−	+	+	+	+	−	−	−	+	+	883	3	0.34%
8	+	+	+	−	−	+	−	−	−	−	+	−	−	+	+	1,180	14	1.19%
9	−	−	−	+	−	−	+	−	+	−	+	+	−	+	+	1,515	17	1.12%
10	+	−	−	+	−	+	−	+	−	+	−	+	−	+	+	883	2	0.23%
11	−	+	−	+	−	+	−	+	+	−	+	−	+	−	+	800	14	1.75%
12	+	+	−	+	−	−	+	−	−	+	−	−	+	−	+	1,180	0	0.00%
13	−	−	+	+	−	+	+	−	−	+	+	−	+	+	−	1,515	9	0.59%
14	+	−	+	+	−	−	−	+	+	−	−	−	+	+	−	883	10	1.13%
15	−	+	+	+	−	−	−	+	−	+	+	+	−	−	−	883	11	1.25%
16	+	+	+	+	−	+	+	−	+	−	−	+	−	−	−	815	9	1.10%
17	−	−	−	−	+	−	−	+	−	+	+	−	+	+	+	1,515	17	1.12%
18	+	−	−	−	+	+	+	−	+	−	−	−	+	+	+	883	7	0.79%
19	−	+	−	−	+	+	+	−	−	+	+	+	−	−	+	690	5	0.72%
20	+	+	−	−	+	−	−	+	+	−	−	+	−	−	+	1,091	16	1.47%
21	−	−	+	−	+	+	−	+	+	−	+	+	−	+	−	1,178	18	1.53%
22	+	−	+	−	+	−	+	−	−	+	−	+	−	+	−	883	3	0.34%
23	−	+	+	−	+	−	+	−	+	−	+	−	+	−	−	883	11	1.25%
24	+	+	+	−	+	+	−	+	−	+	−	−	+	−	−	1,180	4	0.34%
25	−	−	−	+	+	−	+	+	+	+	−	+	+	−	−	1,515	8	0.53%
26	+	−	−	+	+	+	−	−	−	−	+	+	+	−	−	641	7	1.09%
27	−	+	−	+	+	+	−	−	+	+	−	−	−	+	−	883	8	0.91%
28	+	+	−	+	+	−	+	+	−	−	+	−	−	+	−	1,180	13	1.10%
29	−	−	+	+	+	+	+	+	−	−	−	−	−	−	+	1,318	14	1.06%
30	+	−	+	+	+	−	−	−	+	+	+	−	−	−	+	883	9	1.02%
31	−	+	+	+	+	−	−	−	−	−	−	+	+	+	+	883	13	1.47%
32	+	+	+	+	+	+	+	+	+	+	+	+	+	+	+	1,171	8	0.68%

segments) are shown in Table A5.3. Minitab also provides the alias structure for this fractional factorial design. Ignoring 3- and higher-order interactions results in a design that can estimate the 15 main effects of factors A through P, 15 effects each containing seven 2-factor interactions, and one effect containing only 3-factor interactions.

Table A5.3 lists the sample size and response data for each test cell. Since each customer segment was randomly assigned to certain test cells based on the +/− levels in columns A and B, the numbers of customers contacted in each test cell are not the same, and the test is not completed balanced. In addition, after names were assigned to test cells, the final purge/merge (where addresses are double-checked and invalid e-mail addresses are removed) dropped some names from the test. In the end, only 34,060 names were used. Each version of the e-mail was sent to as few as 641 or as many as 1,515 customers.

EXECUTING THE TEST

The creative team—made up of the creative director and a single person who designed every e-mail—was somewhat tentative about the test. The thought of creating so many different e-mails for one drop was daunting. They also didn't know if all required combinations would work from an artistic standpoint.

The consultant worked to minimize their concerns and lighten their workload. First, he helped the team define clear, independent factors that could work together in any combination. Then he sat down with the team to review every required test cell, changing factor definitions to make all test cells essentially simple cut-and-paste combinations. Finally, he worked with the creative team as they developed each version; he checked everything to ensure compliance and consistency and solved any problems as they arose.

Overall, the creative work added two days to the marketing schedule. The team was surprised how smoothly things went once all factors and combinations were clearly defined.

TEST RESULTS

The test dropped on Tuesday, and initial results were analyzed after one week. Since the team wanted to increase the number of orders, the primary metric was the response rate. Average order size was also analyzed to help assess profitability, but this particular analysis is not shown here.

The main effects and 2-factor interactions are shown in the two Pareto charts in Figures A5.1 and A5.2. The effects are calculated by applying the $+/-$ signs to the response rates in the last column of Table A5.3, and dividing the resulting linear combination by 16. Alternatively, the effects can be obtained by regressing the response rate on the design and interaction columns; the only difference in the results is that the size of the effects is cut in half. As discussed below, the significance of the effects is best determined through logistic

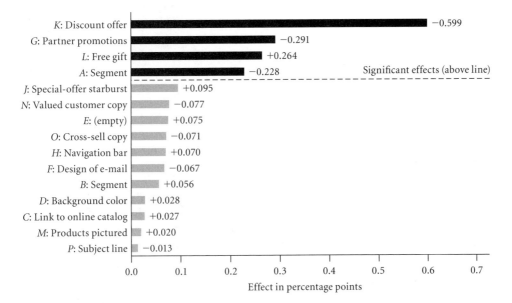

Figure A5.1   Test Results: Main Effects Only

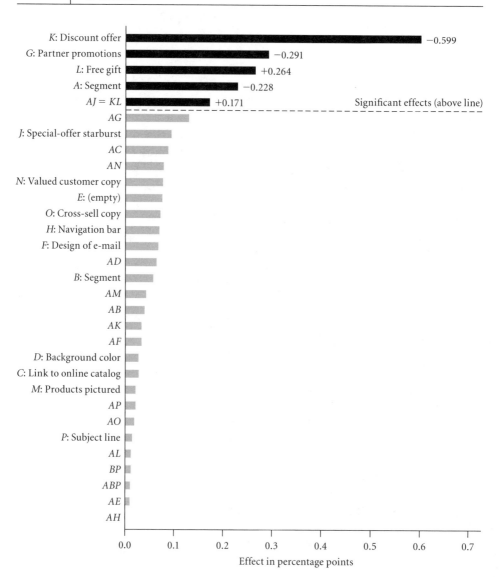

Figure A5.2 Test Results: Main Effects and Interactions

regression. The output of the standard regression and the logistic regression are summarized in Table A5.4. The Minitab statistical software is used for both regressions.

The standard regression with the response rate as the dependent variable has several drawbacks: (1) It ignores the different sample sizes. It treats each response rate as equally precise and analyzes the response rates the same no matter how many e-mails were sent. (2) It uses unweighted averages of the response rates in the estimation of the effects, instead of more appropriate weighted averages that adjust for the unequal precision. (3) The standard errors of the estimates are obtained by pooling smaller insignificant effects into the experimental error term, a somewhat arbitrary decision that can overstate the number of significant terms. *Logistic regression* represents a better approach for analyzing proportions

TABLE A5.4
Standard Regression and Logistic Regression Output of Significant Factors

STANDARD REGRESSION OF RESPONSE RATE ON FACTORS A, B, G, K, L, AB, AND KL

Estimated Effects and Coefficients for Response Rate

Term	Effect	Coef	SE Coef	T	P
Constant		0.9263	0.02823	32.81	0.000
A	-0.2275	-0.1138	0.02823	-4.03	0.000
B	0.0561	0.0280	0.02823	0.99	0.330
G	-0.2913	-0.1457	0.02823	-5.16	0.000
K	-0.5993	-0.2996	0.02823	-10.61	0.000
L	0.2639	0.1319	0.02823	4.67	0.000
AB	-0.0397	-0.0199	0.02823	-0.70	0.488
KL	0.1711	0.0855	0.02823	3.03	0.006

S = 0.159694 R-Sq = 88.68% R-Sq(adj) = 85.38%

BINARY LOGISTIC REGRESSION OF NUMBER OF ORDERS ON FACTORS A, B, G, K, L, AB, AND KL

Link Function: Logit
Response Information

Variable	Value	Count
Orders	Success	312
	Failure	33748
Names	Total	34060

Logistic Regression Table

Predictor	Coef	SE Coef	Z	P	Odds Ratio	95% CI Lower	Upper
Constant	-4.781570	0.0643679	-74.28	0.000			
A	-0.127779	0.0590138	-2.17	0.030	0.88	0.78	0.99
B	0.023847	0.0590880	0.40	0.687	1.02	0.91	1.15
G	-0.163981	0.0577553	-2.84	0.005	0.85	0.76	0.95
K	-0.369910	0.0617801	-5.99	0.000	0.69	0.61	0.78
L	0.196689	0.0618118	3.18	0.001	1.22	1.08	1.37
AB	-0.023241	0.0590546	-0.39	0.694	0.98	0.87	1.10
KL	0.157110	0.0618326	2.54	0.011	1.17	1.04	1.32

Log-Likelihood = -1744.333
Test that all slopes are zero: G = 60.822, DF = 7, P-Value = 0.000

Goodness-of-Fit Tests

Method	Chi-Square	DF	P
Pearson	12.0296	24	0.980
Deviance	15.3150	24	0.911
Hosmer-Lemeshow	2.0619	7	0.956
Brown:			
General Alternative	1.7748	2	0.412
Symmetric Alternative	1.7468	1	0.186

that originate from samples of different sizes. The number of positive responses among the sampled cases in each run is modeled as a binomial random variable, with a success probability that depends on the design variables. Since success probabilities are always between zero and one, logistic regression models the logarithm of the odds (the ratio of the probabilities of success and failure) as a linear function of the design variables. For a detailed discussion of logistic regression and on how to interpret the coefficients in logistic regression, we refer the reader to Chapter 11 in Abraham and Ledolter (2006). Here we use the logistic regression merely to assess the significance of the regression coefficients.

SIGNIFICANT EFFECTS

The average response rate in this test was 0.916%, with just 0–21 orders for each test cell and a total of only 312 orders. This was a small sample size in an unbalanced design with low response rates, and yet the subsequent results were convincing. Significant effects include the following:

K: Discount Offer

The elimination of the 15% discount resulted in a 0.599% reduction in the response rate. The team calculated that the loss of margin from selling the product cheaper is more than covered by the increase in the number of orders.

G: Partner Promotions

The two partner offers in the e-mail reduced the response rate by 0.291%, contrary to what they had expected. The team theorized that the additional offers may have confused the message and given customers too many disjointed offers to choose from.

L: Free Gift

The free pen-and-pencil set increased the response rate by 0.264%. Analyzing profitability, the cost of the gift was easily covered by the increase in orders.

A: Segment

The significance of at least one of the three components responsible for the segment effect (A, B, AB) indicated that the three segments responded differently. The response rates for the three segments are summarized in Table A5.5.

The differences among the three response rates are small and not particularly significant. The 95% confidence intervals in Table A5.5 overlap each other. This finding could suggest that the blocking with respect to the three segments may not have been needed. Nevertheless, the significance of the factor A in the earlier analysis and the summary in Table A5.5 raises the question of why half of segment 1 ($A+B-$) had a response rate lower than any other group, while the other half of segment 1 ($A-B+$) had the highest response rate of all. After some investigation, the problem could be traced to a simple error in the execution of the experiment. The top half of segment 1 (i.e., the best customers) had been placed in the $A-B+$ test cells, while the bottom half were placed in the $A+B-$ test cells. This not only pointed out the risk of a nonrandom assignment of names but also showed that the segmentation model needed some refining; perhaps the best and worst recent buyers should be in different customer segments.

KL Interaction

The final significant effect was the KL 2-factor interaction, with an effect of 0.1711. Before explaining how this interaction affects results, it is worthwhile to take a step back and see where it came from.

Analysis of the data shows 31 independent effects: 15 main effects, 15 strings of 2-factor interactions, and one string of 3-factor interactions. In Minitab, the default is to label the interactions with the first of all confounded interactions. For example, the labeling of the significant interaction in Figure A5.2 starts with AJ because A is the factor that is listed first.

Table A5.5
Response Rates for the Three Customer Segments

Combi-nations	A	B	AB	Segment	Available Names	Average Response	
1	−	−	+	Segment 2	11,586	0.975%	
2	+	−	−	Segment 1	13,508	0.789% }	0.940%
3	−	+	−			1.090% }	
4	+	+	+	Segment 3	8,966	0.803%	

NOTE: The display shows averages and 95% confidence intervals.

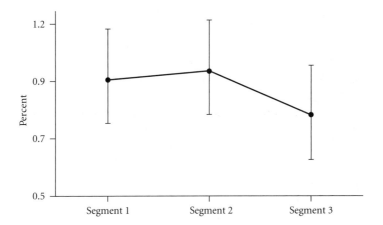

However, this effect could actually be the result of one or more of the seven confounded 2-factor interactions. The list of interactions is given in Minitab, or the interactions can be calculated using the design generators that have been listed earlier in the Test Design section. The seven interactions mixed together in the *AJ* column are *AJ* + *BM* + *CD* + *EP* + *FG* + *KL* + *NO*.

The regression results in Table A5.4 show that this column has a significant effect, but do not identify *which* interaction(s) is most likely. Here is where marketing knowledge and statistical principles come together to help pinpoint the most likely interaction effects. The following two principles can help:

1. Sparsity of effects—Few, if any, interactions are usually significant.
2. Heredity of effects—Large main effects tend to produce interactions.

Even without any understanding of the test factors, these principles imply that few of the seven interactions are likely, and the best-guess interactions are those related to the largest main effects. So, from effect heredity, the most likely interaction candidates would involve factors *K*, *G*, *L*, and *A*. Therefore *AJ* is possible (*J* is not significant but large), *FG* might be possible (though *F* is a smaller effect), and *KL* looks promising (since both main effects are significant).

Limiting the number of interactions to three, the marketing team and consultant could assess which interactions make sense. Though interactions may be completely surprising, often they result from related factors—factors located close together (like two elements on a direct mail envelope) or conceptually related (like price and offer variables).

In this case, the choices are

1. *AJ*   The starburst (*J*) has a different impact depending on the customer segment (*A*)
2. *FG*   The e-mail design (*F*) and partner promotions (*G*) work together to impact the response.
3. *KL*   The 15% discount (*K*) and the free gift (*L*) have different impacts depending on how the other factor is set.

From this one test, there is no way to prove for sure which interaction is present, but *KL* was selected as the most likely interaction. Both *K* and *L* are large significant effects, both are offer-related variables, and an interaction between the two can be logically explained.

The interaction diagram in Figure A5.3 supports the main effects: the 15% discount (*K*−, both points on the left) is always better, and the free gift (*L*+, the top line) increases response over no free gift. However, the 2-factor interaction shows that both together—the 15% discount and the free gift—increase the response less than what can be expected by adding the individual main effects.

The interaction can be understood by comparing both points on the left versus both points on the right. On the right (with no discount, *K*+), offering the free pen-and-pencil set gives a large jump in the response versus offering no free gift—the response more than doubles from 0.41% to 0.84%. In contrast, the points on the left show that, with the 15% discount (*K*−), the free gift increases response only slightly (from 1.18% to 1.27%).

Overall, this interaction shows that the 15% discount is great, the free gift is good, but both together are overkill—the free gift adds little to the benefit of the discount offer. These data helped the company more accurately quantify their return on investment (ROI) on every combination of offers. Also, this gave the marketing team deeper insight into cus-

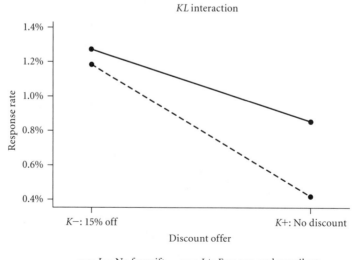

Figure A5.3   Interaction Diagram for Factors *K* and *L*

tomer behavior, showing that one strong incentive is valuable, but additional incentives are probably unnecessary. With these results, the Internet director decided to offer a discount more often, but sometimes switch to a free gift, depending on the e-mail campaign and the profitability of the customer segment.

CONCLUSIONS

The Internet director was amazed by the depth and value of the results of this one test in one drop with just 34,060 names. He learned in one week what would take 6 months using standard techniques of testing one variable at a time. With these results he decided to do the following:

- Consistently offer the 15% discount (testing different discounts in future campaigns).
- Avoid the partner promotions that hurt response.
- Use the special-offer starburst ($J+$), even though it was not quite significant.
- Offer a free gift every few e-mail campaigns to keep the offer fresh and sometimes offer it along with the discount to the highest-value customer segments.
- Improve his segmentation model, adding more variables and splitting apart recent high-value and low-value buyers.

He implemented this strategy in the next campaign. The response jumped to 1.54%, which was somewhat higher than the prediction and much better than the previous performance. The Internet director continued testing offers along with bolder creative changes, eventually achieving response rates consistently between 3% and 5% while adding more names in every drop.

After these results, the marketing team began testing changes in their catalog, retail stores, and regional advertising, continually squeezing greater profit from every marketing dollar. They found few major breakthroughs, but continually uncovered a number of small changes that added up to a big bottom-line impact.

QUESTIONS

Exercise 7 in Chapter 5

CASE 6

MOTHER JONES (PART B)

A random sample of 40,000 persons participated in the test described in Case 3, with the letters mailed on March 15, 2000. The 2^{7-3} design shown below (Table A6.1) was used, resulting in 16 different experimental runs. Each run consisted of a particular combination of factor settings, with each combination sent to 2,500 persons. The response variable shown in Table A6.1 is the net response rate (in %), which is the percentage of people who subscribed and paid (either by cash or credit card). The estimated effects are shown in Table A6.2.

QUESTIONS

1. Analyze the results of the experiment. Which effects are statistically significant at the 5% level? At the 10% level?

2. What settings for the factors would you recommend?

3. What is the regression prediction equation and the predicted response if significant factors are set at their best levels?

TABLE A6.1
2^{7-3} Design with Generators $E = ABC$, $F = BCD$, $G = ACD$, and Responses

Run	A	B	C	D	E	F	G	Response (%)
1	−	−	−	−	−	−	−	2.08
2	+	−	−	−	+	−	+	2.76
3	−	+	−	−	+	+	−	2.36
4	+	+	−	−	−	+	+	3.04
5	−	−	+	−	+	+	+	2.36
6	+	−	+	−	−	+	−	2.52
7	−	+	+	−	−	−	+	2.64
8	+	+	+	−	+	−	−	2.64
9	−	−	−	+	−	+	+	2.40
10	+	−	−	+	+	+	−	2.52
11	−	+	−	+	+	−	+	3.24
12	+	+	−	+	−	−	−	2.12
13	−	−	+	+	+	−	−	2.12
14	+	−	+	+	−	−	+	3.12
15	−	+	+	+	−	+	−	1.96
16	+	+	+	+	+	+	+	3.20

TABLE A6.2
Estimated Effects

Factor	Effect
A	0.345
B	0.165
C	0.005
D	0.035
E	0.165
F	−0.045
G	0.555
AB + CE + FG	−0.145
AC + BE + DG	0.255
AD + CG + EF	−0.035
AE + BC + DF	−0.085
AF + BG + DE	0.205
AG + CD + BF	0.025
BD + CF + EG	−0.075
ABD	−0.105

CASE 7

PEAK ELECTRONICS:

THE BROKEN TENT PROBLEM (PART B)

Peak ran a replicated 2^{5-1} fractional factorial design to solve the problem discussed in Case 4, with each run being a single test panel. The response variable is the number of broken tents. The order of the 32 runs was randomized, with the run sequence shown in parentheses. The results are shown in Table A7.1.

QUESTIONS

1. Analyze the results. Estimate the effects, and obtain their significance by comparing the estimates to their standard error.

2. Which are the significant effects? What are the best settings? How well did Lou Pagentine predict which variables would be significant? What is the regression prediction equation for the number of broken tents on a panel? Estimate by how much Peak would reduce the number of broken tents by using the best settings (as compared to the current settings).

TABLE A7.1
Design and Test Results

A = Lamination Roll Thickness	B = Lamination Exit Temperature	C = Spray Pressure	D = Break Point	E = Hold Time	Response	Average	Standard Deviation
		FACTORS					
−1	−1	−1	−1	1	4 (5) 11 (27)	7.5	4.950
1	−1	−1	−1	−1	6 (9) 5 (23)	5.5	0.707
−1	1	−1	−1	−1	21 (19) 5 (21)	13.0	11.314
1	1	−1	−1	1	2 (7) 3 (30)	2.5	0.707
−1	−1	1	−1	−1	40 (3) 49 (28)	44.5	6.364
1	−1	1	−1	1	29 (11) 15 (32)	22.0	9.899
−1	1	1	−1	1	40 (17) 6 (20)	23.0	24.042
1	1	1	−1	−1	34 (2) 26 (18)	30.0	5.657
−1	−1	−1	1	−1	0 (6) 0 (14)	0.0	0.000
1	−1	−1	1	1	0 (16) 2 (25)	1.0	1.414
−1	1	−1	1	1	1 (1) 1 (15)	1.0	0.000
1	1	−1	1	−1	0 (26) 0 (31)	0.0	0.000
−1	−1	1	1	1	18 (12) 29 (29)	23.5	7.778
1	−1	1	1	−1	6 (8) 3 (24)	4.5	2.121
−1	1	1	1	−1	8 (4) 7 (10)	7.5	0.707
1	1	1	1	1	19 (13) 5 (22)	12.0	9.899

CASE 8

EXPERIMENTS IN RETAIL OPERATIONS: DESIGN ISSUES AND APPLICATION

Gordon H. Bell, Johannes Ledolter, and Arthur J. Swersey

INTRODUCTION

Experimental design methods have long been recognized as an integral part of production and operations management in general and quality management in particular. With its origins in the pioneering work of Sir Ronald Fisher, who published *The Design of Experiments* in 1935, experimental design methods have been widely applied to manufacturing problems, with numerous case studies and examples appearing in the literature.

In the early 1980s, largely in response to competition from Japan, U.S. firms took a renewed interest in these statistical methods, with the Big Three automobile makers at the forefront of these activities. Experimental design was emphasized throughout that decade as an important aspect of statistical process control (SPC) and total quality management (TQM) activities. More recently, Six Sigma programs have gained widespread attention, with experimental design being a prominent part of that methodology. Principles of lean production have been combined with Six Sigma, resulting in an approach that simultaneously focuses on both these methodologies. Over time, the focus of Six Sigma and other quality activities has shifted from its original focus on manufacturing to a broader scope that includes health care and other service areas. Similarly, concepts of lean production have more recently been applied to service operations. For example, patient-focused care in hospitals is designed to increase quality of care by decentralizing many ancillary services and bringing the caregivers to the patient. This approach is very similar to just-in-time production/cell manufacturing in a factory setting.

Bisgaard (1992) provides a notable, historical review of experimental design case studies that includes what he calls "a partial and unsystematic list of articles . . . showing engineering and manufacturing applications of experimental design." This list comprises more than 130 case studies. More recent case studies applying experimental design methods to manufacturing problem are discussed by Lin and Chanada (2003), Cherfi, Bechard, and Boudaoud (2002), Schaub and Montgomery (1997), and Young (1996).

In contrast to work on manufacturing problems, applications of experimental design to service problems, including marketing and retail operations, have been limited, with examples rarely appearing in the academic literature. In searching for service applications of experimental design, we found a few early papers all in the marketing literature, and we found no paper that employed a Plackett-Burman design, which was used in our study. Curhan (1974a) used a 2-level fractional factorial design to test the effects of price, advertising, display space, and display location on sales of fresh fruits and vegetables in supermarkets, while Barclay (1969) used a factorial design to evaluate the effect on profitability of raising the prices of two retail products manufactured by the Quaker Oats Company. Holland and Cravens (1973) presented the essential features of fractional factorial designs and illustrated them with a hypothetical example concerning the effect of advertising and other factors on the sales of candy bars. Wilkinson, Wason, and Paksoy (1982) described a factorial experiment for assessing the impact of price, promotion, and display on the sales of selected items at Piggly Wiggly grocery stores. In addition, marketing researchers have used small experimental designs in survey and conjoint analysis applications (e.g., see Ettenson and Wagner, 1986; Jaffe, Jamieson, and Berger, 1992; Srivastava and Lurie, 2004).

Our purpose in this case study is both to report on a successful retail marketing application of experimental design and to highlight the opportunities that exist for operations management researchers and practitioners to apply these methods to service problems. In general, experimental design in service operations can be used to test the effects on service quality and effectiveness of changes in staffing, training levels, procedures, and service system design. Particular examples in marketing include optimizing the design of Web sites, increasing the effectiveness of direct mail distribution channels for magazines, credit cards, and other products, and various in-store experiments to evaluate changes in factors such as package design, price, and point-of-sale displays.

The traditional approach to experimentation in manufacturing, as well as in retailing and other service areas, is to test one factor at a time while holding the remaining factors at fixed levels. In contrast, in multivariable experimental designs such as factorial, fractional factorial, and Plackett-Burman designs, all factors are tested simultaneously. Because of the orthogonality property of these designs, it is possible to obtain independent estimates of important effects (main effects and interactions), while greatly reducing the required sample size.

In the retail area, which is the focus of this case study, firms typically make very large investments in testing. However, few companies use sophisticated state-of-the-art experimental design techniques for in-store tests, choosing instead to test one variable at a time. Supermarkets offer especially attractive opportunities for experimentation, because of their low profit margins and highly competitive environments.

In designing an in-store experiment, there are many issues that need to be addressed. How many and which factors will be included? How many levels will be tested for each of the factors? What alternative designs should be considered, and which design should be selected? With respect to sample size, how many stores should be included in the experiment, and how should they be chosen? Over how many days or weeks should the test be run to obtain statistically valid and significant results? How should the results be analyzed?

This case study both provides insights into the design issues that are important to decision makers and presents the details and results of an actual application. The product tested was a popular magazine with a very large readership. For proprietary reasons, we do not identify the company or the magazine, and minor changes were made to the data presented. In spite of these modifications, the factors tested and results are essentially unchanged.

Retail testing is ideally suited for the use of experimental design techniques, offering decision makers the opportunity to test numerous variables at a relatively low cost. Dozens of elements can be tested simultaneously with the same sample size as a test of one variable alone.

This case study describes a magazine supermarket test of 10 in-store variables using a 24-run Plackett-Burman experimental design. All 10 factors were tested simultaneously over a 2-week period, with only a fraction of the sample size required for one-variable tests. Results quantified the main effect of each factor and allowed for the analysis of 2-factor interactions.

TEST FACTORS

The supermarket is the final stage of the grocery supply chain. Typically, firms give a great deal of attention to supply-chain management issues that include forecasting, inventory management in stores and warehouses, and transportation and logistics management. Within the supermarket, there is a range of management issues that affect quality and productivity. Some may be addressed with the help of mathematical models, for example, the use of queuing models to schedule the front-end checkout area or computer models for deciding how to allocate shelf space to products.

Designing and implementing in-store experiments offer opportunities to innovate, improve service quality, and increase profits. There are many variables that might be tested, including changes in staffing, training, product location, displays, promotions, and the supermarket environment (store temperature, type of background music, attractiveness, etc.). With respect to the environment, in a supermarket experiment, Milliman (1982) found that slow-tempo music compared to fast tempo music decreased the pace of in-store customer flow and increased daily gross sales.

The focus of this paper is on in-store changes that would increase single-copy magazine sales. The magazine publisher instigated the study, but the variables tested were of general interest to the management of the supermarket chain as well.

Single-copy magazine purchases are often an impulse buy. The cover price paid at a newsstand is usually much higher than the per-copy subscription price; so, loyal customers have a strong incentive to purchase a subscription for its low price and in-home delivery. Publishers invest extensive time and effort on each magazine cover, using experience, focus groups, and one-variable tests to find the right pictures, words, colors, and layouts to attract those impulse buyers who spend just a few seconds selecting a magazine.

In this particular experiment, the magazine itself was not changed. Instead, the project focused on the location, number, and arrangement of magazine racks as well as in-store advertising. Copies of the magazine were primarily displayed near the checkout area. The operations team was particularly interested in the effect on sales of adding additional locations throughout the store. Management of the supermarket chain was also interested in evaluating the effectiveness of these additional sites. These added locations had been unused areas, and because the displays required relatively little space, the magazine was an especially attractive product to test in these additional locations.

The effect of in-store product location on sales has been studied by a number of authors. Dreze, Hoch, and Purk (1994) used a basic test-control experimental approach to assess the

TABLE A8.1
Factors and Their Levels

Factor	(−) Low Level	(+) High Level
A Rack on cooler in produce aisle	No	Yes
B Location on checkout aisle	End cap	Over the belt
C Number of pockets on main racks	Current	More
D Rack by snack foods	No	Yes
E Advertise on grocery dividers	No	Yes
F Distribution of magazines in the store	Random	Even
G Oversized card insert	No	Yes, in 20% of copies
H Clip-on rack advertisement	No	Yes
I Discount on multiple copies	No	Yes
J On-shelf advertisement	No	Yes

sales impact of in-store shelf space management. Changing the location of products among various shelf positions, they found that rearranging products in complementary groups and placing certain products at eye level could increase sales. Placing fabric softener between liquid and powder detergents and moving toothbrushes from a top shelf to a shelf at eye level both increased category sales. They also found that shelf position was more important than the amount of shelf space allocated for a particular product. In earlier research, several authors studied shelf space elasticities, including Brown and Tucker (1961) and Curhan (1974b) while Bultez and Naert (1988) and Bultez et al. (1989) studied space allocation using an attraction model to estimate brand interactions.

In this case, the team wanted to test as many factors as possible, which made sense because most of the cost of the experiment relates to the number of stores included in the test rather than the number of factors. After brainstorming a wide range of new ideas, the team identified the 10 factors in Table A8.1. For each factor, they selected two levels: the low or minus level and the high or plus level. A number of factors related to the number and location of pockets and racks. A *pocket* is one slot in a magazine rack, holding a few copies of the same magazine. A *rack* is the physical display with a few or many pockets. The main magazine rack in one aisle of the supermarket may take up all of the shelf space for 30 feet down the aisle and hold 150 different magazines. A small countertop rack may have just two pockets holding one magazine.

A: Rack on Cooler in Produce Aisle

The team wanted to attract customers as they entered the store. Most supermarkets are designed so that customers begin shopping in the section where produce and other fresh foods are displayed. The team had a new, small rack created with just two pockets. The rack was designed to fit on top of a refrigerated case located in the center of the produce aisle. The display was easy to install and took up little floor space. The team anticipated that a magazine display early in the shopping route would increase the likelihood of purchase.

B: Rack Location on Checkout Aisle

Two different magazine racks were available at the checkout aisles: the end-cap racks that customers see as they approach checkout and the over-the-belt racks above the moving grocery belt, usually with smaller-sized magazines. The team had in the past tried both locations, but never tested one against the other.

C: Number of Pockets on Main Racks

The number and location of pockets on the magazine racks are like shelf position for packaged goods. The publisher already had a number of pockets, but wanted to know if additional pockets would increase incremental sales enough to justify the cost.

D: Rack by Snack Foods

Following the same idea as with factor A, the team felt that an additional rack at the end of the shopping trip might encourage more people to buy a copy. The team developed a similar small rack to place on the shoulder-height snack food shelves. They felt this might also entice buyers on brief shopping excursions who go directly to the beer/snack food aisles.

E: Advertise on Grocery Dividers

With ever-growing alternatives for in-store advertising, from coupon dispensers to floor graphics and public-address announcements, several low-cost options are available for publishers. The operations team brainstormed ideas for something new and narrowed the list to three factors (E, H, and J). One new idea was to advertise on the grocery dividers, the plastic sticks used to separate groceries at the checkout aisle. The team agreed to use the same basic plastic stick, but place short and simple magazine advertisements on each of the four sides. Each test store received new grocery dividers in every checkout aisle where the magazine was sold.

F: Distribution of Magazines Within Each Store

The magazine was normally placed in a number of pockets and racks in the checkout area. The team usually let natural market variation takes its course. Some pockets would empty out completely, while others would remain nearly full. Since customers cannot buy magazines from empty pockets, the team thought that a more even distribution of copies would increase sales. For all $F+$ stores, they would pay people to go around to each store every week and even out the distribution of copies—pulling some copies out of full pockets and placing them in empty pockets.

G: Oversized Card Insert

The one test factor that related to the magazine itself was a card insert similar to subscription cards in many magazines. The insert was made taller than the magazine, so its promotional message extended above the top of the magazine. This approach let the team promote the magazine without paying the supermarket for in-store advertising. The card was added to about 20% of the copies used in the test stores, so it would stand out among all copies on the newsstand. This factor was implemented by the distributor who made special deliveries of magazines with oversized inserts to all $G+$ stores.

H: Clip-on Rack Advertisement

Another new strategy for in-store advertising was the addition of a small, plastic, clip-on sign with a promotional message about the magazine. These clip-ons were designed for the wire racks at checkout and were added to about 50% of all of the pockets around the checkout aisles.

I: Discount on Multiple Copies

The team wanted to test the effect of a special promotion. They printed stickers for the front of each pocket, promoting a discount on the second copy purchased at the same time.

Cash registers were programmed to register the discount when two or more copies were purchased.

J: On-Shelf Advertisement

The final in-store advertising factor the team tested was an on-shelf "billboard." These small signs in plastic frames were attached to the edge of shelves so they stick out into the aisle. These on-shelf signs were placed in a few of the nonmagazine supermarket aisles.

TEST DESIGN

With 10 factors, there are several alternative designs that can be considered. One possibility is a 32-run 2^{10-5} fractional factorial design of resolution IV. In this design, main effects are confounded with 3- and 4-factor interactions, whereas pairs of 2-factor interactions are confounded with each other, except for one contrast in which four 2-factor interactions are confounded. For a discussion of design resolution and confounding patterns, see Box, Hunter, and Hunter (2005). Assuming that the 3- and 4-factor interactions are negligible, this design provides clear estimates of all main effects. In addition, with proper labeling of the factors, it may be possible to anticipate which 2-factor interactions are likely to be negligible and thereby estimate 2-factor interactions as well. A second alternative is a 16-run 2^{10-6} fractional factorial design. But this design is resolution III, with main effects confounded with 2-factor interactions.

A third alternative is to choose a Plackett-Burman design (Plackett and Burman, 1946). Plackett-Burman designs are a class of orthogonal designs for factors with two levels, with the number of runs N a multiple of 4 (i.e., 4, 8, 12, 16, 20, and so on). If N is a power of 2 (i.e., 4, 8, 16, 32, 64, . . .), these designs coincide with the fractional factorial designs. The orthogonal Plackett-Burman designs with $N = 12$, $N = 20$, $N = 24$ runs are important in practice because they result in uncorrelated estimates of main effects of a large number of factors in very few runs. For 2-level (fractional) factorials the run size must be 4, 8, 16, 32, and so forth. This leaves large gaps in the run sizes. In our case, with 10 factors, a minimum of 16 runs would be needed, while the next-highest run size would be 32 runs, as noted above.

Orthogonality of the design implies that the main effect of one factor can be calculated independently of the main effects of all others. The main effect of a factor is the difference between the response averages at the high (plus) and low (minus) levels of that factor. Plackett-Burman designs have fairly complex confounding schemes. In contrast to the fractional factorial designs where main effects and interactions are either not confounded or "fully aliased," Plackett-Burman designs leave main and interaction effects "partially aliased." This means that the absolute values of the alias coefficients are strictly less than one. The literature refers to designs that lead to partial aliasing as nonregular designs; see Wu and Hamada (2000).

The authors selected the 12-run reflected Plackett-Burman design in Table A8.2, consisting of a total of 24 runs. This design was chosen to increase resolution while minimizing the number of treatment combinations, or "test cells." The 12-run reflected test design can include up to 11 factors. With only 10 factors, the 11th column, K, was simply left empty. Note that although factor columns may be left empty, test cells (rows in the matrix) may not be eliminated. In an empty column, the resulting effect is simply a measure of experimen-

TABLE A8.2

The Reflected Plackett-Burman Design in 24 Runs

Test Cell	Rack on Cooler in Produce Aisle	Location on Checkout Aisle	Number of Pockets on Main Racks	Rack by Snack Foods	Advertisement on Grocery Dividers	Distribution of Magazines in the Store	Oversized Card Insert	Clip-on Rack Advertisement	Discount on Multiple Copies	On-Shelf Advertisement	(empty)
	A	B	C	D	E	F	G	H	I	J	K
1	+	+	−	+	+	+	−	−	−	+	−
2	+	−	+	+	+	−	−	−	+	−	+
3	−	+	+	+	−	−	−	+	−	+	+
4	+	+	+	−	−	−	+	−	+	+	−
5	+	+	−	−	−	+	−	+	+	−	+
6	+	−	−	−	+	−	+	+	−	+	+
7	−	−	−	+	−	+	+	−	+	+	+
8	−	−	+	−	+	+	−	+	+	+	−
9	−	+	−	+	+	−	+	+	+	−	−
10	+	−	+	+	−	+	+	+	−	−	−
11	−	+	+	−	+	+	+	−	−	−	+
12	−	−	−	−	−	−	−	−	−	−	−
13	−	−	+	−	−	−	+	+	+	−	+
14	−	+	−	−	−	+	+	+	−	+	−
15	+	−	−	−	+	+	+	−	+	−	−
16	−	−	−	+	+	+	−	+	−	−	+
17	−	−	+	+	+	−	+	−	−	+	−
18	−	+	+	+	−	+	−	−	+	−	−
19	+	+	+	−	+	−	−	+	−	−	−
20	+	+	−	+	−	−	+	−	−	−	+
21	+	−	+	−	−	+	−	−	−	+	+
22	−	+	−	−	+	−	−	−	+	+	+
23	+	−	−	+	−	−	−	+	+	+	−
24	+	+	+	+	+	+	+	+	+	+	+

tal error and/or interactions. The removal of test cells, on the other hand, creates a nonorthogonal test design destroying the independence of the main effects.

In a resolution III Plackett-Burman design, each main effect is confounded with all 2-factor interactions that do not include the main effect, but it is unconfounded with 2-factor interactions that include it. Plackett-Burman designs are nonregular designs with confounding (alias) coefficients strictly less than one in absolute value. In the 12-run Plackett-Burman design, for example, the alias coefficients are either $+1/3$ or $-1/3$.

A complete foldover (or "reflection") of a Plackett-Burman design, such as the one used in Table A8.2, leads to a resolution IV design where main effects are unconfounded with all 2-factor interactions. The term *reflection* is used because 12 additional test cells are run with every plus and minus switched, somewhat like holding a mirror up to the original design. For example, test cell 1 is: $A+, B+, C-, D+, E+, F+, G-, H-, I-, J+, K-$. For the first reflected test cell, 13, all signs are reversed to become: $A-, B-, C+, D-, E-, F-, G+, H+,$ $I+, J-, K+$. Though main effects are independent of all 2-factor interactions, each 2-factor interaction is confounded with many other 2-factor interactions. A reflected Plackett-Burman design provides more accurate estimates of the main effects of a large number of factors, but it creates challenges in trying to identify significant 2-factor interactions.

Reflected designs can show the presence of 2-factor interactions, but it is difficult to quantify individual interactions. A significant difference between effects calculated from the 12 original Plackett-Burman runs and effects from the 12 reflected runs is due to one or more interactions, because the interactions switch signs from the original design to the reflection. However, the group of interactions confounded within each column cannot be separated mathematically. Experience and general statistical principles, like effect heredity, can lead to the subjective selection of likely interactions, but selective analyses can only offer clues to potential interactions. If important interactions seem to be present, the best course of action is to run a higher-resolution follow-up experiment where all interactions can be clearly quantified.

Because of the time and cost of producing many test cells, reflected designs are seldom used in direct mail, print advertising, or even Internet applications. But for retail testing, additional test cells add little, if any, further cost. Each store needs to be set up and monitored individually, so more stores require more effort, but the number of unique test cells does not make a difference. The statistical benefits far outweigh the cost of implementation. The only constraint is the number of test units available (i.e., the number of stores that can be used for the test).

In this case, a larger 32-run fractional factorial design with less confounding would have been preferable. But the company chose to limit the number of stores used in the test, so the larger design was not possible.

DEFINITION OF KEY METRICS AND SELECTION OF TEST UNITS

The key metric for this test was unit sales. The team wanted to uncover any factors that increased the number of magazine copies sold throughout the supermarkets. After analyzing sales data, the team could then calculate profitability based on sales and the cost of each new pocket, rack, and advertisement.

Unit sales were easily and reliably measured using scanner data from each store in the test. However, unit sales were not directly comparable among stores because each store had a different historical sales level. For example, a large supermarket may sell 100 copies per week, while a small store sells only 50. These store-to-store differences would likely overshadow any differences due to the test factors. Therefore, sales data during the test were standardized based on the historical sales volume of each store. The actual key metric was the percent change in sales relative to the historical baseline: 100(actual units sold − baseline units)/(baseline units).

Calculating the baseline sales level for each store can be complicated and potentially a large source of error. If stores vary widely in sales levels, then they should not be grouped together in the same test, because our confidence in a 10% change in sales is much different for a store that sells 10 magazines one week and 11 the next, as compared to a store selling 100 magazines one week and 110 the next.

Initially, the authors suggested a minimum of 96 stores for the test. With a resolution IV 32-run fractional factorial design, this would have given three stores—three replicates—per test cell. However, analyzing test costs, management set the limit at 50 stores, all from a single supermarket chain. At this point, not wanting to risk having just a single store in some test cells, the authors changed the test to the 12-run reflected design (described earlier) with just two replicates in each of the 24 test cells.

With 24 unique combinations in the 12-run reflected test design, at least 24 stores must be selected as test units for the experiment. Two or three times the minimum number of stores is often better for three reasons:

1. *Larger sample size.* More stores offer greater sales volume per week, so the test can be completed more quickly.

2. *Variability analysis.* Within-cell variation can be used as a measure of experimental error or stores can be combined together to reduce total variability.

3. *Identification of outliers.* With three or mores stores per test cell, a store with surprisingly high or low sales can be identified, scrutinized, and, if appropriate, eliminated from the analysis of test results.

Selection of a Retail Partner

The first step is the selection of a retail chain that is used for the test. In this case, the publisher selected a grocery chain known for its excellent cooperation with previous test campaigns. Many of the chain's supermarkets were located in close proximity, had strong magazine sales, and had a fairly standard store layout. There were strong reasons to expect that the test results would transfer to other chains.

After approving all test factors, the retail partner agreed to run the test and share scanner sales data for all stores during the course of the test. The chain's management team was supportive and agreed to arrange meetings with store managers so that the team could explain and manage the execution of the test.

Analysis of Available Stores and Selection/Matching of Test Units

The grocery chain had nearly 100 stores from which the team could select the final 48. The first step in this selection was to analyze the past sales of all stores and eliminate outliers. New stores, highly seasonal locations, and stores with dramatic rates of growth (or decline) were eliminated first. Then stores with low sales volumes were removed.

Control charts (individuals, X, and moving range, MR, charts) of weekly sales data were created for all remaining stores—plotting unit sales per week and adding control limits to quantify variability and identify special causes (see Montgomery, 2004, for a discussion of control charts for individual measurements).

The authors selected 48 stores with high sales volumes, low variability, and stable sales over time. The next step was simply to match smaller stores with larger stores so that the average sales volume per test cell would be relatively constant.

The final baseline sales numbers were calculated after stores were matched and placed in each test cell. Each pair of stores was considered one test unit, and 24 new control charts were created. All pairs were similar in showing minimal sales growth over the previous few weeks, with average sales consistent with the long-term average over the last few months. A couple of special causes—identifiable sources of variation—affected previous weeks. A holiday 6 weeks before caused a large jump in sales, and a special issue of the magazine before that also caused a shift in sales. Therefore, average sales over the previous 5 weeks were selected as a baseline for the test.

The average sale over the 5 previous weeks for each two-store test unit was selected because it gave a valid, easily understood baseline for comparison. More complex options could have been used instead of the 5-week average. For example, a regression model based on past performance—including seasonality and growth rates—could have been used to predict future sales. Covariates could be added to the model based on information about competitor pricing, promotions, and special offers. With sufficient and accurate data, a regression model may work well. However, historical results do not always predict future performance, and numerous predictor variables can potentially create additional sources of error. Also, in this case competitive data were not available, and recent sales were fairly consistent among all test stores. Therefore, the 5-week average sales level gave a clear and simple method for standardizing all test units without undue complexity or potential error.

Minimizing and Measuring Experimental Error

Since two stores were combined into one test unit, store-to-store differences were not used as measure of experimental error. To get greater consistency among test units, large stores were matched with small stores, potentially creating higher within-test cell variation. Therefore, week-to-week variation of each pair of stores over time was used to calculate experimental error for the test. With the same combination of factors run in the same stores over a number of weeks, the weekly difference in sales paralleled the natural market variation. Each additional week provided an additional replicate for each test cell.

Sample Size

The power of the test was determined by the overall sample size. This number depended on the number of the stores, plus how long the test was to run. Once the number of stores was set, the only way to obtain more data and increase power was to run the test for a longer period of time. Sample size is an important issue in planning the testing schedule. Company executives were concerned about the cost of testing, while the team wanted to run the test long enough to identify small effects.

Sample size calculations require a reliable estimate of the variance of the key metric, in this case the percent change in sales. This variance must come from the test units as defined for the test. In this case, the important number was the variance in weekly sales for each *pair* of stores used for each test cell. An estimate of the variance was obtained by pooling the information from the control charts of all 24 pairs of stores. An average of 125 copies of the magazine was sold in each pair of stores every week, with a standard deviation of about 12 copies, or 10%.

The authors recommended running the test for at least 5 weeks, assuming the standard deviation during the test would remain at 10%. With a total sample size of 120 test units (24 pairs of stores $\times$ 5 weeks) and standard deviation of 10%, the team would have an 80% chance of detecting any factors that impacted sales by 5% or more.

The overall sample size in factorial-type experiments where factors are changed simultaneously has a different meaning from the sample size in the experiments that test one variable at a time. With one-variable tests, each individual comparison represents a separate statistical test requiring a certain sample size. A test of one factor alone would require the same 120 test units recommended for this 10-factor Plackett-Burman design, or a total of 50 weeks for a series of 10, one-variable tests within the same 48 stores.

As the launch date approached, company executives felt the need to speed up the project and reduce costs, so they limited the test to just 4 weeks. Then, just before the test began, further delays reduced the run length to only 2 weeks.

TEST RESULTS

The 12-run reflected Plackett-Burman test matrix and the resulting percent changes for weeks 1 and 2 are shown in Table A8.3.

The main effects are obtained by applying the plus and minus signs in the design columns to the averages in the last column of Table A8.3, and dividing the resulting sum by 12 (the number of plus signs). Alternatively, one can regress the response (the averages in the last column of Table A8.3) on the design vectors. The only difference with the regression is the definition of the effects, which are cut in half when using regression.

We treat the changes in weeks 1 and 2 as independent replications and calculate for each run an estimate of the variance of individual measurements. For example, for the first run, the variance estimate is $[(12.5 - 17.9)^2 + (23.3 - 17.9)^2]/1 = 58.32$. We pool the 24 variances to obtain an overall estimate $s^2 = 93.75$. The variance of each run average (average for weeks 1 and 2) that goes into the main effects calculation is given by $s^2/2$. The variance of an effect is $var(\text{effect}) = 2(s^2/2)/12 = s^2/12 = 93.75/12 = 7.81$, and the standard error is $\sqrt{7.81} = 2.79$. Effects that are larger than 1.96 times the standard error (5.47) are considered significant. The effects are displayed graphically in Figure A8.1.

Three effects are statistically significant:

A+: Rack on Cooler in Produce Aisle
The display on top of the refrigerated case in the produce section increased sales by 10.8%. This identified the one most profitable new location to sell the magazine and supported the idea that attracting the customer early increases sales. This was a major change, placing magazines far from their usual locations.

F−: Not Adjusting the Number of Magazines in the Pockets
Sales dropped 10.6% when workers adjusted the number of magazines among the pockets. This result was completely surprising, but saved a great deal of money on unnecessary effort. After seeing these results, the team thought that empty pockets might create the perception of greater demand—perhaps customers are thinking, "if everyone else is buying this issue of the magazine, then it must be worth reading." Another explanation was that an uneven distribution—with more copies in some pockets and just one or two in others—might be more eye-catching, adding "texture" to the numerous rows and columns of magazines. Of course, there was also the possibility that these results were due to chance. But even if the apparent negative effect was random, it was clear that there was no benefit to evening out the distribution of magazines across pockets.

TABLE A8.3
Test Results

Test Cell	Rack on Cooler in Produce Aisle A	Location on Checkout Aisle B	Number of Pockets on Main Racks C	Rack by Snack Foods D	Advertisement on Grocery Dividers E	Distribution of Magazines in the Store F	Oversized Card Insert G	Clip-on Rack Advertisement H	Discount on Multiple Copies I	On-Shelf Advertisement J	(empty) K	PERCENT CHANGE IN SALES (2 STORES/TEST CELL)		
												Week 1	Week 2	Average
1	+	+	−	+	+	+	−	−	−	+	−	12.5	23.3	17.90
2	+	−	+	+	+	−	−	−	+	−	+	29.1	26.0	27.55
3	−	+	+	+	−	−	−	+	−	+	+	−5.8	4.2	−0.80
4	+	+	+	−	−	+	+	−	+	+	−	36.8	1.8	19.30
5	+	+	−	−	−	+	−	+	+	−	+	−4.9	−3.6	−4.25
6	+	−	−	−	+	−	+	+	−	+	+	17.8	12.8	15.30
7	−	−	−	+	−	+	+	−	+	+	+	−3.7	0.0	−1.85
8	−	−	+	−	+	+	−	+	+	+	−	11.0	−8.8	1.10
9	−	+	−	+	+	−	+	+	+	−	−	11.0	1.5	6.25
10	+	−	+	+	−	+	+	+	−	−	−	−1.3	33.1	15.90
11	−	+	+	−	+	+	+	−	−	−	+	−6.0	−11.9	−8.95
12	−	−	−	−	−	−	−	−	−	−	−	−5.3	−8.7	−7.00
13	−	−	+	−	−	−	+	+	+	−	+	17.5	8.4	12.95
14	−	+	−	−	−	+	+	+	−	+	−	−14.6	3.9	−5.35
15	+	−	−	−	+	+	+	−	+	−	−	7.7	−6.4	0.65
16	−	−	−	+	+	+	−	+	−	−	+	28.9	11.4	20.15
17	−	−	+	+	+	−	+	−	−	+	−	17.1	6.7	11.90
18	−	+	+	+	−	+	−	+	−	−	−	−6.6	−5.1	−5.85
19	+	+	+	−	+	−	−	+	−	−	−	16.8	23.2	20.00
20	+	+	−	+	−	−	+	−	−	−	+	25.0	35.0	30.00
21	+	−	+	−	−	+	−	−	+	+	+	12.5	8.4	10.45
22	−	+	−	−	+	−	−	+	+	+	+	8.9	18.3	13.60
23	+	−	−	+	−	−	+	+	+	+	−	14.1	17.1	15.60
24	+	+	+	+	+	+	+	+	+	+	+	−3.2	−2.4	−2.80

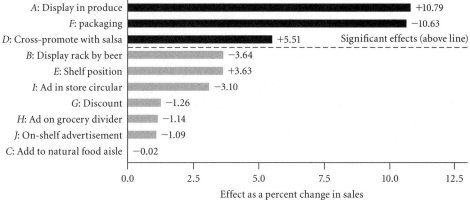

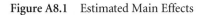

Figure A8.1   Estimated Main Effects

D+: Rack by Snack Foods

The second new display that worked well was a magazine rack at the other end of the store next to the snack foods and beer, increasing sales by 5.5%. Once again, a small rack in a completely new location was beneficial.

The average sales increase during the test was 8.4%. Adding these three significant effects (calculated as the overall average plus one-half of each effect) resulted in a sales increase of 21.8% as compared to the 5-week baseline. Profitability analyses (not shown here) showed that these three remained valuable when taking into account the cost of the two additional racks.

Confounded 2-factor interactions were also analyzed by comparing the effects of the original 12 runs with the effects calculated from the reflected 12 runs. No significant difference in effects was found, so interaction columns were not analyzed further.

The nonsignificant effects were also very valuable. With such a brief test, it can be risky to assume that nonsignificant effects have no impact on sales, but these results can signal where the company can achieve significant savings. All of the in-store advertising had no impact, so the publisher avoided continuing with the on-shelf, clip-on, and grocery divider ads. The location on the checkout aisle was not statistically significant, but because the estimated effect was negative, the team decided to keep the end-cap displays. The second-copy discount and oversized card insert had no impact as well and were eliminated. Surprisingly, more pockets on the main racks had no impact on sales. The team not only avoided the additional cost of more pockets, but also realized that the incremental benefit of additional pockets has some point of diminishing returns.

FINAL COMMENTS

In 48 stores over 2 weeks, the team learned more than they could have learned from months of testing one variable at a time. Two new rack locations were the significant winners among the four factors that related to the number of pockets and location of racks. The team avoided unnecessary operating costs after all five in-store advertising factors showed no effect. Finally, the common perception that redistributing copies was a worthwhile investment proved to be a significant misconception.

The focus of this case study is on increasing magazine sales in a retail setting, but the methods we have presented and discussed apply to retail products in general. In testing such products, decision makers are interested in a range of factors, including price, package design, location, and advertising. The experimental design methodology described here can be used to test specific options, for example, one package design versus another, or it can focus on providing more general insights about the effectiveness of factors such as advertising and product location. More generally, the experimental design approach has applicability to a wide range of problems involving service operations and marketing programs. These statistical tools offer an efficient methodology for future studies aimed at improving the quality and effectiveness of service systems.

QUESTIONS

Exercise 1 in Chapter 6

CASE 9

EXPERIMENTAL DESIGN ON THE FRONT LINES
OF MARKETING: TESTING NEW IDEAS
TO INCREASE DIRECT MAIL SALES

Gordon H. Bell, Johannes Ledolter, and Arthur J. Swersey

This case is reprinted from the *International Journal of Research in Marketing*, Vol. 23 (2006), pp. 309–319 with permission of the Elsevier Publishing Company.

INTRODUCTION

"Test everything" has been a rallying cry in the marketing and advertising industry throughout the 20th century. Industry experts like Hopkins (1923), Caples (1974), Ogilvy (1983), and Stone and Jacobs (2001) have stressed the importance of testing new ideas in the marketplace. But as statisticians developed and refined sophisticated experimental design techniques, most marketers held firm to the approach of changing one factor at a time, often called "split-run testing" (also referred to as *A/B* splits, test-control, or champion-challenger testing). Only in the last few years have marketing leaders begun to embrace advanced techniques for real-world testing.

The financial industry—including insurance, investment, credit card, and banking firms—was among the first to use experimental design techniques for marketing testing. The project described here is from a leading Fortune 500 financial products and services firm. The company name and proprietary details have been removed, but the test strategy, designs, results, and insights are accurate. Tests were run within two direct-mail campaigns that focused on increasing the number and profitability of new customers. The initial experiment, a Plackett-Burman screening design of 19 factors in 20 runs, was followed by a 4-factor 16-run full-factorial experiment.

Although factorial, fractional factorial, and related methods of experimental design have been widely applied to manufacturing problems, there have been few applications to direct mail, Internet, retail, and other market-testing programs, and we found no papers that apply Plackett-Burman designs to these problems. For in-market testing, in an early paper Curhan (1974) used a fractional factorial design to examine the effects of price, advertising, display space, and display location on the sales of fresh fruits and vegetables in a supermarket, while Barclay (1969) used a factorial design to evaluate the effect on profitability of raising the prices of two retail products manufactured by the Quaker Oats Company. Holland and Cravens (1973) presented the essential features of fractional factorial designs and illustrated them with a hypothetical example concerning the effect of advertising and other factors on the sales of candy bars. Wilkinson, Wason, and Paksoy (1982) described a factorial experiment for assessing the impact of price, promotion, and display on the sales of selected items at Piggly Wiggly grocery stores.

Although the market-testing literature is sparse on the use of experimental design models with many factors, 1- or 2-factor experiments have been common. For example, Lodish et al. (1995a) analyzed the results of 389 television advertising experiments to determine the effect of advertising on sales. Their data set included three types of tests: comparing two different versions of advertising copy, comparing two different levels of exposure, and testing copy and exposure simultaneously using a factorial design. In a related paper, Lodish et al. (1995b) examined the carryover effect of television-advertising exposure by tracking sales for an additional two years beyond the original one-year test period.

Factorial and fractional factorial designs are well known and have been widely used in behavioral marketing experiments in laboratory settings (see e.g. Jaffe, Jamieson, and Berger, 1992, Srivastava and Lurie, 2004, and Ettenson and Wagner, 1986) as well as in conjoint analysis applications. Green, Krieger, and Wind (2001) described a credit-card study that illustrates how fractional factorial designs may be used in conjoint analysis. Their design consisted of 12 attributes relating to potential credit-card services, each having two to six levels. For example, annual price (six alternatives), retail purchase insurance (no, yes), rental car insurance (no, yes), and airport club admission (no admission, $5 fee per visit, $2 fee per visit). Using a fractional factorial design, 64 profiles were created out of a total of 186,624 possible attribute-level combinations. The 64 profiles were partitioned into "blocks" of eight profiles each, with all profiles in a given block being presented to each respondent. For each profile of credit card services, the respondent was asked to indicate the likelihood of purchase on a 0–100 point scale. This blocked fractional design provided independent (uncorrelated) estimates of main effects.

Green, Carroll, and Carmone (1978) provided an excellent overview and discussion of the key elements in fractional factorial designs, while Green and Srinivasan (1978, 1990) and Green, Krieger, and Wind (2001) provided notable reviews of the extensive literature on conjoint analysis. Bradlow (2005) discussed current issues in conjoint analysis and the need for future research; Wittink and Cattin (1989) and Wittink, Vriens, and Burhenne (1994) documented the widespread commercial use of conjoint models. Although Green, Carroll, and Carmone (1978) briefly discussed Plackett-Burman designs, we found no papers that used these designs in conjoint and discrete choice models.

Our Plackett-Burman design is a main-effects model that, as we will show, may provide evidence of likely 2-factor interactions under some circumstances. The fractional designs used in conjoint analysis are typically main effects models as well, confounding main effects and 2-factor interactions. Carmone and Green (1981) showed how selected 2-factor interactions can be included in fractional main-effects designs. Plackett-Burman and fractional factorial models are orthogonal designs, which means that effects are estimated independently and with minimum variance. Orthogonal designs may be prohibitively large in situations with many factors, including some at more than two levels, and in cases where interactions are important. For these circumstances, nonorthogonal designs are available and may be generated using statistical software. Kuhfeld, Tobias, and Garratt (1994) discussed such nonorthogonal designs and their use in conjoint and discrete choice studies.

Our review of the literature shows that fractional designs and related orthogonal designs have been used extensively in conjoint and discrete choice studies. As we have noted, there have also been a few papers on market tests involving relatively few factors that use factorial or fractional factorial designs. However, it has been our experience that until recently the great majority of market-testing practitioners relied on the traditional approach of testing one factor at a time. In this case we show the benefits of statistical methods that simultaneously test many factors and also demonstrate the usefulness of Plackett-Burman designs, an important class of experimental design models.

THE EXPERIMENT

The Factors

The firm's marketing group regularly mailed out credit-card offers and wanted to find new ways of increasing the effectiveness of its direct mail program. The 19 factors shown in Table A9.1 were thought to influence a customer's decision to sign up for the advertised product. Factors *A–E* were approaches aimed at getting more people to look inside the envelope, while the remaining factors related to the offer inside. Factor *G* (sticker) refers to the peel-off sticker at the top of the letter to be applied by the customer to the order form. The firm's marketing staff believed that a sticker increases involvement and is likely to increase the number of orders. Factor *N* (product selection) refers to the number of different credit card images that a customer could chose from, while the term "buckslip" (factors *Q* and *R*) describes a small separate sheet of paper that highlights product information.

A Plackett-Burman Design for 19 Factors

With so many factors, we chose a 2-level design. By doing so, we could keep the number of runs relatively low and avoid more complicated and possibly nonorthogonal designs. Two-level screening designs are common in the field of experimental design; see Box,

TABLE A9.1
The 19 Test Factors and Their Low and High Levels

Factor	(−) Control	(+) New Idea
A Envelope teaser	General offer	Product-specific offer
B Return address	Blind	Add company name
C "Official" ink-stamp on envelope	Yes	No
D Postage	Preprinted	Stamp
E Additional graphic on envelope	Yes	No
F Price graphic on letter	Small	Large
G Sticker	Yes	No
H Personalize letter copy	No	Yes
I Copy message	Targeted	Generic
J Letter headline	Headline 1	Headline 2
K List of benefits	Standard layout	Creative layout
L Postscript on letter	Control version	New postscript
M Signature	Manager	Senior executive
N Product selection	Many	Few
O Value of free gift	High	Low
P Reply envelope	Control	New style
Q Information on buckslip	Product info	Free gift info
R Second buckslip	No	Yes
S Interest rate	Low	High

TABLE A9.2

Response Rates in the 20-Run Plackett-Burman Design

Test Cell	A	B	C	D	E	F	G	H	I	J	K	L	M	N	O	P	Q	R	S	Orders	Response Rate
	Envelope Teaser	Return Address	"Official" Ink-stamp on Envelope	Postage	Additional Graphic on Envelope	Price Graphic on Letter	Sticker	Personalize Letter Copy	Copy Message	Letter Headline	List of Benefits	Postscript on Letter	Signature	Product Selection	Value of Free Gift	Reply Envelope	Information on Buckslip	Second Buckslip	Interest Rate		
1	+	+	−	−	+	+	+	+	−	+	−	+	−	−	−	−	+	+	−	52	1.04%
2	−	+	+	−	−	+	+	+	+	−	+	−	+	−	−	−	−	+	+	38	0.76%
3	+	−	+	+	−	−	+	+	+	+	+	−	+	−	+	−	−	−	+	42	0.84%
4	+	+	−	+	+	−	−	+	+	+	+	−	+	−	+	−	−	−	−	134	2.68%
5	−	+	+	−	+	+	−	−	+	+	+	+	−	+	−	+	−	−	−	104	2.08%
6	−	−	+	+	−	+	+	−	−	+	+	+	+	−	+	−	+	−	−	60	1.20%
7	−	−	−	+	+	−	+	+	−	−	+	+	+	+	−	+	−	+	−	61	1.22%
8	−	−	−	−	+	+	−	+	+	−	−	+	+	+	+	−	+	−	+	68	1.36%
9	+	−	−	−	−	+	+	−	+	+	−	−	+	+	+	+	−	+	−	57	1.14%
10	−	+	−	−	−	−	+	+	−	+	+	−	−	+	+	+	+	−	+	30	0.60%
11	+	−	+	−	−	−	−	+	+	−	+	+	−	−	+	+	+	+	−	108	2.16%
12	−	+	−	+	−	−	−	−	+	+	−	+	+	−	−	+	+	+	+	39	0.78%
13	+	−	+	−	+	−	−	−	−	+	+	−	+	+	−	−	+	+	+	40	0.80%
14	+	+	−	+	−	+	−	−	−	−	+	+	−	+	+	−	−	+	+	49	0.98%
15	+	+	+	−	+	−	+	−	−	−	−	+	+	−	+	+	−	−	+	37	0.74%
16	+	+	+	+	−	+	−	+	−	−	−	−	+	+	−	+	−	−	−	99	1.98%
17		+	+	+	+	−	+	−	+	−	−	−	−	+	+	−	+	+	−	86	1.72%
18	−	−	+	+	+	+	−	+	−	+	−	−	−	−	+	+	−	+	+	43	0.86%
19	+	−	−	+	+	+	+	−	+	−	+	−	−	−	−	+	+	−	+	47	0.94%
20	−	−	−	−	−	−	−	−	−	−	−	−	−	−	−	−	−	−	−	104	2.08%

Hunter, and Hunter (2005). Our philosophy in testing many factors, each at two levels, was to identify which factors were active—that is, which factors had a significant effect on the response. Once these active factors were identified, it would be possible (if needed) to test each of them at more than two levels while still maintaining an orthogonal design.

With 19 factors, we created the 20-run *Plackett-Burman main effects design* shown in Table A9.2. Plackett-Burman designs are orthogonal designs for factors that have two levels each, with the number of runs N given by a multiple of 4 (see Plackett and Burman, 1946). For 2-level fractional factorials, the run size N must be a power of 2, leaving large gaps in the run sizes. For example, a minimum of 32 runs is required in a fractional factorial design involving 19 factors. The Plackett-Burman design, on the other hand, can study 19 factors in just 20 runs. This is why these designs are useful in situations where the number of runs is critical.

In a Plackett-Burman design each pair of factors (columns) is orthogonal, which by definition means that each of the four factor-level combinations $[(--), (-+), (+-), (++)]$ appears in the same number of runs. In the 20-run design (Table A9.2), for every pair of columns, each of the four combinations appears five times. As a consequence of orthogo-

TABLE A9.3

A Fractional Factorial Design with Generators E = AB, F = AC, G = AD, H = BC, I = BD, J = CD, K = ABC, L = ABD, M = ACD, N = BCD, and O = ABCD

	FACTOR														
Run	*A*	*B*	*C*	*D*	*E*	*F*	*G*	*H*	*I*	*J*	*K*	*L*	*M*	*N*	*O*
1	−	−	−	−	+	+	+	+	+	+	−	−	−	−	+
2	+	−	−	−	−	−	−	+	+	+	+	+	+	−	−
3	−	+	−	−	−	+	+	−	−	+	+	+	−	+	−
4	+	+	−	−	+	−	−	−	−	+	−	−	+	+	+
5	−	−	+	−	+	−	+	−	+	−	+	−	+	+	−
6	+	−	+	−	−	+	−	−	+	−	−	+	−	+	+
7	−	+	+	−	−	−	+	+	−	−	−	+	+	−	+
8	+	+	+	−	+	+	−	+	−	−	+	−	−	−	−
9	−	−	−	+	+	+	−	+	−	−	−	+	+	+	−
10	+	−	−	+	−	−	+	+	−	−	+	−	−	+	+
11	−	+	−	+	−	+	−	−	+	−	+	−	+	−	+
12	+	+	−	+	+	−	+	−	+	−	−	+	−	−	−
13	−	−	+	+	+	−	−	−	−	+	+	+	−	−	+
14	+	−	+	+	−	+	+	−	−	+	−	−	+	−	−
15	−	+	+	+	−	−	−	+	+	+	−	−	−	+	−
16	+	+	+	+	+	+	+	+	+	+	+	+	+	+	+

nality, the main effect of one factor can be calculated independently of the main effect of all others. Plackett and Burman showed that the complete design can be generated from the first row of +'s and −'s. In Table A9.2, the last entry in row 1 (−) is placed in the first position of row 2. The other entries in row 1 fill in the remainder of row 2, by each moving one position to the right. The third row is generated from the second row using the same method, and the process continues until the next to the last row is filled in. A row of −'s is then added to complete the design.

In what follows, we will assume that 3-factor and higher-order interactions are negligible and can therefore be ignored. The main effect of a factor is the difference between the response averages at the high (plus) and low (minus) levels of that factor. Both fractional factorial designs and Plackett-Burman designs are orthogonal, but the natures of their confounding patterns differ. Consider a fractional factorial design in which main effects are confounded with 2-factor interactions; for example, the saturated design for 15 factors in 16 runs shown in Table A9.3. The design matrix is constructed by first writing columns of signs for a full factorial design in four factors (columns *A*–*D*). The signs for the remaining columns are determined from 11 generators that use all interaction columns in the full factorial design. For example, consider the generator *K* = *ABC*. Multiplying the signs in columns *A*, *B*, and *C*, row by row, results in the column of signs for factor *K*. There are 15 main effects and 105 2-factor interactions (15!/2!13!). Each interaction belongs to a single set of seven 2-factor interactions, and each main effect is confounded with one of these sets. For example, we find that *A* is confounded with *BE*, *CF*, *DG*, *HK*, *IL*, *JM*, and *NO*. The factor *A* does not appear as a letter in any of the seven interactions, and no two interactions include the same factor. The column of signs for factor *A* is identical to the column of signs for each of the 2-factor interactions that are confounded with the main effect of *A*. Hence there is perfect correlation ($\rho = 1$) between the column of signs of *A* and the column of signs for each of its confounded 2-factor interactions. For example, multiplying the signs in columns

B and *E* row by row to obtain a column representing the *BE* interaction results in a column of signs that is identical to the column of signs for factor *A*. Because of this perfect correlation, estimating the main effect by taking the difference between the response averages at the high (plus) and low (minus) levels of a particular factor actually gives an estimate of the main effect of that factor *plus* the sum of the seven 2-factor interactions that are confounded with that main effect. If all of these interactions are negligible, then the result will be a clear estimate of the main effect. If one or more of the interactions are significantly different from zero, the estimate of the main effect will be biased. The books by Berger and Maurer (2002), Box, Hunter, and Hunter (1978), and Ledolter and Burrill (1999) discuss fractional factorial designs, confounding, and the analysis of experimental results.

Plackett-Burman designs have more complex confounding patterns. Each main effect is confounded with all 2-factor interactions except those that involve that main effect. In our 19-factor design in Table A9.2, the main effect for each factor is confounded with all 2-factor interactions involving the other 18 factors for a total of 153 interactions (18!/2!16!). But in contrast to the fractional factorial design shown in Table A9.3, the column of signs for each main effect is not identical to the column of signs for each of its confounded 2-factor interactions. Although not identical and thus not perfectly correlated, these columns of signs are correlated. That is, the correlation between the signs in a main-effect column and the signs in each 2-factor interaction column that is confounded with that main effect is strictly less than 1 in absolute value ($|\rho| < 1$). As a consequence, it can be shown (see Chapter 6) that estimating the main effect of a particular factor by taking the difference between the high (plus) and low (minus) levels for that factor actually provides an estimate of the main effect plus the *weighted* sum of the 2-factor interactions that are confounded with that main effect.

The weight associated with each 2-factor interaction is the correlation between that 2-factor interaction and the main effect; see Barrentine (1996) for a discussion of the structure of confounding patterns in Plackett-Burman designs. Enumerating all correlations among factor columns and interaction columns reveals that, for the 20-run Plackett-Burman design in Table A9.2, the weights (correlations) are either -0.2, $+0.2$ or -0.6. Of the 153 interactions confounded with each main effect, 144 have weights of -0.2 or $+0.2$ while 9 have weights of -0.6. A particular 2-factor interaction will appear in the confounding pattern of 17 main effects. For 16 of these main effects, the weight associated with this interaction will be -0.2 or $+0.2$, while for a single main effect, the weight associated with this interaction will be -0.6. For example, consider the main effect of factor *R* and the *SG* interaction. We use $+1$ and -1 to represent the column signs and multiply the entries in columns *S* and *G* to obtain the entries in column *SG*. Writing each column as a row to save space and listing the run numbers above the entries, we obtain

Run	1	2	3	4	5	6	7	8	9	10	11	12	13	14	15	16	17	18	19	20
Column *R*	+1	+1	−1	−1	−1	−1	+1	−1	+1	−1	+1	+1	+1	+1	−1	−1	+1	+1	−1	−1
Column *SG*	−1	+1	+1	+1	+1	−1	−1	−1	−1	+1	+1	−1	−1	−1	+1	+1	−1	−1	+1	+1

Both columns have 10 plus signs and 10 minus signs, and the entries in each column add to zero. Furthermore, the sum of the squares of the entries in each column is 20, the number of runs *N*. The columns are correlated. In 4 of the 20 runs the signs match, while in 16 runs the signs are opposite. The correlation between these two mean-zero columns (call them *x* and *z*) is given by

$$\rho = \frac{\sum x_i z_i}{\sqrt{\sum x_i^2}\sqrt{\sum z_i^2}} = \frac{-12}{20} = -0.6$$

For simplicity, suppose a single 2-factor interaction confounded with a particular main effect is important. A total of 17 main effects will be confounded with that interaction. For each of these main effects, taking the difference between the high (plus) and low (minus) levels for that factor provides an estimate of the main effect plus α times the magnitude of the confounded 2-factor interaction. As noted previously, for 16 main effects the fraction α will be -0.2 or 0.2, and the bias in our estimate of each main effect will be relatively small: plus or minus 0.2 times the magnitude of the interaction. For a single main effect, α will be -0.6 and the bias will be -0.6 times the magnitude of the interaction.

Given the complex confounding patterns of Plackett-Burman designs, it may seem at first glance that they would not provide any useful information about 2-factor interactions. In fact, traditionally they have been used as main-effects designs. More recently, however, Plackett-Burman designs have received much greater attention from researchers because of what Box, Hunter, and Hunter (2005) call "their remarkable projective properties." In analyzing the results of our experiment in the remainder of this case, we will discuss these projective properties and show how they can be used in certain circumstances to estimate one or more 2-factor interactions from the results of a Plackett-Burman experiment.

The Results

The focus of the experiment was on increasing response rate: the fraction of people who respond to the offer. A large mailing list of potential customers was available for the test. The overall sample size (the number of people to receive test mailings) was determined according to statistical and marketing considerations. The chief marketing executive wanted to limit the number of names in order to minimize the cost of test mailings that performed worse than the control (especially when testing a higher interest rate) and also to reduce postage costs. Of the 500,000 packages that were mailed, 400,000 names received the control mailing (that was run in parallel to the test) while 100,000 were used for the test itself. Therefore, each of the 20 test cells in Table A9.2 was sent to 5,000 people, resulting in the response rates listed in the last column.

For each factor in the experiment, 50,000 people received a mailing with the factor at the plus level and 50,000 people received a mailing with the factor at the minus level. Each main effect is obtained by comparing average responses from these two independent samples of 50,000 each. Because the design is orthogonal, the same 100,000 people are used to obtain independent estimates of each main effect.

The marketing team regularly used 25,000–50,000 names for each split-run test, so the sample size of 100,000 for the designed experiment was not much different from what had been done in the past. Power calculations using the Minitab software convinced the authors that this sample size was large enough to detect meaningful differences. Determining the statistical significance of each main effect is equivalent to the standard statistical test for comparing two independent sample proportions, in this case of size 50,000 each. The firm estimated an average response rate of 1% and wanted to be quite confident of detecting a change of 0.2% (either an increase from 1% to 1.2% or a decrease from 1% to 0.8%). At the 5% significance level with a sample of 100,000, the detection probability (statistical power)

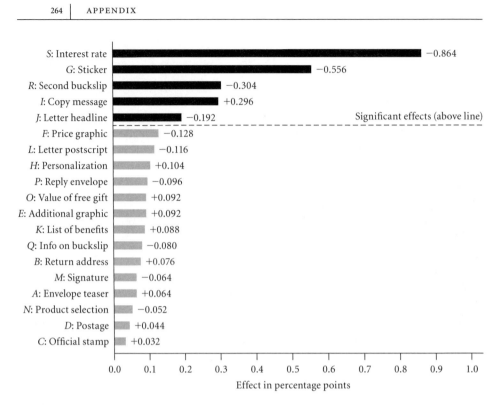

S: Interest rate −0.864
G: Sticker −0.556
R: Second buckslip −0.304
I: Copy message +0.296
J: Letter headline −0.192   Significant effects (above line)
F: Price graphic −0.128
L: Letter postscript −0.116
H: Personalization +0.104
P: Reply envelope −0.096
O: Value of free gift +0.092
E: Additional graphic +0.092
K: List of benefits +0.088
Q: Info on buckslip −0.080
B: Return address +0.076
M: Signature −0.064
A: Envelope teaser +0.064
N: Product selection −0.052
D: Postage +0.044
C: Official stamp +0.032

Effect in percentage points

Figure A9.1   Main Effects Estimates: Plackett-Burman Design

was found to be 0.86 for a change to 1.2% and 0.92 for a change to 0.8%. Thus, with a sample size of 100,000, the authors and marketing team were confident of being able to detect very small yet economically meaningful differences.

Testing all factors simultaneously has large sample-size advantages compared to testing each of the 19 factors one at a time. Suppose we kept the total sample size at 100,000. Then a sample of 5,263 persons would be used for each of the 19 tests of one factor at a time. Because the control package was already being sent to 400,000 people, the group of 5,263 people would receive a mailing of the control with one factor changed. The two sample proportions would then be compared to determine the effect of changing that one factor. Using Minitab, we calculated the power of such a test using the assumptions just described—a 5% significance level and an average response rate of 1%. The statistical power (detection probability) is 0.32 for a change to 1.2% and 0.29 for a change to 0.8%, compared to 0.86 and 0.92 (respectively) for the Plackett-Burman design calculated previously. Again using Minitab, we found that to obtain the same statistical power as the Plackett-Burman design would require a sample size of about 25,000 for each of the 19 tests of one factor at a time; this would yield a total sample size of 475,000 people, an increase of 375,000 persons.

Initial Analysis of the Results

The estimated effects, which are differences between average responses at the plus and minus levels of the factor columns, are shown in Figure A9.1. In the figure, effects are ordered from the largest (at top) to the smallest (at bottom) in terms of their absolute values.

The sign of each effect shows which level is better: for positive effects, the "+" level increases response; for negative effects, the "−" level decreases response.

Significance of the effects was determined by comparing the estimated effects with their standard errors. The result of each experimental run is the proportion of customers who respond to the offer. Each proportion is an average of $n = 5{,}000$ individual binary responses; its standard deviation is given by $\sigma = \sqrt{\pi(1 - \pi)/n}$, where π is the underlying true proportion. Each estimated effect is the difference of two averages of $N/2 = 10$ such proportions. Hence its standard deviation is

$$StdDev(\text{effect}) = \sqrt{\frac{2}{N}\frac{\pi(1 - \pi)}{n} + \frac{2}{N}\frac{\pi(1 - \pi)}{n}} = \sqrt{4/N}\sqrt{\frac{\pi(1 - \pi)}{n}}$$

Replacing the unknown proportion π by the overall success proportion (averaged over all runs and samples), $\bar{p} = (\#Purchases)/(nN) = 1{,}298/100{,}000 = 0.01298$, leads to the standard error of an estimated effect,

$$StdError(\text{effect}) = \sqrt{4/20}\sqrt{\frac{(0.01298)(0.98702)}{5{,}000}} = 0.00072$$

The standard error is 0.072 if effects are expressed in percentage terms. Significance (at the 5% level) is determined by comparing the estimated effect with 1.96 times its standard error, $\pm 1.96(0.072) = \pm 0.141$. The dashed line in Figure A9.1 separates significant and insignificant effects.

The following five factors had a significant effect on the response rate.

S− or Low Interest Rate

Increasing the credit-card interest rate reduces the response by 0.864 percentage points. In addition, it was very clear based on the firm's financial models that the gain from the higher rate would be much less than the loss due to the decrease in the number of customers.

G− or Sticker

The sticker ($G-$) increases the response by 0.556 percentage points, resulting in a gain much greater than the cost of the sticker.

R− or No Second Buckslip

A main effect interpretation shows that adding another buckslip reduces the number of buyers by 0.304 percentage points. One explanation offered for this surprising result was that the buckslip added unnecessary information and obscured the simple "buy now" offer. A more compelling explanation, which we discuss in the next section, is that the significant effect is due not to the main effect of factor R but rather to an interaction between two other factors.

I+ or Generic Copy Message

The targeted message ($I-$) emphasized that a person could choose a credit card design that reflected his or her interests, while the generic message ($I+$) focused on the value of the offer. The creative team was certain that appealing to a person's interests would increase the response, but they were wrong. The generic message increased the response by 0.296 percentage points.

J— or Letter Headline #1

The result showed that all "good" headlines were not equal. The best wording increased the response by 0.192 percentage points.

The response rate from the 400,000 control mailings was 2.1%, while the average response for the test was 1.298%. The predicted response rate for the implied best strategy, starting with the overall average and adding half of each significant effect, amounted to 2.40%. This represented a 15% predicted increase over the response rate of the "control."

Further Analysis of the Results

The confounding of main effects and interactions introduces some uncertainty into our interpretation of the results. A straightforward approach for obtaining unconfounded main effects is a "foldover" of the original Plackett-Burman design. In such a foldover design, the 20-run Plackett-Burman design would be augmented by an additional 20 runs in which the signs of each of the 19 design columns are switched. The combination of a Plackett-Burman design and its complete foldover creates a design in which main effects are no longer confounded with 2-factor interactions. In our experiment, a foldover was not carried out (with 40 runs it would have greatly increased the operational complexity of the mailing), and we cannot be certain which combinations of main effects and interactions are responsible for the significant estimates in Figure A9.1.

The use of our Plackett-Burman design is supported by empirical experimental design principles. Effect sparsity (Box and Meyer, 1986) means that the number of important effects is typically small; hierarchical ordering means that important interactions are usually fewer in number, and smaller in magnitude, than main effects (Wu and Hamada, 2000). In addition, on the basis of effect heredity (Hamada and Wu, 1992)—the principle that significant interactions are likely to involve factors with significant main effects—it is possible in some circumstances to identify likely 2-factor interactions.

Factors S (interest rate) and G (presence of a sticker) are by far the largest effects in Figure A9.1. The correlation between the main effect of R (second buckslip) and the SG interaction is -0.6. Hence, a significant SG interaction would bias the estimate of the main effect of R by -0.6 times the value of the interaction. This suggests that it may not be the main effect of factor R that is important, but the 2-factor interaction between S and G. This interpretation is supported by the principle of effect heredity, since the main effects of S and G are the most important factors. As one might expect, at the high interest rate the effect of having a sticker is small (a change from 0.776% to 0.956% is implied by the results in Table A9.2); at the low interest rate, however, the effect of having the sticker is much larger (a change from 1.264% to 2.024%). The sticker is most effective when the customer receives a more attractive offer.

Box and Tyssedal (1996) showed that the 20-run Plackett-Burman design produces, for any three factors, a complete factorial arrangement with some combinations replicated. The design is said to have "projectivity" 3. In contrast, fractional factorial designs that confound main effects with 2-factor interactions, such as the one shown in Table A9.3, fail to produce a complete factorial for some sets of three factors and hence only have projectivity 2. We use this projectivity idea to provide more evidence that the apparent main effect of R (second buckslip) is actually a consequence of the bias created by the SG interaction.

Consider the three factors S, G, and R. Of the 20 runs in Table A9.2, there is at least one run at each of the eight factor-level combinations of these three factors. In specifying each combination, we let the first sign indicate the level of S, the second sign the level of G, and the last sign the level of R. There are four runs at each of the four combinations $(- - -)$, $(- + +)$, $(+ + -)$, $(+ - +)$ and one run at each of the remaining four combinations. Because we have at least one response at each combination, we have a full factorial arrangement in factors S, G, and R (ignoring the other factors). Because the number of runs at each combination is not the same, we must use regression to estimate the effects. Doing so, we find that the three significant effects are S, G, and SG, confirming that it is the SG interaction and not the main effect of R that is significant.

Table A9.4(a) shows the results when regressing the response rate on the main and interaction effects of the three factors S, G, and R. The standard errors of the estimated regression coefficients use the pooled variance from the eight factor-level combinations, assuming that the other factors have no effect on the response. The t-ratios and the probability

TABLE A9.4

Regression Results for Models Relating the Response Rate to Factors S (Interest Rate), G (Sticker), R (Second Buckslip), I (Copy Message), and J (Letter Headline)

(A) REGRESSION OF RESPONSE RATE ON S, G, R, AND THEIR INTERACTIONS

Rate = $1.325 - (0.386)S - (0.320)G - (0.061)R + 0.151(SG) - (0.070)SR + (0.076)GR + (0.045)SGR$; $R^2 = 0.902$

Predictor	Coefficients	StdError	t-ratio	P-value
Constant	1.325	0.066	20.07	0.000
S	-0.386	0.066	-5.85	0.000
G	-0.320	0.066	-4.85	0.000
R	-0.061	0.066	-0.93	0.372
SG	0.151	0.066	2.29	0.041
SR	-0.070	0.066	-1.06	0.310
GR	0.076	0.066	1.16	0.271
SGR	0.045	0.066	0.68	0.508

(B) REGRESSION OF RESPONSE RATE ON S, G, AND SG

Rate = $1.298 - (0.432)S - (0.278)G + (0.188)SG$; $R^2 = 0.872$

Predictor	Coefficients	StdError	t-ratio	P-value
Constant	1.298	0.052	24.75	0.000
S	-0.432	0.052	-8.24	0.000
G	-0.278	0.052	-5.30	0.000
SG	0.188	0.052	3.58	0.002

(C) REGRESSION OF RESPONSE RATE ON S, G, SG, I, AND J

Rate = $1.298 - (0.432)S - (0.278)G + (0.151)SG + (0.118)I - (0.066)J$; $R^2 = 0.921$

Predictor	Coefficients	StdError	t-ratio	P-value
Constant	1.298	0.044	29.46	0.000
S	-0.432	0.044	-9.80	0.000
G	-0.278	0.044	-6.31	0.000
SG	0.151	0.046	3.29	0.005
I	0.118	0.045	2.62	0.020
J	-0.066	0.045	-1.46	0.166

values of the regression coefficients listed in this table indicate that S, G, and SG are significant whereas all other effects (including the main effect of factor R) are insignificant. Table A9.4(b) lists the results of the regression on the significant effects S, G, and SG. The regression explains 87.2% of the variability in the response rate.

Cheng (1995) showed that in the 20-run Plackett-Burman design, for any four factors, estimates of the four main effects and the six 2-factor interactions involving these four factors can be obtained when their higher-order (3- and 4-factor) interactions are assumed to be negligible. Having eliminated factor R, we apply Cheng's finding and consider a model that includes the four factors that were significant in our initial main effects analysis: S, G, I, and J, together with their six 2-factor interactions. The result of this regression shows that all 2-factor interactions except SG are insignificant, leading to a model with the four main effects and the SG interaction. The fitting results for the model with S, G, SG, and the two main effects of I and J are shown in Table A9.4(c). These five effects explain 92.1% of the variation, a rather modest improvement over the 87.2% that is explained by S, G, and SG. It is clear that factors S (interest rate) and G (sticker) and their interaction SG are the main drivers of the response rate.

A FOLLOW-UP EXPERIMENT

Full Factorial Design in Four Factors

In light of the positive Plackett-Burman test results, the chief marketing executive wanted to continue testing. Since the long-term interest rate was such an important factor in the first test, he decided to focus on a smaller test of just interest rates and fees. In the first test, the introductory interest rate was fixed. Now, he wanted to test changes in both introductory and long-term rates as well as the effects of adding an account-opening fee and lowering the annual fee. The four factors are shown in Table A9.5. Although the account-opening fee was likely to reduce response, one manager thought the fee would give an impression of exclusivity that would mitigate the magnitude of the response decline. The team also wanted once again to test the effect of a small increase in the long-term interest rate. At the same time they wanted to test the effect of two alternative initial interest rates, both lower than the long-term rate.

Each of the factors affected the cost to the customer, so it was expected that 2-factor interactions might well exist. In order to study these interactions along with all main effects, the authors recommended a full-factorial design. The marketing team used columns A–D of the test matrix in Table A9.6 to create the 16 mail packages. The $+/-$ combinations in the 11 interaction (product) columns are used solely for the statistical analysis of the results. All pairs of columns in Table A9.6 are orthogonal. All 15 effects (4 main ef-

TABLE A9.5
Factors and Their Low and High Levels in the Follow-up Experiment

Factor	($-$) Control	($+$) New Idea
A Annual fee	Current	Lower
B Account-opening fee	No	Yes
C Initial interest rate	Current	Lower
D Long-term interest rate	Low	High

TABLE A9.6
Results of the Follow-up Experiment

Test Cell	Annual Fee — A	Account Opening Fee — B	Initial Interest Rate — C	Long-Term Interest Rate — D	Interactions AB	AC	AD	BC	BD	CD	ABC	ABD	ACD	BCD	ABCD	Orders	Response Rate
1	−	−	−	−	+	+	+	+	+	+	−	−	−	−	+	184	2.45%
2	+	−	−	−	−	−	−	+	+	+	+	+	+	−	−	252	3.36%
3	−	+	−	−	−	+	+	−	−	+	+	+	−	+	−	162	2.16%
4	+	+	−	−	+	−	−	−	−	+	−	−	+	+	+	172	2.29%
5	−	−	+	−	+	−	+	−	+	−	+	−	+	+	−	187	2.49%
6	+	−	+	−	−	+	−	−	+	−	−	+	−	+	+	254	3.39%
7	−	+	+	−	−	−	+	+	−	−	−	+	+	−	+	174	2.32%
8	+	+	+	−	+	+	−	+	−	−	+	−	−	−	−	183	2.44%
9	−	−	−	+	+	+	−	+	−	−	−	+	+	+	−	138	1.84%
10	+	−	−	+	−	−	+	+	−	−	+	−	−	+	+	168	2.24%
11	−	+	−	+	−	+	−	−	+	−	+	−	+	−	+	127	1.69%
12	+	+	−	+	+	−	+	−	+	−	−	+	−	−	−	140	1.87%
13	−	−	+	+	+	−	−	−	−	+	+	+	−	−	+	172	2.29%
14	+	−	+	+	−	+	+	−	−	+	−	−	+	−	−	219	2.92%
15	−	+	+	+	−	−	−	+	+	+	−	−	−	+	−	153	2.04%
16	+	+	+	+	+	+	+	+	+	+	+	+	+	+	+	152	2.03%

fects and 11 interactions) can be analyzed independently, and none of these effects are confounded.

The Results

Each of the $N = 16$ test cells was mailed to $n = 7,500$ potential customers. A total of 2,837 customers, or $100(2,837)/(16)(7500) = 2.364\%$, responded to the offer and placed an order. Main and interaction effects were calculated by applying the plus and minus signs to the response column and dividing the weighted sum by $N/2 = 8$. The results are shown in Figure A9.2. Standard errors of the effects (expressed in percentage changes) are obtained by substituting $\bar{p} = 0.02364$ into

$$StdError(\text{effect}) = 100\sqrt{4/N}\sqrt{\frac{\bar{p}(1 - \bar{p})}{n}} = 0.0877$$

Effects outside $\pm 1.96(0.0877) = \pm 0.172$ are statistically significant at the 5% level.

As shown in Figure A9.2, all four main effects as well as one or two (the AB and the CD) interactions are significant. Note that the CD interaction is just slightly smaller than 1.96 times the standard error.

B− or No Account-Opening Fee

Although one manager had thought that charging an initial fee would give the impression of exclusivity, this fee had the largest negative effect, reducing the response rate by 0.518 percentage points.

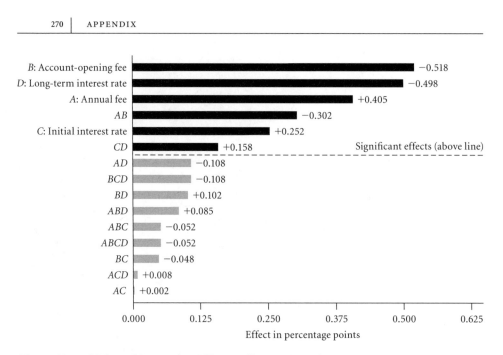

Figure A9.2 Main and Interaction Effects: Follow-up Experiment

D− or Low Long-Term Interest Rate

Another attempt to slightly increase the interest rate showed, once again, that the long-term interest rate had to stay low. Raising the interest rate reduced response on average by 0.498 percentage points.

A+ or Lower Annual Fee

The annual fee was not charged until the end of the first year, but the fee was stated in the mailing. It was not surprising that, as with the other charges, a lower fee was better—here it increased the response by 0.405 percentage points.

C+ or Lower Initial Interest Rate

Reducing the introductory interest rate increased response by 0.252 percentage points.

The main effects are quite strong. However, the significant interactions (*AB* and *CD*) imply that one needs to look at the effects of *A* and *B* and of *C* and *D* jointly. The diagrams in Figure A9.3 show the nature of the interactions. The *AB* interaction supports both main effects, but provides additional important insights. With an account-opening fee (*B*+), the lower annual fee results in only a small increase in response from 2.05% to 2.16%, but with no account-opening fee (*B*−), a lower annual fee results in a large increase in response from 2.27% to 2.98%. The estimated response of 2.98% is highest for the combination *A*+*B*−, the lower annual fee and no account-opening fee. The *AB* interaction expresses that *A*+ and *B*− together increase the response rate beyond what can be expected by either of the two factors separately. This may result from positive synergies or may be due to the negative impact of the account-opening fee, which for some customers may cause an immediate rejection of the offer. The nature of this 2-factor interaction provides extremely valuable infor-

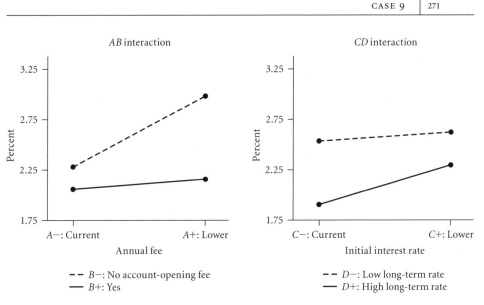

Figure A9.3   Interaction Plots: Follow-up Experiment

mation. Using its financial models, the company found that the increase in response result-ing from no account-opening fee and a lower annual fee $(A+B-)$ was much greater than the loss in revenue that would result from eliminating these fees.

The CD interaction shows that when the long-term rate is low $(D-)$, the effect of a lower initial rate is small and not statistically significant (a change in response from 2.57% to 2.66%). It is clear that offering the lower initial rate would not be profitable if the lower long-term rate were also offered. However, if the long-term rate is high $(D+)$, then the lower initial rate has a large impact, with the response changing from 1.91% to 2.32%. The interaction shows that, for persons receiving both lower rates, the increase in response is considerably less than the sum of the two main effects. This customer behavior is consistent with the concave value function used by Thaler (1985) and based on the earlier work of Kahneman and Tversky (1979). In contrast to the main effects that suggest both interest rates should be low, these results followed by additional analysis using the company's finan-cial models showed that a lower long-term rate coupled with the current (higher) initial rate was the most profitable.

FINAL COMMENTS

After these two mailings—one with a 19-factor Plackett-Burman screening test and the other with the 4-factor full-factorial follow-up test—the marketing team learned more than they had ever before when using the simple technique of testing one variable at a time. The specific findings of these experiments led to immediate and substantial improvements: increased re-sponse rates, lower costs, and higher profits. But the longer-term benefits have been even more substantial. This study introduced the company to the use of formal experimental de-sign methods. Since then the firm has continued to experiment, increasing the speed and profitability of its testing programs and becoming a leader in the application of these tools to direct marketing. Testing has given the company the ability to quickly prove what sells and to greatly improve its performance in the highly competitive financial services marketplace.

Although the focus of this case has been on direct marketing, the potential applications of experimental design approaches are widespread. Website design, on-line advertising, telemarketing, catalog design, and retail tests are fertile areas for multivariable experiments. As marketing applications of large fractional factorial and Plackett-Burman designs are more widely disseminated, the real-world use of these powerful techniques should become more commonplace.

QUESTIONS

Exercise 2 in Chapter 6

CASE 10

PIGGLY WIGGLY

Wilkinson et al. [Wilkinson, J. B., Wason, J. B, and Paksoy, C. H.: "Assessing the Impact of Short-Term Supermarket Strategy Variables," *Journal of Marketing Research*, Vol. 19 (1982), 72–86] described the results of an experiment that assesses the impact of price, promotion, and display on the sales of several grocery items.

THE EXPERIMENT

They considered three price levels:

- Regular price, which is the recommended retail price to customers as listed in the regional warehouse price manual
- Cost price, which is the cost to the supermarket
- Reduced price, which is the price halfway between the recommended retail price and the cost to the supermarket

They studied three display choices:

- Normal display space as determined at the beginning of the experiment on the stock manager's recommendation
- Expanded display, which amounts to twice the normal display area
- Special display, which is normal display, plus some type of additional display alternative such as special display at another location in the store

They considered two levels of advertising:

- Advertising, which means including the product and its price in the supermarket chain's Wednesday advertisement
- No advertising

Detailed operational definitions of the factor levels are given in Wilkinson, Wason, and Paksoy (1982).

The authors used a single, carefully selected Piggly Wiggly supermarket for their experiment: a medium-size store with a loyal, varied, and stable clientele. Four products were studied: Camay soap (bath size), White House apple juice (32 oz), Mahatma rice (1 lb), and Piggly Wiggly frozen pie shells. Sales of these products were stable without trends, and they exhibited limited seasonality.

A complete factorial experiment was carried out. With three price levels, three display options, and two advertising options, the design called for 18 factor-level combinations (treatments). Since the design was replicated once, 36 weeks were needed. Furthermore, each week was preceded and followed by a base week (which is a week where all four products are priced at regular price, displayed at normal shelf position, and not advertised). For such a time arrangement and because holiday weeks were not used, the experiment spanned roughly 80 consecutive weeks. The response was the number of units sold between Wednesday noon and Sunday 9 p.m. of each experimental week.

Trend and seasonality were not considered serious factors because products were stable with minimal seasonality. Furthermore, prior studies showed that the customer flow varies little throughout time.

The precise schedule of the treatment weeks, and a detailed discussion of the necessary preparations and logistic problems that are associated with running such an experiment are given in Wilkinson et al. (1982).

THE DATA AND THE ANALYSIS

Here, we consider Mahatma rice and White House apple juice. The ANOVA tables (for the model with the three main effects, three 2-factor interactions, and the 3-factor interaction) are shown below. We also list the cell averages for the various factor-level combinations.

ANOVA for Mahatma Rice

Source	Sum of Squares
Main effects	
Price	4,376
Display	7,430
Advertising	900
2-factor interactions	
Price × Display	1,068
Price × Advert	107
Display × Advert	345
3-factor interaction	
$P \times D \times A$	99
Error	1,070
Total	15,395

ANOVA for White House Apple Juice

Source	Sum of Squares
Main effects	
Price	2,843
Display	3,983
Advertising	261
2-factor interactions	
Price × Display	428
Price × Advert	74
Display × Advert	208
3-factor interaction	
$P \times D \times A$	624
Error	1,440
Total	9,861

Average Unit Sales for Treatments: Mahatma Rice

	NO ADVERTISING			ADVERTISING		
	Regular Price	Reduced Price	Cost Price	Regular Price	Reduced Price	Cost Price
Regular display	28.0	35.5	32.5	22.5	37.5	43.0
Expanded display	21.0	42.0	46.0	32.5	51.5	55.5
Special display	38.0	60.0	76.5	53.0	73.0	101.0

Average Unit Sales for Treatments: White House Apple Juice

	NO ADVERTISING			ADVERTISING		
	Regular Price	Reduced Price	Cost Price	Regular Price	Reduced Price	Cost Price
Regular display	19.0	23.5	31.5	26.5	46.0	38.0
Expanded display	26.0	36.0	41.0	22.5	44.5	43.5
Special display	37.0	65.0	61.0	38.0	51.5	78.0

COMMENT

In Section 7.2, we discussed the analysis of a general *2-factor* factorial experiment. Here we face a *3-factor* factorial experiment. However, extending the ANOVA table to this situation is straightforward. Now we have a total of *abcn* responses,

$$y_{ijkl} \quad i = 1, 2, \ldots, a \text{ (factor } A); j = 1, 2, \ldots, b \text{ (factor } B); k = 1, 2, \ldots, c \text{ (factor } C),$$

$$l = 1, 2, \ldots, n \text{ (replications)}$$

In our case, A = Price, B = Display, C = Advertising, and $a = 3, b = 3, c = 2, n = 2$. For each of the *abc* factor-level combinations (groups), we obtain the sum of the squared deviations from the group mean. The sum of these sums of squares across the *abc* groups is the error sum of squares, and it has $abc(n - 1)$ degrees of freedom. The other sum of squares entries in the ANOVA table are for (i) main effects of A, B, and C with $a - 1$, $b - 1$, and $c - 1$ degrees of freedom; (ii) 2-factor interactions: AB, AC, and BC with $(a - 1)(b - 1)$, $(a - 1)(c - 1)$, and $(b - 1)(c - 1)$ degrees of freedom; and (iii) the 3-factor interaction ABC with $(a - 1)(b - 1)(c - 1)$ degrees of freedom. The significance of each effect is tested through an F-ratio that compares the mean square of the effect to the mean square error. The degrees of freedom in the numerator and denominator become the degrees of freedom of the F-distribution that is used to test the significance. Remember to always test the highest-order interaction first, because lower-order interactions and main effects make sense only if this interaction is negligible.

Standard statistical software will give you the ANOVA table. For example, you can use the Minitab command "Stat > ANOVA > General Linear Model."

QUESTIONS

Exercise 2 in Chapter 7

CASE 11

UNITED DAIRY INDUSTRIES

This case is adapted from D. G. Clarke [*Marketing Analysis and Decision Making*, The Scientific Press (1987)].

Researchers at the United Dairy Industry Association (UDIA) were evaluating the results of a recent field experiment that tested the impact of varying levels of advertising on the sales of cheese. The principal objective of the study was to measure the retail sales response (pounds of cheese sold) to varying levels of advertising. Eight markets were selected for the experiment—two from each of the four geographic regions: Northeast, Midwest, Southwest, and Southeast. Two markets with similar monthly sales patterns were selected from each geographic region in a way that minimized overlap of local television and newspaper coverage. Within each geographic region, the two markets were designated as test or control market on a random basis.

Executives determined the levels of advertising to be tested in the experiment. It was believed that the levels should be distinct enough to generate measurable differences in the results. They decided to tests the impact of four levels of advertising: 0 cents (level A), 3 cents (B), 6 cents (C), and 9 cents (D), all expressed on a per-capita basis. The 6 cents per-capita level represents a national campaign costing approximately 12 million dollars (in 1973). The principal medium for advertising was television, with point-of-purchase display materials in stores and newspaper ads playing a secondary role. Each of the four levels of advertising was implemented within each test market during one of four 3-month periods between May 1972 and April 1973. The sequence in which the advertising levels were tested was selected so that each advertising level was tested in only one test market during any one time period. Such an arrangement is referred to as a *Latin square* design. You can check that each letter in the Table A11.2 (A, B, C, D) appears only once in each column and each row.

Within each market, UDIA executives obtained the cooperation of approximately 30 supermarkets in obtaining quarterly audits of cheese sales. Average cheese sales (in pounds) per store in each test market across the four 3-month periods between May 1972 and April 1973 are listed in Table A11.3.

TABLE A11.1
Test and Control Markets

Region	Test Market	Control Market
Northeast	Binghamton, NY	Utica-Rome, NY
Midwest	Rockford, IL	Fort Wayne, IN
Southwest	Albuquerque, NM	El Paso, TX
Southeast	Chattanooga, TN	Montgomery, AL

TABLE A11.2
Test Design

	TEST MARKET				CONTROL MARKET			
Time	Bing-hamton	Rock-ford	Albu-querque	Chatta-nooga	Utica-Rome	Fort Wayne	El Paso	Mont-gomery
May–July 72	A	B	C	D	A	A	A	A
Aug–Oct 72	B	D	A	C	A	A	A	A
Nov–Jan 73	C	A	D	B	A	A	A	A
Feb–Apr 73	D	C	B	A	A	A	A	A

TABLE A11.3
Results

	TEST MARKETS			
Time Period	Binghamton	Rockford	Albuquerque	Chattanooga
May–Jul 1972	7,360 (A)	11,258 (B)	11,800 (C)	7,776 (D)
Aug–Oct 1972	7,364 (B)	13,147 (D)	11,852 (A)	8,501 (C)
Nov–Jan 1973	8,049 (C)	13,153 (A)	11,450 (D)	7,900 (B)
Feb–Apr 1973	9,010 (D)	13,880 (C)	12,089 (B)	7,557 (A)

	CONTROL MARKETS (ADVERTISING LEVEL A)			
Time Period	Utica-Rome	Fort Wayne	El Paso	Montgomery
May–Jul 1972	7,166	10,970	11,706	7,441
Aug–Oct 1972	7,489	12,718	11,495	8,250
Nov–Jan 1973	7,679	12,902	11.753	7,853
Feb–Apr 1973	8,536	13,826	12,008	7,768

ANALYSIS OF THE DATA

Part 1 (Test Markets)

The 16 test-market responses in the Latin square allow for the estimation of the main effects of the three factors: location, time, and advertising. We are most interested in the effect of advertising, after adjusting the analysis for possible location and time effects. The ANOVA shown below indicates that there is no strong evidence for an advertising effect. The F-statistic for testing the significance of an advertising effect is $639,139/330,445 = 1.93$; its probability value $P[F(3, 6) \geq 1.93] = 0.225$ is larger than the significance level 0.05. Time is also not significant ($F = 2.42$, with probability value 0.164). The only significant factor is location, with Rockford and Albuquerque having considerably higher cheese sales.

Results for Test Markets (Latin Square Design)

ANOVA

Source	DF	SS	MS	F	P
Advert	3	1917416	639139	1.93	0.225
Cities	3	79308210	26436070	80.00	0.000
Time	3	2398201	799400	2.42	0.164
Error	6	1982671	330445		
Total	15	85606498			

AVERAGES AND STANDARD ERRORS

Advertising
1 (0 cents) 9981 287.4
2 (3 cents) 9653 287.4
3 (6 cents) 10558 287.4
4 (9 cents) 10346 287.4

Cities
1 (Binghamton)   7946 287.4
2 (Rockford)   12859 287.4
3 (Albuquerque) 11798 287.4
4 (Chattanooga) 7934 287.4

Time
1 (May-Jul 72) 9549 287.4
2 (Aug-Oct 72) 10216 287.4
3 (Nov-Jan 73) 10138 287.4
4 (Feb-Apr 73) 10634 287.4

Part 2 (Control Markets)

The 16 control market responses (all under zero cent advertising) originate from a factorial experiment with two factors: location and time. The ANOVA table allows us to test whether there are time and location effects. There is some indication for a time effect, but the evidence is weak (probability value 0.075); the location effect is very significant, with Fort Wayne and El Paso having considerably higher cheese sales. The weak time effect and the significant location effect confirm the findings of Part 1.

Results for Control Markets (Factorial Design)

ANOVA

Source	DF	SS	MS	F	P
Cities	3	78938086	26312695	85.41	0.000
Time	3	2985501	995167	3.23	0.075
Error	9	2772578	308064		
Total	15	84696166			

AVERAGES AND STANDARD ERRORS

Cities
1 (Utica-Rome)   7718 277.5
2 (Fort Wayne) 12604 277.5
3 (El Paso)   11741 277.5
4 (Montgomery)   7828 277.5

Time
1 (May-Jul 72) 9321 277.5
2 (Aug-Oct 72) 9988 277.5
3 (Nov-Jan 73) 10047 277.5
4 (Feb-Apr 73) 10535 277.5

Part 3 (Combining Test and Control Markets)

We combine the observations from the test and control markets and fit a regression model that includes variables for the four levels of advertising, the eight different locations, and the four time periods. The ANOVA table for this model is shown below. This is no longer an orthogonal design, because the different factor-level combinations do not have the same number of runs; for example, there are no observations for control cities and advertising at levels B–D. A consequence of a nonorthogonal design is that sequential and adjusted sums of squares are no longer the same. We are interested in adjusted sums of squares because they tell us about the regression contribution of each factor, on top of all other factors that are part of the analysis. We find that there is not a huge benefit to increased advertising. The test statistic $F = 2.45$ is not significant at the 0.05 level. There is evidence for a time effect, with sales increasing linearly with time. The effect of location is quite strong, with higher sales for Rockford, Albuquerque, Fort Wayne, and El Paso. The main effects plots in Figure A11.1 illustrate these relationships graphically.

Results for Test and Control Markets (Combined Analysis)

ANOVA

Source	DF	Seq SS	Adj SS	Adj MS	F	P
Advert	3	2126193	1917416	639139	2.40	0.101
Cities	7	158246501	158246501	22606643	84.94	0.000
Time	3	5348522	5348522	1782841	6.70	0.003
Error	18	4790430	4790430	266135		
Total	31	170511645				

AVERAGES AND STANDARD ERRORS

Advertising
1 (0 cents) 9977 144.2
2 (3 cents) 9649 295.5
3 (6 cents) 10554 295.5
4 (9 cents) 10342 295.5

Cities
1 (Binghamton)   7946 257.9
2 (Rockford)   12860 257.9
3 (Albuquerque) 11798 257.9
4 (Chattanooga) 7933 257.9
5 (Utica-Rome)   7871 341.2
6 (Fort Wayne) 12758 341.2
7 (El Paso)   11894 341.2
8 (Montgomery)   7982 341.2

Time
1 (May-Jul 72) 9511 213.9
2 (Aug-Oct 72) 10179 213.9
3 (Nov-Jan 73) 10169 213.9
4 (Feb-Apr 73) 10661 213.9

QUESTIONS

Exercise 3 in Chapter 7

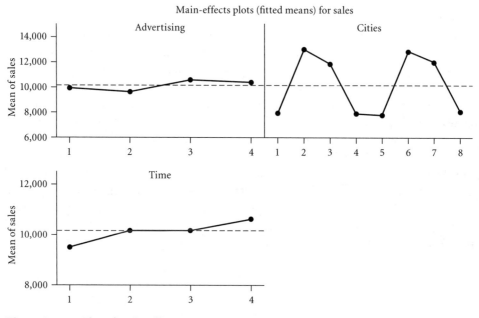

Figure A11.1   Plot of Main Effects

Eric Almquist and Gordon Wyner ["Boost Your Marketing ROI with Experimental Design," *Harvard Business Review* (October 2001), 135–141] make a convincing argument why experimental design can speed up the learning curve of marketing research. Direct marketers have used simple techniques such as split mailings to compare consumer reactions to different prices or promotional offers. However, such traditional testing techniques that change one factor at a time become prohibitively expensive if more than just a couple of advertising techniques need to be evaluated. Changing factors simultaneously and changing the factors according to a well-constructed experimental plan is the key to efficiently learning which of many factors have an influence. Almquist and Wyner discussed two examples.

EXAMPLE 1

The first example describes how a company called BizWare tests the sales response to a campaign that varies three factors: *Price* at four levels ($150, $160, $170, and $180), two different *messages* (one emphasizing speed, the other power), and two *promotion* strategies (one involving a free trial period, the other a free gift). With 3 factors—two at 2 levels and one at 4 levels—the $2^2 4^1$ factorial involves the 16 experiments listed in Table A12.1. The last column in this table indicates the orthogonal half-fraction that is suggested by the design software JMP as an 8-run screening design.

A fraction of a mixed-level design with one factor at 4 levels is easy to generate. One writes down an 8-run design in seven 2-level factors (see Table A12.2 given below) and uses two columns and their interaction to assign the 4 levels of the 4-level factor. This procedure generates an orthogonal design with one factor at 4 levels, and up to 4 factors at 2 levels each. Here, we let the first two columns represent the levels of the two 2-level factors. The columns 3, 12, and their product $(3)(12) = 123$ are used to determine the levels of the 4-level factor. These are the columns in boldface. Level 1 of the 4-level factor is associated with $(-1, 1, -1)$; level 2 with $(-1, -1, 1)$; level 3 with $(1, -1, -1)$; and level 4 with $(1, 1, 1)$. This leads to the 8 runs that are indicated in Table A12.1. Note that the unused columns la-

TABLE A12.1
The 16 Runs of the $2^2 4^1$ Design, and Its Half-Fraction

| | FACTOR | | |
| | | | Run Included |
Message	Promotion	Price	in Fraction?
−1	−1	1	Yes
−1	−1	2	
−1	−1	3	
−1	−1	4	Yes
1	−1	1	
1	−1	2	Yes
1	−1	3	Yes
1	−1	4	
−1	1	1	
−1	1	2	Yes
−1	1	3	Yes
−1	1	4	
1	1	1	Yes
1	1	2	
1	1	3	
1	1	4	Yes

TABLE A12.2
Construction of the Half-Fraction of the Mixed $2^2 4^1$ Design

1 = Message	2 = Promotion	3	12	13	23	123	Price	Response
−1	−1	−1	1	1	1	−1	1	$y_1 = 0.14$
1	−1	−1	−1	−1	1	1	2	$y_2 = 0.09$
−1	1	−1	−1	1	−1	1	2	$y_3 = 0.13$
1	1	−1	1	−1	−1	−1	1	$y_4 = 0.40$
−1	−1	1	1	−1	−1	1	4	$y_5 = 0.01$
1	−1	1	−1	1	−1	−1	3	$y_6 = 0.06$
−1	1	1	−1	−1	1	−1	3	$y_7 = 0.10$
1	1	1	1	1	1	1	4	$y_8 = 0.07$

beled 13 and 23 could have been used for two additional 2-level factors, resulting in an orthogonal fraction of a 5-factor $2^4 4^1$ factorial design.

The design for the factors Message, Promotion, and Price in Table A12.2 is balanced and orthogonal. The design is balanced as the factor levels of each factor occur in the same number of runs. It is orthogonal as the factor-level combinations for each pair of factors appear in the same number of runs. Because of orthogonality, the main effects of the three factors can be estimated independently of each other. The effect of message can be obtained by averaging over the other two factors; the same is true for the effects of promotion and price.

The orthogonal half-fraction was carried out and the results (proportions of sampled individuals responding to the offer) are shown in the last column of Table A12.2. The estimated main effects are

Message: Ave(Message at −1) = 0.095; Ave(Message at +1) = 0.155;

Main effect of Message = 0.155 − 0.095 = 0.06

Promotion: Ave(Promotion at −1) = 0.075; Ave(Promotion at +1) = 0.175;

Main effect of Promotion = 0.175 − 0.075 = 0.10

Price: Ave(Price at 1) = 0.27; Ave(Price at 2) = 0.11;

Ave(Price at 3) = 0.08; Ave(Price at 4) = 0.04;

The main effects of Message, Promotion, and Price are confounded by two factor interactions. The interaction between the two 2-level factors, Message and Promotion, does not affect the main effects of Message and Promotion, but it does confound the main effect of Price.

Price is a continuous factor with 4 levels ($150, $160, $170, $180), and it makes sense to partition its effect into three orthogonal components: a linear, a quadratic, and a cubic component. We can express the main effects of the three factors through a regression of the response on the six columns listed in Table A12.3.

The second and third columns reflect the levels of Message and Promotion; the next three columns representing the linear, quadratic, and cubic components of price (see Appendix 7.1).

The regression output is shown in Table A12.4. The regression coefficients for Message and Promotion are one-half of the main effects that have been listed previously. The standard errors of the regression estimates in Table A12.4 pool the effects of the two unused columns (13 and 23) in Table A12.2. The type of promotion and the price matter most. There

TABLE A12.3
Regression Formulation of the Half-Fraction of the $2^2 4^1$ Design

	REGRESSOR COLUMNS					
Constant	Message	Promotion	Price(lin)	Price(qua)	Price(cubic)	Response
1	−1	−1	−3	1	−1	$y_1 = 0.14$
1	1	−1	−1	−1	3	$y_2 = 0.09$
1	−1	1	−1	−1	3	$y_3 = 0.13$
1	1	1	−3	1	−1	$y_4 = 0.40$
1	−1	−1	3	1	1	$y_5 = 0.01$
1	1	−1	1	−1	−3	$y_6 = 0.06$
1	−1	1	1	−1	−3	$y_7 = 0.10$
1	1	1	3	1	1	$y_8 = 0.07$

TABLE A12.4
Regression Output of the Main Effects Model with Orthogonal Trend Components

```
The regression equation is
Response = 0.125 + 0.0300 Message + 0.0500 Promotion - 0.0360 PriceLin
           + 0.0300 PriceQua - 0.0070 PriceCub

Predictor      Coef   SE Coef      T       P
Constant    0.12500   0.02500    5.00   0.038
Message     0.03000   0.02500    1.20   0.353
Promotion   0.05000   0.02500    2.00   0.184
PriceLin   -0.03600   0.01118   -3.22   0.084
PriceQua    0.03000   0.02500    1.20   0.353
PriceCub   -0.00700   0.01118   -0.63   0.595

S = 0.0707107   R-Sq = 89.8%   R-Sq(adj) = 64.4%

Analysis of Variance

Source          DF        SS        MS      F      P
Regression       5  0.088200  0.017640   3.53  0.235
Residual Error   2  0.010000  0.005000
Total            7  0.098200
```

is a strong linear component to the price effect. This pattern was seen earlier in the averages, which decreased with increasing price. Of course, these results come from a very small study, and they should be confirmed by additional experiments.

EXAMPLE 2

In a second example, Almquist and Wyner discussed the launch of a creative arts and activities Internet portal for Crayola, the maker of colored markers and crayons. The goal was to design a letter marketing campaign that attracts target customers to the site and converts browsers into buyers. In their letter to potential customers, Crayola varied several levels of the following 5 factors: (1) two different *subject lines*; (2) three *salutations*; (3) two calls to *action*; (4) three *promotions*; and (5) two different *closings*. A full $2^3 3^2$ factorial design that includes all possible factor-level combinations requires 72 different letters. Constructing and sending each one of 72 letters to a reasonably large sample of potential customers and monitoring their performance is a challenging task. It would be preferable to reduce the number of different letters by considering suitably chosen fractions. The discussion of fractional experiments in Chapter 5 has shown that while fractions of factorial experiments confound effects, they can provide much useful information about the importance of the studied factors.

It is straightforward to construct an orthogonal half-fraction of 36 runs by combining the 9 runs in the full 3^2 with a half fraction 2^{3-1} for the 2-level factors. Such a design confounds the main and interaction effects of the 2-level factors, but it does not confound the main effects of the 3-level factors. However, 36 runs may still be too many, and one may want to look for designs with fewer runs. Almquist and Wyner mention running a 16-run design, but they do not specify how they selected these runs. A balanced and orthogonal design in 16 runs is not possible. The 16 runs cannot be divided evenly among the 3 levels; hence, the design cannot be balanced. Furthermore, there is no arrangement that achieves the same number of runs at all factor-level combinations of each pair of factors; hence the design cannot be orthogonal.

The design software JMP was used to obtain the 16-run *D-optimal* design in Table A12.5. A D-optimal design minimizes the determinant of the covariance matrix of the main-effects estimates; it maximizes the precision of the parameter estimates. Note that this design is quite close to being an orthogonal design.

QUESTIONS

Exercise 1 in Chapter 8

The 16-Run D-Optimal Design for the Crayola Marketing Campaign

Subject	Action	Closing	Salutation	Promotion
−1	1	1	0	0
1	1	−1	−1	−1
1	−1	1	1	0
1	1	1	0	−1
1	−1	−1	0	0
−1	−1	1	−1	1
1	−1	1	1	−1
−1	1	−1	1	1
−1	−1	−1	1	0
1	−1	1	−1	1
1	1	−1	1	1
−1	−1	−1	−1	−1
−1	−1	−1	0	1
1	1	−1	−1	0
−1	1	1	−1	0
−1	1	1	1	−1

CASE 13

PHONEHOG

This case continues our discussion in Section 8.2. PhoneHog recorded the number of distinct visitors to the PhoneHog site (VISITS), the number of times visitors click on the subsequent page to obtain additional information (CLICKS), and the number of actions of actually completing the subscription agreement (ACTIONS). The click-through rate, CTR = CLICKS/VISITS and the action rate AR = ACTIONS/VISITS measure the success of the creatives. The results are shown in Table A13.1.

QUESTIONS

Exercise 2 in Chapter 8

Table A13.1

Description of the 45 Creatives and the Resulting VISITS, CLICKS, ACTIONS, CTR = CLICKS/VISITS, and AR = ACTIONS/VISITS

Creatives (runs)	A-bottom	A-top	B	C	D	E	F	G	H	I	Visitors	Clicks	Actions	CTR (%)	AR (%)
1	6	3	3	2	2	2	4	3	2	1	2,177	405	239	18.60	10.97
2	10	3	6	1	1	4	2	4	4	2	2,150	420	212	19.53	9.86
3	10	4	1	2	4	1	5	3	4	1	1,988	376	203	18.91	10.21
4	10	3	6	6	6	2	1	1	2	1	2,163	412	232	19.04	10.72
5	6	2	3	6	1	3	7	1	4	1	2,239	391	231	17.46	10.31
6	6	4	6	1	6	2	2	2	1	1	2,204	404	249	18.33	11.29
7	9	2	3	1	6	4	1	3	2	2	2,119	410	213	19.34	10.05
8	9	1	6	2	1	2	5	5	1	2	2,124	434	183	20.43	8.61
9	6	4	1	5	4	4	7	4	2	1	2,088	360	196	17.24	9.38
10	9	2	1	1	4	2	1	2	4	2	2,208	452	245	20.47	11.09
11	6	1	6	1	4	1	4	4	3	2	2,131	413	215	19.38	10.08
12	4	2	1	5	5	2	4	2	4	2	2,262	479	252	21.17	11.14
13	8	3	1	2	6	4	7	1	3	2	2,151	420	202	19.52	9.39
14	10	4	1	1	1	3	4	3	2	1	2,242	393	214	17.52	9.54
15	10	2	6	5	3	1	2	5	3	1	2,134	391	214	18.32	10.02
16	6	3	1	6	3	3	5	2	3	2	2,101	356	186	16.94	8.85
17	9	2	1	2	2	3	2	4	1	1	2,089	353	188	16.89	8.99
18	10	1	1	2	6	3	6	5	4	1	2,144	385	226	17.95	10.54
19	8	2	6	6	2	1	6	3	3	1	2,090	360	202	17.22	9.66
20	4	4	3	6	6	1	5	4	4	2	2,068	413	196	19.97	9.47
21	6	4	6	2	5	4	1	5	1	2	2,054	379	180	18.45	8.76
22	10	1	6	2	3	3	7	2	2	2	2,202	453	230	20.57	10.44
23	4	3	6	1	2	3	1	5	4	1	2,087	374	216	17.92	10.34
24	10	1	3	5	3	2	1	4	3	1	2,087	393	216	18.83	10.34
25	1	3	1	6	5	2	6	4	1	2	2,123	433	208	20.39	9.79
26	8	2	6	5	5	3	5	4	2	1	2,077	347	188	16.70	9.05
27	9	4	6	5	3	4	6	1	4	1	2,059	339	183	16.46	8.88
28	9	3	3	6	5	1	2	2	2	1	2,211	413	242	18.67	10.94
29	1	2	3	2	3	1	4	1	1	2	2,123	426	211	20.06	9.93
30	1	2	1	6	4	4	2	5	2	2	2,162	406	201	18.77	9.29
31	4	1	1	6	3	4	2	3	1	1	1,649	289	161	17.52	9.76
32	8	1	3	5	5	2	2	3	4	2	2,257	513	252	22.72	11.16
33	4	1	1	5	2	1	6	1	2	2	2,123	384	177	18.08	8.33
34	4	3	3	1	4	1	7	5	1	1	2,188	389	227	17.77	10.37
35	10	1	3	5	2	4	5	2	1	2	2,244	406	186	18.09	8.28
36	8	4	1	5	1	1	1	2	1	2	2,202	434	206	19.70	9.35
37	1	3	6	5	4	3	6	3	1	2	2,241	448	223	19.99	9.95
38	4	4	3	2	4	3	2	1	3	2	2,185	454	241	20.77	11.02
39	4	2	3	2	1	4	6	2	3	1	2,166	479	242	22.11	11.17
40	1	4	6	6	2	2	7	3	3	2	2,194	451	243	20.55	11.07
41	1	1	1	1	5	4	5	1	3	1	2,094	364	217	17.38	10.36
42	8	1	6	6	4	4	4	2	4	1	2,214	397	214	17.93	9.66
43	9	4	3	5	6	3	4	5	3	2	2,061	390	210	18.92	10.18
44	8	4	3	1	3	2	6	5	2	2	2,072	441	222	21.28	10.71
45	1	1	1	1	1	1	1	1	1	1	7,698	1,282	741	16.65	9.62

NOTE: The numbers under areas *A*-bottom, *A*-top, *B* through *I* refer to the available levels in each of the 10 test areas. Level 1 represents the baseline level.

REFERENCES

CHAPTER 1

Box, George E. P., Hunter, William G., and Hunter, J. Stuart: *Statistics for Experimenters: Design, Innovation, and Discovery.* New York: Wiley, 1978 (2nd ed., 2005).

Deming, W. Edwards: *Out of the Crisis.* Cambridge, MA: MIT Press, 1982.

Fisher, Ronald A.: *The Design of Experiments.* Edinburgh: Oliver & Boyd, 1935 (and various later editions).

Fisher Box, Joan: *R. A. Fisher, The Life of a Scientist.* New York: Wiley, 1978.

Pande, Peter S., Neuman, Robert P., and Cavanagh, Ronald R.: *The Six Sigma Way: How GE, Motorola, and Other Top Companies Are Honing Their Performance.* New York: McGraw-Hill, 2000.

Salsburg, David: *The Lady Tasting Tea: How Statistics Revolutionized Science in the Twentieth Century.* New York: W. H. Freeman, 2001.

CHAPTER 2

Clarke, D. G.: *Marketing Analysis and Decision Making.* Redwood City, CA: The Scientific Press, 1987.

Hald, Anders: *A History of Probability and Statistics and Their Applications Before 1750.* New York: Wiley, 1986.

Shewhart, W. A.: *Economic Control of Quality of Manufactured Product.* New York: Van Nostrand, 1931.

Stigler, Stephen M.: *Statistics on the Table.* Cambridge, MA: Harvard University Press, 1999.

Welch, B. L.: "The significance of the difference between two means when the population variances are unequal." *Biometrika,* Vol. 29 (1937), 350–362.

CHAPTER 3

Box, George E. P., Hunter, William G., and Hunter, J. Stuart: *Statistics for Experimenters: Design, Innovation, and Discovery.* New York: Wiley, 1978 (2nd ed., 2005).

Clarke, D. G.: *Marketing Analysis and Decision Making.* Redwood City, CA: The Scientific Press, 1987.

Fisher, R. A.: *Statistical Methods for Research Workers.* Edinburgh: Oliver & Boyd, 1925.

CHAPTER 4

Fisher, Ronald A.: *The Design of Experiments*. Edinburgh: Oliver & Boyd, 1935 (and various later editions).

Lenth, R. V.: "Quick and easy analysis of unreplicated factorials." *Technometrics*, Vol. 31 (1989), 469–473.

Montgomery, D. C.: *Introduction to Statistical Quality Control* (3rd ed.). New York: Wiley, 1996.

Yin, G. Z., and Jillie, D. W.: "Orthogonal design for process optimization and its application in plasma etching." *Solid State Technology* (May 1987), 127–132.

CHAPTER 5

Abraham, B., and Ledolter, J.: *Introduction to Regression Modeling*. Belmont, CA: Duxbury Press, 2006.

Eibl, S., Kess, U., and Pukelsheim, F.: "Achieving a target value for a manufacturing process: A case study," *Journal of Quality Technology*, Vol. 24 (1992), 22–26.

Ledolter, J., and Swersey, A.: "Dorian Shainin's variables search procedure: A critical assessment," *Journal of Quality Technology*, Vol. 29 (1997), 237–247.

CHAPTER 6

Abraham, B., and Ledolter, J.: *Introduction to Regression Modeling*. Belmont, CA: Duxbury Press, 2006.

Box, G. E. P., and Tyssedal, J.: "Projective properties of certain orthogonal arrays." *Biometrika*, Vol. 83 (1996), 950–955.

Cheng, C. S.: "Some projection properties of orthogonal arrays." *Annals of Statistics*, Vol. 23 (1995), 1223–1233.

Draper, N. R.: "Plackett and Burman designs." *Encyclopedia of Statistical Sciences*. New York: Wiley, 1985, 754–758.

Draper, N. R., and Smith, H.: *Applied Regression Analysis* (2nd ed.). New York: Wiley, 1981.

Margolin, B. H.: "Orthogonal main-effect $2^n 3^m$ designs and two-factor interaction aliasing." *Technometrics*, Vol. 10 (1968), 559–573.

Plackett, R. L., and Burman, J. P.: "The design of optimum multifactorial experiments." *Biometrika*, Vol. 33 (1946), 305–325.

CHAPTER 7

Box, G. E. P., Hunter, William G., and Hunter, J. Stuart: *Statistics for Experimenters: Design, Innovation, and Discovery*. New York: Wiley, 1978 (2nd ed., 2005).

John, P. W. M.: *Statistical Methods in Engineering and Quality Assurance*. New York: Wiley, 1990.

Montgomery, D. C.: *Design and Analysis of Experiments* (6th ed.). New York: Wiley, 2005.

CHAPTER 8

Box, G. E. P., and Draper, N. R.: *Empirical Model Building and Response Surfaces*. New York: Wiley, 1987.

John, P. W. M.: *Statistical Design and Analysis of Experiments*. New York: Macmillan, 1971.

Kuhfeld, W. F., and Tobias, R. D.: "Large factorial designs for product engineering and marketing research applications." *Technometrics*, Vol. 47 (2005), 132–141.

Kuhfeld, W. F., Tobias, R. D., and Garratt, M.: "Efficient experimental design with marketing research applications." *Journal of Marketing Research*, Vol. 41 (1994), 545–557.

Meyer, R. K., and Nachtsheim, C. J.: "The coordinate-exchange algorithm for constructing exact optimal experimental designs." *Technometrics*, Vol. 37 (1995), 60–69.

APPENDIX

Case 4

Abraham, B., and Ledolter, J.: *Introduction to Regression Modeling*. Belmont, CA: Duxbury Press, 2006.

Case 8

Barclay, W. D.: "Factorial design in a pricing experiment." *Journal of Marketing Research*, Vol. 6 (1969), 427–429.

Bisgaard, S.: "Industrial use of statistically designed experiments: Case study references and some historical anecdotes." *Quality Engineering*, Vol. 4 (1992), 547–562.

Box, G. E. P., Hunter, W. G., and Hunter, J. S.: *Statistics for Experimenters*. New York: Wiley, 1978 (2nd ed., 2005).

Brown, W., and Tucker, W. T.: "The marketing center: Vanishing shelf space." *Atlanta Economic Review*, Vol. 46 (1961), 9–13.

Bultez, A., Gijsbrechts, E., Naert, P., and Vanden Abeele, P.: "Asymmetric cannibalism in retail assortments." *Journal of Retailing*, Vol. 65 (1989), 153–192.

Bultez, A., and Naert, P.: "S.H.A.R.P.: Shelf allocation for retailer's profit." *Marketing Science*, Vol. 7 (1988), 211–231.

Cherfi, Z., Bechard, B., and Boudaoud, N.: "Case study: Color control in the automotive industry." *Quality Engineering*, Vol. 15 (2002), 161–170.

Curhan, R. C.: "The effects of merchandising and temporary promotional activities on the sales of fresh fruits and vegetables in supermarkets." *Journal of Marketing Research*, Vol. 11 (1974a), 286–294.

Curhan, R. C.: "The relationship between shelf space and unit sales." *Journal of Marketing Research*, Vol. 9 (1974b), 406–412.

Dreze, X., Hoch, S. J., and Purk, M. E.: "Shelf management and space elasticity." *Journal of Retailing*, Vol. 70 (1994), 301–326.

Ettenson, R., and Wagner, J.: "Retail buyers' saleability judgments: A comparison of information use across three levels of experience." *Journal of Retailing*, Vol. 62 (1986), 41–63.

Fisher, R. A.: *The Design of Experiments* (8th ed.). New York: Hafner, 1966.

Holland, C. W., and Cravens, D. W.: "Fractional factorial designs in marketing research." *Journal of Marketing Research*, Vol. 10 (1973), 270–276.

Jaffe, L. J., Jamieson, L. F., and Berger, P. D.: "Impact of comprehension, positioning, and segmentation on advertising response." *Journal of Advertising Research*, Vol. 32 (1992), 24–33.

Lin, T., and Chanada, B.: "Quality improvement of an injection-molded product using design of experiments: A case study." *Quality Engineering*, Vol. 16 (2003), 99–104.

Milliman, R. E.: "Using background music to affect the behavior of supermarket shoppers." *Journal of Marketing*, Vol. 46 (1982), 86–91.

Montgomery, D. C.: *Introduction to Statistical Process Control* (5th ed.). New York: Wiley, 2004.

Plackett, R. L., and Burman, J. P.: "The design of optimum multifactorial experiments." *Biometrika*, Vol. 33 (1946), 305–325.

Schaub, D. A., and Montgomery, D. C.: "Using experimental design to optimize the stereolithography process." *Quality Engineering*, Vol. 9 (1997), 575–585.

Srivastava, J., and Lurie, N.: "Price-matching guarantees as signals of low store prices: Survey and experimental evidence." *Journal of Retailing*, Vol. 80 (2004), 117–128.

Wilkinson, J. B., Wason, J. B., and Paksoy, C. H.: "Assessing the impact of short-term supermarket strategy variables." *Journal of Marketing Research*, Vol. 19 (1982), 72–86.

Wu, C. F. J., and Hamada, M.: *Experiments: Planning, Analysis, and Parameter Design Optimization*. New York: Wiley, 2000.

Young, J. C.: "Blocking, replication, and randomization—the key to effective experimentation: A case study." *Quality Engineering*, Vol. 9 (1996), 269–277.

Case 9

Barclay, W. D.: "Factorial design in a pricing experiment." *Journal of Marketing Research*, Vol. 6 (1969), 427–429.

Barrentine, L. B.: "Illustration of confounding in Plackett-Burman designs." *Quality Engineering*, Vol. 9 (1996), 11–20.

Berger, P. D., and Maurer, R. E.: *Experimental Design with Applications in Management, Engineering and the Sciences*. Belmont, CA: Duxbury Press, 2002.

Box, G. E. P., Hunter, W. G., and Hunter, J. S.: *Statistics for Experimenters*. New York: Wiley, 1978 (2nd ed., 2005).

Box, G. E. P., and Meyer, R. D.: "Dispersion effects from fractional designs." *Technometrics*, Vol. 28 (1986), 19–27.

Box, G. E. P., and Tyssedal, J.: "Projective properties of certain orthogonal arrays." *Biometrika*, Vol. 83 (1996), 950–955.

Bradlow, E. T.: "Current issues and a wish list for conjoint analysis." *Applied Stochastic Models in Business and Industry*, Vol. 21 (2005), 319–323.

Caples, J.: *Tested Advertising Methods* (4th ed.). Englewood Cliffs, NJ: Prentice-Hall, 1974.

Carmone, F. J., and Green, P. E.: "Model misspecification in multiattribute parameter estimation." *Journal of Marketing Research*, Vol. 18 (February 1981), 87–93.

Cheng, C. S.: "Some Projection Properties of Orthogonal Arrays." *Annals of Statistics*, Vol. 23 (1995), 1223–1233.

Curhan, R. C.: "The effects of merchandising and temporary promotional activities on the sales of fresh fruits and vegetables in supermarkets." *Journal of Marketing Research*, Vol. 11 (August 1974), 286–294.

Ettenson, R., and Wagner, J.: "Retail buyers' saleability judgments: A comparison of information use across three levels of experience." *Journal of Retailing*, Vol. 62 (Spring 1986), 41–63.

Green, P. E., Carroll, J. D., and Carmone, F. J.: "Some new types of fractional factorial designs for marketing experiments." *Research in Marketing*, J. N. Sheth (ed.), Vol. 1 (1978), 99–122.

Green, P. E., Krieger, A. M., and Wind, Y.: "Thirty years of conjoint analysis: Reflections and prospects." *Interfaces*, Vol. 31, Issue 3 (May/June 2001), S56–S73.

Green, P. E., and Srinivasan, V.: "Conjoint analysis in consumer research: Issues and outlook." *Journal of Consumer Research*, Vol. 5 (September 1978), 103–123.

Green, P. E., and Srinivasan, V.: "Conjoint analysis in marketing: New developments with implications for research and practice." *Journal of Marketing*, Vol. 54, No. 1 (1990), 3–19.

Hamada, M., and Wu, C. F. J.: "Analysis of designed experiments with complex aliasing." *Journal of Quality Technology*, Vol. 24 (1992), 130–137.

Holland, C. W., and Cravens, D. W.: "Fractional factorial designs in marketing research." *Journal of Marketing Research*, Vol. 10 (1973), 270–276.

Hopkins, C. C.: *Scientific Advertising*. New York: Lord & Thomas, 1923. (Reprinted by NTC Business Books, Chicago, 1966.)

Jaffe, L. J., Jamieson, L. F., and Berger, P. D.: "Impact of comprehension, positioning, and segmentation on advertising response." *Journal of Advertising Research*, Vol. 32 (1992), 24–33.

Kahneman, D., and Tversky, A.: "Prospect theory: An analysis of decision under risk." *Econometrica*, Vol. 47, No. 2 (1979), 263–292.

Kuhfeld, W. F., Tobias, R. D., and Garratt, M.: "Efficient experimental design with marketing research applications." *Journal of Marketing Research*, Vol. 31 (November 1994), 545–557.

Ledolter, J., and Burrill, C. W.: *Statistical Quality Control: Strategies and Tools for Continual Improvement*. New York: Wiley, 1999.

Lodish, L. M., Abraham, M. M., Livelsberger, J., Lubetkin, B., Richardson, B., and Stevens, M. E.: "A summary of fifty-five in-market experimental estimates of the long-term effect of TV advertising." *Marketing Science*, Vol. 14, No. 3, part 2 of 2 (1995a), G133–G140.

Lodish, L. M., Abraham, M. M., Livelsberger, J., Lubetkin, B., Richardson, B., and Stevens, M. E.: "How TV advertising works: A meta-analysis of 389 real world split cable TV advertising experiments." *Journal of Marketing Research*, Vol. 32, No. 2 (1995b), 125–139.

Ogilvy, D.: *Ogilvy on Advertising*. New York: Random House, 1983.

Plackett, R. L., and Burman, J. P.: "The design of optimum multifactorial experiments." *Biometrika*, Vol. 33 (1946), 305–325.

Srivastava, J., and Lurie, N.: "Price-matching guarantees as signals of low store prices: Survey and experimental evidence." *Journal of Retailing*, Vol. 80 (2004), 117–128.

Stone, B., and Jacobs, R.: *Successful Direct Marketing Methods* (7th ed.). New York: McGraw-Hill, 2001.

Thaler, R.: "Mental accounting and consumer choice." *Marketing Science*, Vol. 4, No. 3 (1985), 199–214.

Wilkinson, J. B., Wason, J. B., and Paksoy, C. H.: "Assessing the impact of short-term supermarket strategy variables." *Journal of Marketing Research*, Vol. 19 (1982), 72–86.

Wittink, D. R., and Cattin, P.: "Commercial use of conjoint analysis: An update." *Journal of Marketing*, Vol. 53, No. 3 (1989), 91–96.

Wittink, D. R., Vriens, M., and Burhenne, W.: "Commercial use of conjoint in Europe: Results and critical reflections." *International Journal of Research in Marketing*, Vol. 11, No. 1 (1994), 41–52.

Wu, C. F. J., and Hamada, M.: Experiments: Planning, Analysis, and Parameter Design Optimization. New York: Wiley, 2000.

Case 10

Wilkinson, J. B., Wason, J. B., and Paksoy, C. H.: "Assessing the impact of short-term supermarket strategy variables." *Journal of Marketing Research*, Vol. 19 (1982), 72–86.

Case 11

Clarke, D. G.: *Marketing Analysis and Decision Making*. Redwood City, CA: The Scientific Press, 1987.

Case 12

Almquist, E., and Wyner, G.: "Boost your marketing ROI with experimental design." *Harvard Business Review* (October 2001), 135–141.

INDEX

Page numbers in italics refer to figures and tables.